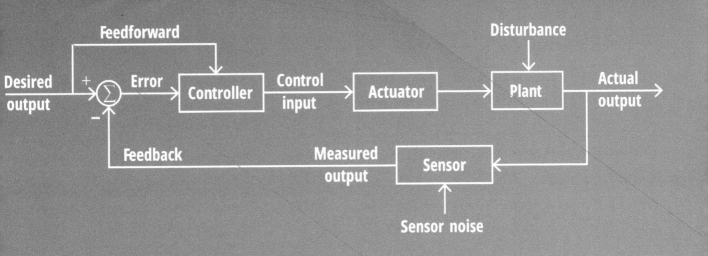

CONTROL SYSTEMS: AN INTRODUCTION

Hassan K. Khalil
Michigan State University

Copyright © 2023 Hassan K. Khalil

This book is published by Michigan Publishing under an agreement with the author. It is made available free of charge in electronic form to any student or instructor interested in the subject matter.

Published in the United States of America by Michigan Publishing.
Manufactured in the United States of America.

ISBN 978-1-60785-826-3 (print)
ISBN 978-1-60785-827-0 (OA)

The Free ECE Textbook Initiative is sponsored by the ECE Department at the University of Michigan.

Preface

I wrote this book in response to the initiative established by Professor Fawwaz Ulaby of the University of Michigan to have textbooks available to students as free PDF files or reasonably priced printed versions. The book is based on my experience teaching control courses at Michigan State University for 42 years. I am grateful to many people. I start with my students and colleagues at Michigan State University who played an important role in shaping the content of the book. Professor Vaibhav Srivastava did me a great favor by preserving my ECE 313 class notes, which I didn't keep at the time of my retirement.

I am grateful to Professor Ulaby, not only for his initiative but also for providing help throughout the project. I thank Professor Necmiye Ozay of the University of Michigan for reviewing the book and providing feedback, Andrew Yagle for his language editing, and Richard Carnes for his book compositing and design.

The book was typeset using LaTeX. Computations were done using MATLAB. The figures were generated using MATLAB, the graphics tool of LaTeX, or TikZ.

HASSAN KHALIL

Contents

Preface		iv
1	**Introduction**	**1**
2	**Transfer Function Models**	**9**
	2-1 Linear Systems	10
	2-2 Examples	14
	2-2.1 Electric Circuits	14
	2-2.2 Mechanical Systems	24
	2-2.3 Electromechanical Systems	37
	2-3 Block Diagrams	40
	2-4 Linearization	50
3	**Stability and Dynamic Response**	**64**
	3-1 BIBO Stability	65
	3-2 Routh-Hurwitz Criterion	70
	3-3 Dynamic Response	78
	3-4 Step Response	81
	3-4.1 First-Order Transfer Functions	81
	3-4.2 Second-Order Transfer Functions	83
	3-4.3 Effect of Zeros	90
	3-4.4 Approximation of Higher-Order Transfer Functions	93
4	**Feedback Control Systems**	**107**
	4-1 Stability	108
	4-2 The Root Locus Method	110
	4-2.1 Rules for Constructing the Root Locus	111
	4-2.2 Refinement of the Root Locus	122
	4-3 Steady-State Error	126
	4-3.1 Unity Feedback System	127
	4-3.2 Non-Unity Feedback System	131
	4-4 Effect of Disturbance	133
	4-5 Plant Parameter Uncertainty	136
	4-6 Effect of Measurement Noise	139

	4-7 Control Constraints	140
	4-8 Why Use Closed-Loop Instead of Open-Loop?	142

5 Time-Domain Design — 154
- 5-1 Design Specifications . . . 155
- 5-2 Design by Gain Adjustment . . . 156
- 5-3 Design by Cascade Compensation . . . 163
- 5-4 Lead and PD Controller . . . 165
  - 5-4.1 Lead Compensator . . . 165
  - 5-4.2 Proportional Derivative (PD) Controller . . . 173
- 5-5 Lag and PI Controller . . . 179
  - 5-5.1 Lag Compensator . . . 179
  - 5-5.2 Proportional-Integral (PI) Controller . . . 181
- 5-6 Lead-Lag and PID Controller . . . 186
  - 5-6.1 Lead-Lag Compensator . . . 187
  - 5-6.2 Proportional-Integral-Derivative (PID) Controller . . . 190
- 5-7 Tuning PID Controllers and Anti-Windup Schemes . . . 194

6 Frequency Response Methods — 210
- 6-1 Polar Plot . . . 211
- 6-2 Bode Plots . . . 217
  - 6-2.1 Asymptotic Bode Plots . . . 220
  - 6-2.2 Underdamped Second-Order Transfer Functions . . . 228
  - 6-2.3 Error Constants from Bode Plots . . . 231
- 6-3 Time-Delay Systems . . . 233
- 6-4 Nyquist Criterion . . . 234
  - 6-4.1 Contour Mapping . . . 234
  - 6-4.2 Cauchy's Principle of the Argument . . . 235
  - 6-4.3 Stability of Feedback Systems . . . 238
  - 6-4.4 Application of the Nyquist Criterion with Gain Adjustment . . . 240
  - 6-4.5 The Case When $GH(s)$ Has Poles on the Imaginary Axis . . . 243
- 6-5 Relative Stability . . . 249
  - 6-5.1 Stability Margins . . . 250
  - 6-5.2 Stability Margins from Bode Plots . . . 251
- 6-6 Relations between Frequency-Domain and Time-Domain Responses . . . 256

7 Frequency-Domain Design — 277
- 7-1 Design Specifications . . . 278
- 7-2 Design by Gain Adjustment . . . 280
  - 7-2.1 Gain Adjustment to Achieve a Desired Crossover Frequency . . . 280
  - 7-2.2 Gain Adjustment to Achieve a Desired Phase Margin . . . 281
  - 7-2.3 Gain Adjustment to Achieve a Desired K_p, K_v, or K_a . . . 282
  - 7-2.4 Gain Adjustment to Shape the Low Frequency Gain . . . 284
- 7-3 Lead Compensation . . . 287

7-3.1 Design of a Lead Compensator to Increase the Phase Margin at a Desired Crossover Frequency 290
7-3.2 Design of a Lead Compensator to Increase the Phase Margin without Changing the Low-Frequency Gain 292
7-4 Lag Compensation 295
7-4.1 Design of a Lag Compensator to Increase the Low-Frequency Gain with Minimal Effect on the Phase Margin of the Uncompensated System 297
7-5 PI Compensation 299
7-5.1 Design of a PI Controller to Increase the Type of the Feedback System and Achieve a Desired Phase Margin 299
7-6 Lead-Lag Compensation 302
7-7 Advantages of the Frequency-Domain Approach 306

8 State-Space Methods 317
8-1 State-Space Models 318
    8-1.1 Examples 321
    8-1.2 Linearization 324
8-2 Solution of the State Equation 327
8-3 Stability 330
8-4 Controllability and Observability 333
    8-4.1 Controllability 333
    8-4.2 Observability 338
    8-4.3 Duality 340
8-5 Transfer Function of a State-Space Model 340
8-6 State-Space Model of a Transfer Function 344
    8-6.1 Controllable Canonical Form 344
    8-6.2 Observable Canonical Form 347
    8-6.3 Parallel Realization 349
8-7 State-Space Design 352
    8-7.1 State Feedback Control 353
    8-7.2 Observers and Output Feedback Control 358
    8-7.3 Set-Point Regulation 361

9 Digital Control 377
9-1 Digital Control Systems 378
9-2 Discrete-Time Systems 381
9-3 Discrete-Time Equivalent of Digital Control System 387
9-4 Stability of Digital Control Systems 393
    9-4.1 Calculation of the Closed-Loop Poles 394
    9-4.2 Root Locus Method 394
    9-4.3 Routh-Hurwitz Criterion 397
    9-4.4 Nyquist Criterion 398
9-5 Steady-State Error 403
9-6 Sample-Frequency Selection 405
    9-6.1 Sampling Theory 405

		9-6.2	Antialiasing Filter (Prefilter) 407

         9-6.2   Antialiasing Filter (Prefilter) 407
         9-6.3   Methods for Choosing ω_s 408
  9-7   Design by Emulation 411
  9-8   Discrete-Time Design 416
         9-8.1   Root Locus Design 416
         9-8.2   Frequency-Domain Design 419
  9-9   Quantization 422

A   Review of Laplace Transform   438

B   Elements of Matrix Analysis   442

C   Laplace and *z*-Transform Tables   444

D   MATLAB Tutorial   446
  D-1   Transfer Function Definition and Manipulation 446
  D-2   Root Locus Analysis 447
  D-3   Error Constants 447
  D-4   Frequency Response 448
  D-5   State-Space Calculations 448
  D-6   Digital Control 449

E   Symbols   451

    Bibliography   453

    Index   456

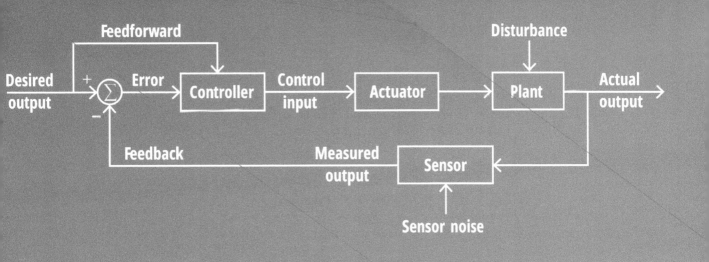

Chapter 1

Introduction

What do rockets, airplanes, automobiles, robots, autonomous vehicles, power grids, power stations, electric motor drives, manufacturing processes, CD drives, and artificial pancreas systems have in common?

Each of these systems uses one or more automatic control systems. In an automatic control system, certain variables are controlled to behave in a desired way without the intervention of a human operator. The system to be controlled is called the *process* or *plant*. It has input and output variables. The output variables are the variables we can monitor or measure, some or all of which are the variables that we control to behave in a desired way. Examples include:

- controlling a car to have a certain speed or to stay in a specified lane,

- maintaining the temperature of a room at a desired set point,

- having the position or velocity of an electric motor follow a certain pattern,

- maintaining the frequency of the electrical signal on the power grid at a specified frequency, and

- having a rocket follow a desired path.

The input variables determine the behavior of the system. There are two types of inputs: those we can manipulate and those we cannot. The inputs that we can manipulate to shape the desired behavior of the system are called *control inputs*, while the inputs that affect the system but are beyond our control are called *disturbance inputs*. For example, in maintaining the temperature of a house, we can control the on and off times of the furnace to determine the heat flow into the house, but we cannot control the heat loss due to heat transfer from inside the house to the outside environment.

A system may have multiple inputs and multiple outputs. However, to focus on the basic elements of control systems, this book is limited to the study of single-input–single-output systems; that is, systems with one control input and one controlled/measured output.

Control systems are classified into

- **Open-loop control systems** and

- **Closed-loop control systems**.

In open-loop control, a mathematical model of the system is used to determine the control input that causes the output to follow a desired response. No feedback information is used to adjust the control input.

In closed-loop control, the output variable is measured and fed back to the controller to be compared with the desired output and the control signal changes in a way that drives the output towards its desired value.

The feedback concept is not alien to us. Feedback is used in our lives whether or not we realize it. When you drive a car to stay in a specified lane, your vision provides feedback about the orientation of the car and you, the driver, determine the action needed to correct the car's orientation if such correction is needed. Using the steering wheel you apply appropriate control action. Our body utilizes feedback to regulate various variables. Examples include thermoregulation (if body temperature changes, mechanisms are induced to restore normal levels) and blood sugar regulation (insulin lowers blood glucose when levels are high; glucagon raises blood glucose when levels are low).

Example 1-1: Cruise Control

Let us use a familiar example to introduce the components of a control system and illustrate the difference between open-loop and closed-loop control. We expect that most readers are familiar with cruise control in automobiles.

The starting point is to develop a mathematical model of the vehicle for which we wish to design the control. Figure 1-1 shows the forces acting on a vehicle moving on a road that makes an angle θ with the horizontal. There are three forces tangential to the road. T is the thrust (in newtons) produced by the engine, which moves the vehicle forward with speed v. It is given by $T = au$, where u is the throttle angle (in degrees) (or accelerator pedal) and a is the engine gain (newtons/degree). F is the force due to road friction, which is assumed to be proportional to the speed v so that $F = bv$. W is a disturbance force that opposes the motion of the vehicle, and it has two components: a force due to the road angle, $mg\sin\theta$, where m is the vehicle's mass and g is the acceleration due to gravity, and a force due to air drag, D; that is, $W = mg\sin\theta + D$. By Newton's second law of motion, the equation of motion is

$$m\dot{v} = T - F - W = au - bv - mg\sin\theta - D.$$

In this model, the angle u is the control input, the velocity v is the controlled output, and $\dot{v} = dv/dt$. The only parameter of the model that is known is the gain a. On the other hand, the mass m, the friction coefficient b, and the disturbance W are uncertain. The mass m depends on the number of passengers and their masses. The friction coefficient b depends on the road condition, which may change from one segment of the road to another. The disturbance W changes with the road angle and the air drag. Both b and W could be changing with time during the motion of the vehicle.

Open-loop control is designed assuming $W = 0$ and using nominal values $\hat{m}$ and $\hat{b}$ for m and b, respectively. This leads to the nominal model

$$\hat{m}\dot{v} = au - \hat{b}v. \tag{1.1}$$

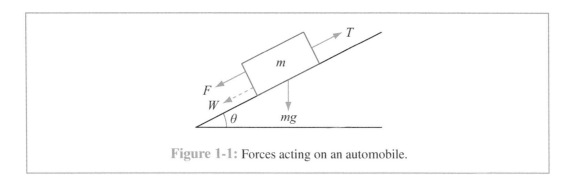

Figure 1-1: Forces acting on an automobile.

The solution of the differential equation has a transient part during which $v(t)$ changes from its initial value to its constant steady-state value v_{ss} at which $\dot{v} = 0$. At steady state

$$0 = au - \hat{b}v_{ss} \quad \Longrightarrow \quad v_{ss} = \frac{au}{\hat{b}}.$$

Focusing on the steady-state behavior, for the purpose of this example, u is designed as a constant control that makes $v_{ss} = v_{des}$, where v_{des} is the desired speed:

$$u = \frac{\hat{b}v_{des}}{a}. \tag{1.2}$$

When this control is applied to the actual system with the true parameters m and b and in the presence of the disturbance W, the system is given by

$$m\dot{v} = \hat{b}v_{des} - bv - W.$$

Assuming W is constant, the steady-state speed v_{ss} is given by

$$v_{ss} = \frac{\hat{b}v_{des} - W}{b}.$$

and the steady-state speed error is given by

$$v_{des} - v_{ss} = \left(\frac{b - \hat{b}}{b}\right)v_{des} + \frac{1}{b}W. \tag{1.3}$$

In **closed-loop control**, the speed v is measured by a speedometer and the measurement is fed back to the controller, which defines the control as

$$u = \frac{\hat{b}v_{des}}{a} + K(v_{des} - v). \tag{1.4}$$

The control has two components. The first component $\hat{b}v_{des}/a$, which is the same as the open-loop control, is called the *feedforward control*. The second component $K(v_{des} - v)$ is called the *feedback control*. This component corrects the control action depending on the speed error. The constant K is the *feedback gain*. Under this control, the closed-loop system is given by

$$m\dot{v} = \hat{b}v_{des} + aK(v_{des} - v) - bv - W = -(b + aK)v + (\hat{b} + aK)v_{des} - W.$$

When W is constant, the steady-state speed error is given by

$$v_{des} - v_{ss} = \left(\frac{b - \hat{b}}{b + aK}\right)v_{des} + \frac{1}{b + aK}W. \tag{1.5}$$

Comparison of Eqs. (1.3) and (1.5) shows that the denominator term b in Eq. (1.3) is replaced by the denominator term $(b + aK)$ in Eq. (1.5). If K is designed to be large, the steady-state closed-loop error will be much smaller than the steady-state open-loop error.

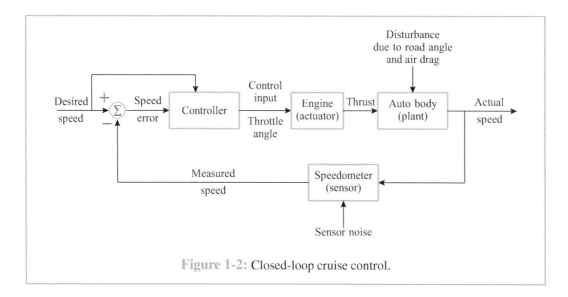

Figure 1-2: Closed-loop cruise control.

Closed-loop systems that are designed with large feedback gain are called *high-gain feedback* systems and the system is said to have a *tight feedback loop*. High-gain feedback is one of the tools used to achieve the goals of closed-loop control, but it is not without limitations. First, the feedback gain cannot be made arbitrarily large because the control input (the throttle angle in this example) has a limited range. Second, the feedback gain affects the transient response of the system and the gain is limited by the design specifications on the transient response. Third, and more importantly, increasing the feedback gain could make the system unstable. This does not happen in this cruise control example but it could happen in systems with more complicated dynamics. Have you been in a room with a public address (PA) system and heard the squeal of oscillation when the volume is increased? That is an example of instability caused by high-gain feedback.

The closed-loop cruise control system is represented by the block diagram of Fig. 1-2. The diagram shows the four basic components of a feedback control system: *plant*, *actuator*, *sensor*, and *controller*.

The speed of the vehicle is measured by a speedometer, which is the **sensor** component of the system. It is represented by a block whose input is the actual speed and whose output is the measured speed. Sensors are not perfect; the measured speed is not exactly equal to the actual speed. The error is the sensor noise, also called measurement noise. The sensor noise depends on the quality of the sensor. A high-quality sensor (typically more expensive) will have less significant sensor noise, while a low-quality sensor (typically cheaper) will have more significant sensor noise.

The measured speed is fed back to a summation node that calculates the speed error by subtracting the measured speed from the desired speed. The summation node is represented by a circle with the summation sign inside the circle. Arrows coming into the circle have positive or negative signs that indicate whether the incoming signal is added or subtracted.

The **controller** is represented by a block that receives the speed error as well as the desired speed and computes the throttle angle, which is the control input.

The control input signal is sent to the **actuator**, which produces the actual physical signal that affects the plant. In our example, the actuator is the vehicle's engine that translates the throttle angle into the thrust force that moves the vehicle. The actuator is represented by a block whose input is the throttle angle and whose output is the thrust.

The body of the automobile is the **plant**. The word "plant" is used in general to describe the system whose output is to be controlled. The inputs to the plant are the control and the disturbance signals. The output is the vehicle's speed, which is regulated to the desired speed.

What are the features of a well-designed control system?

- The system must be stable. We have not defined stability yet, but you can think of it as a requirement mandating that the output of the system should be well behaved when its input is well behaved.

- The output must track the desired output (also called the reference or command) with zero or small steady-state error.

- The transient response of the system must be acceptable.

- The effect of disturbance and measurement noise on the system's response should be made as small as possible.

- The design should be robust to model uncertainties.

- The design should meet any constraints associated with the control input.

The design of a control system uses a mathematical model of the system. Deriving the mathematical model is crucial and challenging. It is crucial because if you start with a bad model you cannot expect to end up with a good control design. It is challenging because a detailed model that captures every aspect of the behavior of the system will most likely be too complicated for analysis and design. Such complicated models are suitable for computer simulation because there are well-developed numerical algorithms that can be used with such models. For analysis and design we need to simplify the model. How to simplify the model while still capturing the essential phenomena of the system is challenging. It requires deep understanding of the physics or chemistry of the system, as well as good understanding of the control techniques that will use the model.

Developing a mathematical model can be done from

- first principles of physics, chemistry or biology,
- identification experiments, or
- combination of both.

We shall see examples of deriving mathematical models from physical principles in Chapter 2 where we model electrical, mechanical, and electromechanical systems. Even in that case we may need to perform experiments to estimate the parameters of the model. Identification is a branch of system science where a model of a dynamic system is determined from experiments in which inputs are applied to the system and the outputs are measured. A mathematical model is obtained from the input–output data.[1]

Overview of the Book

- Chapter 2 introduces differential-equation and transfer-function models of linear systems and provides examples of deriving transfer functions of electrical, mechanical, and electromechanical systems. It introduces block-diagram representations of systems with rules to simplify them. It concludes with a brief discussion of linearization.

- Chapter 3 defines the stability criterion for linear systems, together with the Routh-Hurwitz criterion to check stability. The chapter examines the response of the system to various inputs, with a focus on step inputs.

- Chapter 4 outlines the features that characterize the performance of the feedback control system. It starts with stability and the root-locus method, followed by the calculation of steady-state tracking errors. It describes the effect of disturbance, plant parameter uncertainty, measurement noise, and control constraints on the performance of the system. It concludes with a comparison between open-loop and closed-loop control.

- Chapter 5 presents methods to design the controller to meet time-domain design specifications. It demonstrates how to tune PID (Proportional-Integral-Derivative) controllers.

- Chapter 6 introduces frequency-domain methods as an alternative way to the analysis and design of feedback control systems. Polar and Bode plots are described, leading to the Nyquist stability criterion and stability margins that characterize the robustness of the system. The chapter describes the relations between time-domain and frequency-domain responses.

- Chapter 7 presents methods to design the controller to meet frequency-domain design specifications. It concludes with a discussion on why to use the frequency-domain approach.

- Chapter 8 describes state-space methods for the analysis and design of control systems, using a different approach from the approach used in the previous chapters (which were based on transfer-function models). The chapter introduces state-space models with physical examples. It describes the stability, controllability, and observability properties of such models. It shows how to go back and forth between state-space models and transfer-function models. Finally, the chapter presents the design of feedback control systems using state-space models.

- Chapter 9 deals with the implementation of controllers using digital computers. It describes digital control systems. It reviews discrete-time systems and the z-transform.

---

[1] To read about identification and parameter estimation, see [20] or [26]. A brief introduction can be found in [4] and [12].

A digital control system has continuous-time and discrete-time components. The chapter shows how to represent the whole system as a discrete-time system. This leads to digital characterization of stability and the steady-state error. Rules for selecting the sampling frequency are discussed, followed by the design of digital controllers.

Chapters 2 through 7 would form the core of a first course on control systems. In a semester system, they can be covered together with either Chapter 8 or Chapter 9. It will probably be a stretch to cover all chapters in one semester, unless some parts of Chapters 2 through 7 are skipped or covered in a previous systems course.

This book is based on class notes developed by the author while teaching control courses at Michigan State University for many years. Chapter 2 to 7 are covered in a junior-level course, while Chapters 8 and 9 are included in a senior-level course. Over the years several control books were used as textbooks or references. This includes the books by Nise [28], Dorf and Bishop [8], Franklin, Powell and Emami-Naeini [11], and Ogata [29] for Chapters 2 through 7, the books by Rugh [32] and Anstaklis and Michel [2] for Chapter 8, and the books by Franklin, Powell and Workman [12], Philips, Nagle and Chakrabortty [31], Fadali and Visioli [9], and Astrom and Wittenmark [4] for Chapter 9.[2] Naturally, the contents of this book overlap with the contents of these references, but it is expected that the concise and pedagogical treatment in this book will appeal to instructors and students.

---

[2]The version of a book used in the course might not be the more recent edition listed here.

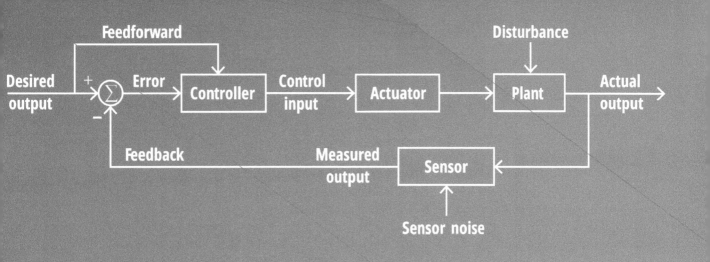

Chapter 2
Transfer Function Models

Chapter Contents

     Overview, 10
2-1 Linear Systems, 10
2-2 Examples, 14
2-3 Block Diagrams, 40
2-4 Linearization, 50
     Chapter Summary, 52
     Problems, 54

Objectives

Upon learning the material presented in this chapter, you should be able to:

1. Represent linear systems by differential equations or transfer functions.
2. Obtain a transfer function model from a differential equation model and vice versa.
3. Derive transfer functions of electric circuits formed of resistors, inductors, capacitors, and operational amplifiers.
4. Derive transfer functions of translation mechanical systems formed of masses, springs, and viscous dampers.
5. Derive transfer functions of rotational mechanical systems formed of rotating masses, torsional springs, viscous dampers, and gears.
6. Derive the transfer function of an armature-controlled dc motor.
7. Represent dynamical systems by block diagrams.
8. Simplify block diagrams.
9. Linearize a nonlinear system to obtain a linear model.

Overview

The first step in the analysis and design of a control system is the development of a mathematical model of the system. There are two forms of mathematical models of linear systems: transfer functions and state-space models. In all chapters except Chapter 8, we use transfer function models, which are the models used in classical control. State-space models are briefly introduced in Chapter 8. A mathematical model of a system can be derived from physical principles, or by performing experiments where certain inputs are applied to the system and then used together with the measured outputs to fit a model to the input-output data. The latter approach is known as *system identification*.[1] Even when the model is derived from physical principles, identification techniques might be used to estimate the parameters of the system.

The focus of this chapter is on transfer function models. In Section 2-1, we present the general form of differential equation models of linear systems and show how such models are converted into transfer functions using the Laplace transform. Section 2-2 gives examples of transfer functions derived from physical principles. Section 2-2.1 deals with electric circuits, Section 2-2.2 with mechanical systems, both translational and rotational, and Section 2-2.3 with electromechanical systems. A control system is typically formed of the connection of several components. A transfer function of the whole system can be derived from the transfer functions of its components using block diagram representation and reduction rules, which are discussed in Section 2-3. While our focus in this book is on linear systems, it is recognized that mathematical models of many physical systems are nonlinear. The justification for basing the analysis and design of control systems on linear models stems from the fact that many nonlinear systems can be described by linear models in a restricted regime of operation. In such cases, linear models are derived using *linearization*, which is briefly discussed in Section 2-4.

2-1 Linear Systems

A large class of systems can be represented by the linear constant-coefficient differential equation

$$y^{(n)}(t) + a_{n-1}y^{(n-1)}(t) + \cdots + a_1 y^{(1)}(t) + a_0 y(t) =$$
$$b_m u^{(m)}(t) + b_{m-1} u^{(m-1)}(t) + \cdots + b_1 u^{(1)}(t) + b_0 u(t) , \quad (2.1)$$

where $u(t)$ and $y(t)$ are the input and output of the system, respectively, $y^{(i)}(t)$ denotes the ith derivative of $y(t)$ with respect to the time variable t; that is, $y^{(i)}(t) = d^i y/dt^i$, n is a positive integer, m is a nonnegative integer with $m \leq n$, and a_0 to a_{n-1} and b_0 to b_m are constant coefficients. Later in the chapter we shall see several examples of physical systems that can be represented in the form Eq. (2.1). Because the coefficients a_0 to a_{n-1} and b_0 to b_m are constant; that is, independent of t, the system is time-invariant. Had some of these coefficients been dependent on t, the system would have been time-varying. We shall not consider time-varying systems in this book. A time-invariant system has the property that changing the

---

[1] See, for example, [26].

2-1. LINEAR SYSTEMS

time variable from t to $\tau = t + t_0$, for some constant t_0, will still lead to Eq. (2.1) except that the variables u, y, and their derivatives would be functions of τ. Therefore in studying time-invariant systems we can, without loss of generality, take the initial time to be zero.

The solution, $y(t)$, of Eq. (2.1) depends on the input $u(\tau)$, for $0 \leq \tau \leq t$, and the initial conditions $y(0)$ to $y^{(n-1)}(0)$. To focus on the input-output behavior of the system, the initial conditions are set to zero. The solution, $y(t)$, will then represent the response of the system to the input excitation $u(t)$. Note that the output y at time t depends on the input u up to time t. It does not depend on the input at future time greater than t. A system having this property is defined to be *causal*. In a noncausal system, the output at time t might depend on the input at time greater than t. The causality of Eq. (2.1) is a consequence of the fact that $m \leq n$. Systems that run in real time are causal by nature because in real time the future input will not be known.

An efficient way to represent the system defined by Eq. (2.1) is to use the Laplace transform to convert Eq. (2.1) into a transfer function. Taking the Laplace transform of both sides of Eq. (2.1), assuming zero initial conditions, yields

$$s^n Y(s) + a_{n-1} s^{n-1} Y(s) + \cdots + a_1 s Y(s) + a_0 Y(s) =$$
$$b_m s^m U(s) + b_{m-1} s^{m-1} U(s) + \cdots + b_1 s U(s) + b_0 U(s) \,. \tag{2.2}$$

Recognizing $Y(s)$ as a common factor on the left side and $U(s)$ as a common factor on the right side, we arrive at

$$(s^n + a_{n-1} s^{n-1} + \cdots + a_1 s + a_0) Y(s) = (b_m s^m + b_{m-1} s^{m-1} + \cdots + b_1 s + b_0) U(s) \,.$$

Hence,

$$\frac{Y(s)}{U(s)} = \frac{b_m s^m + b_{m-1} s^{m-1} + \cdots + b_1 s + b_0}{s^n + a_{n-1} s^{n-1} + \cdots + a_1 s + a_0} \,.$$

Next, we define the polynomials

$$N(s) = b_m s^m + b_{m-1} s^{m-1} + \cdots + b_1 s + b_0 \,, \tag{2.3a}$$
$$D(s) = s^n + a_{n-1} s^{n-1} + \cdots + a_1 s + a_0 \,, \tag{2.3b}$$

and their ratio as

$$G(s) = \frac{N(s)}{D(s)} \,.$$

Hence,

$$\frac{Y(s)}{U(s)} = G(s) \,. \tag{2.4}$$

$G(s)$ is called the *transfer function* of the system. It is the ratio of the Laplace transform of the output to the Laplace transform of the input when the initial conditions are zero. The transfer function $G(s)$ is equivalent to the differential equation Eq. (2.1). It is a rational function of s because it is the ratio of two polynomials. The order of the transfer function is the degree of its denominator polynomial. So if $n = 1$ we have a first-order transfer function; if $n = 2$

second-order, and so on. Because $m \leq n$, $G(s)$ is defined to be **proper**. If $m < n$, it is **strictly proper**. Defining
$$G(\infty) = \lim_{s \to \infty} G(s),$$
we see that $G(\infty) = 0$ for a strictly proper transfer function and $G(\infty) \neq 0$ if the transfer function is proper but not strictly proper. In the latter case with $m = n$, $G(\infty) = b_n$, and we can write $G(s)$ as
$$G(s) = G(\infty) + G_{\text{sp}}(s),$$
where the strictly proper transfer function $G_{\text{sp}}(s) = G(s) - G(\infty)$.

Example 2-1: Transfer Function of a System Represented by a Differential Equation

---

Find the transfer function of the system represented by the differential equation
$$\dddot{y} + 6\ddot{y} + 11\dot{y} + 6y = 2\dddot{u} + 16\ddot{u} + 22\dot{u} - 40u,$$
where $\dot{u} = u^{(1)}$, $\ddot{u} = u^{(2)}$, and so on. Determine if it is proper or strictly proper. If it is not strictly proper, find the strictly proper part.

Solution: The transfer function can be written by inspection as
$$G(s) = \frac{2s^3 + 16s^2 + 22s - 40}{s^3 + 6s^2 + 11s + 6}.$$

The third-order transfer function $G(s)$ is proper but not strictly proper because the numerator and denominator polynomials have the same degree. Also, $G(\infty) = 2$. Hence,
$$G(s) - G(\infty) = \frac{2s^3 + 16s^2 + 22s - 40}{s^3 + 6s^2 + 11s + 6} - 2$$
$$= \frac{2s^3 + 16s^2 + 22s - 40 - 2(s^3 + 6s^2 + 11s + 6)}{s^3 + 6s^2 + 11s + 6}$$
$$= \frac{4s^2 - 52}{s^3 + 6s^2 + 11s + 6} \stackrel{\text{def}}{=} G_{\text{sp}}(s).$$

$G_{\text{sp}}(s)$ is strictly proper because the degree of the numerator is smaller than the degree of the denominator. Hence, $G(s)$ can be expressed as
$$G(s) = 2 + \frac{4s^2 - 52}{s^3 + 6s^2 + 11s + 6}.$$

The denominator polynomial given by Eq. (2.3b), namely
$$s^n + a_{n-1}s^{n-1} + \cdots + a_1 s + a_0,$$

can be factored as

$$s^n + a_{n-1}s^{n-1} + \cdots + a_1 s + a_0 = (s - p_1)(s - p_2) \cdots (s - p_n) . \tag{2.5a}$$

Similarly, the numerator polynomial in Eq. (2.3a) can be factored as

$$b_m s^m + b_{m-1} s^{m-1} + \cdots + b_1 s + b_0 = b_m (s - z_1)(s - z_2) \cdots (s - z_m) . \tag{2.5b}$$

Thus, the transfer function $G(s)$ can be written as

$$G(s) = K \frac{(s - z_1)(s - z_2) \cdots (s - z_m)}{(s - p_1)(s - p_2) \cdots (s - p_n)} , \tag{2.6}$$

where $K = b_m$. The roots of the denominator polynomial p_1 to p_n are the **poles** of $G(s)$ and the roots of the numerator polynomial z_1 to z_m are its **zeros**. The roots of a polynomial with real coefficients can be complex, but complex roots will be in conjugate pairs. We shall see later on that the poles and zeros of a transfer function shape its response to various inputs.

Example 2-2: Poles and Zeros of a Transfer Function

Find the poles and zeros of the third-order transfer function of the previous example.

Solution: By factoring the numerator and denominator polynomials of $G(s)$ in Example 2-1, the transfer function can be written as

$$G(s) = 2 \frac{(s+4)(s+5)(s-1)}{(s+1)(s+2)(s+3)} .$$

The poles of $G(s)$ are located at -1, -2, and -3, while its zeros are located at -4, -5, and 1.

Time-Delay and Padé Approximations

Signals may be delayed, for example, due to communication time. A time-delay unit is an element whose output, $y(t)$, is delayed from its input, $u(t)$, by a delay time τ; that is,

$$y(t) = u(t - \tau) .$$

Taking the Laplace transform of both sides yields

$$\mathcal{L}\{y(t)\} = \mathcal{L}\{u(t - \tau)\} \quad \longrightarrow \quad Y(s) = e^{-\tau s} U(s) . \tag{2.7}$$

Therefore, the transfer function of the time-delay unit is $e^{-\tau s}$. This transfer function is not a rational function of s; that is, it is not a ratio of two polynomials in s. Time-domain analysis and design deal with rational functions of s. We shall see later on in Section 6-3 that frequency-domain analysis and design can deal directly with the transfer function $e^{-\tau s}$. For time domain analysis, the transfer function $e^{-\tau s}$ is approximated by a rational function of s. Frequently used approximations include the multiple versions of the Padé approximation:

- First-order Padé approximation:

$$e^{-\tau s} \approx \frac{1 - \tau s/2}{1 + \tau s/2} \, . \tag{2.8a}$$

- Second-order Padé approximation:

$$e^{-\tau s} \approx \frac{1 - \tau s/2 + (\tau s)^2/12}{1 + \tau s/2 + (\tau s)^2/12} \, . \tag{2.8b}$$

- Second-order lag:

$$e^{-\tau s} \approx \frac{1}{1 + \tau s + (\tau s)^2/2} \, . \tag{2.8c}$$

2-2 Examples

2-2.1 Electric Circuits

Deriving a mathematical model for an RLC (Resistor-Inductor-Capacitor) electric circuit starts from knowing the mathematical model of each component of the circuit and then using the laws that govern the connection of these components. **Figure 2-1** shows the basic components of the circuit.

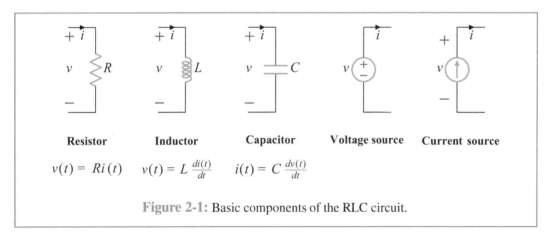

Figure 2-1: Basic components of the RLC circuit.

For each component, we have the voltage across the component, v, and the current through it, i. For the resistor, the mathematical model is given by Ohm's law, $v(t) = Ri(t)$, where R is the resistance. For the inductor, $v(t) = L(di(t)/dt)$, where L is the inductance. For the capacitor, $i(t) = C(dv(t)/dt)$, where C is the capacitance. It is important to distinguish between the resistor on one hand and the inductor and capacitor on the other hand. The relationship between v and i for the resistor is an algebraic one; that is, given the voltage $v(t)$ at any time t determines the current $i(t)$, and vice versa. There is no need to know the past history of v or i. This is different from the inductor and capacitor. Integrating the equation $v(t) = L[di(t)/dt]$

2-2. EXAMPLES

of the inductor over the period $[t_0, t_1]$ yields

$$L[i(t_1) - i(t_0)] = \int_{t_0}^{t_1} v(\tau) \, d\tau . \tag{2.9}$$

To determine the current at time t_1, it is not enough to know the voltage at t_1. We also need to know the initial current $i(t_0)$ and the voltage over the period $[t_0, t_1]$. Therefore, the inductor is said to be a dynamic component or a component with memory. Similarly, for the capacitor

$$C[v(t_1) - v(t_0)] = \int_{t_0}^{t_1} i(\tau) \, d\tau , \tag{2.10}$$

so the capacitor also is a dynamic component or component with memory. In contrast, the resistor is said to be a memoryless or static component. The remaining two components are the voltage and current sources. The voltage v of the voltage source is constant and independent of the circuit connection, which determines its current. The current i of the current source is constant and independent of the circuit connection, which determines its voltage.

The connection of the components of an electric circuit is governed by *Kirchhoff's voltage law* and *Kirchhoff's current law*. Both laws are illustrated in Fig. 2-2.

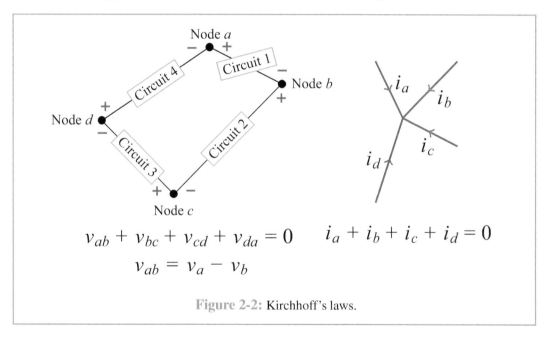

Figure 2-2: Kirchhoff's laws.

Kirchhoff's voltage law states that the sum of the voltage drops around a loop is zero, while the current law states that the sum of the currents coming into a node is zero. Both laws are algebraic equations where the signs of terms are changed if the variables are reversed. For example, if in the equation

$$v_{ab} + v_{bc} + v_{cd} + v_{da} = 0$$

the term $v_{da} = v_d - v_a$ is replaced by $v_{ad} = v_a - v_d$ the equation is written as

$$v_{ab} + v_{bc} + v_{cd} - v_{ad} = 0 .$$

Similarly, in the current law, if the direction of i_b is taken to be leaving the node, the equation is written as
$$i_a - i_b + i_c + i_d = 0 \ .$$

Example 2-3: Parallel RLC Circuit

Find the transfer function V/I of the circuit in Fig. 2-3. Assume $v(t) = 0$ for $t < 0$.

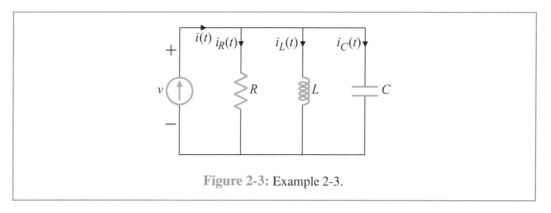

Figure 2-3: Example 2-3.

Solution: The voltage across all components is the same. Application of Kirchhoff's current law results in
$$i(t) = i_R(t) + i_L(t) + i_C(t) = \frac{v(t)}{R} + \frac{1}{L}\int_0^t v(\tau)\,d\tau + i_L(0) + C\frac{dv(t)}{dt} \ .$$

To cast this equation as a differential equation, differentiate both sides with respect to time:
$$\frac{di(t)}{dt} = \frac{1}{R}\frac{dv(t)}{dt} + \frac{1}{L}v(t) + C\frac{d^2v(t)}{dt^2} \ .$$

This equation can be put in the form of Eq. (2.1) with input i and output v by dividing by C:
$$\frac{d^2v(t)}{dt^2} + \frac{1}{CR}\frac{dv(t)}{dt} + \frac{1}{CL}v(t) = \frac{1}{C}\frac{di(t)}{dt} \ .$$

To find the transfer function, we take the Laplace transform assuming zero initial conditions, which gives
$$s^2 V(s) + \frac{1}{CR}sV(s) + \frac{1}{CL}V(s) = \frac{1}{C}sI(s) \ ,$$

leading to
$$G(s) = \frac{V(s)}{I(s)} = \frac{\frac{1}{C}s}{s^2 + \frac{1}{CR}s + \frac{1}{CL}} \ . \tag{2.11}$$

2-2. EXAMPLES

> Example 2-4: Series RLC Circuit

Find the transfer function I/V of the circuit in Fig. 2-4. Assume that no current was flowing through the loop prior to $t = 0$.

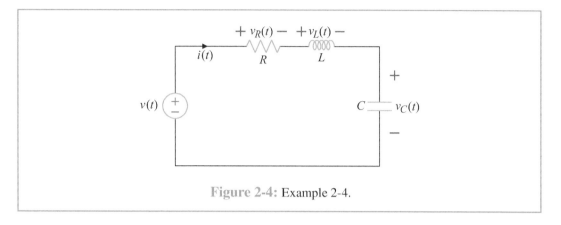

Figure 2-4: Example 2-4.

Solution: The current through all components is the same. Application of Kirchhoff's voltage law yields

$$v(t) = v_R(t) + v_L(t) + v_C(t)$$
$$= Ri(t) + L\frac{di(t)}{dt} + \frac{1}{C}\int_0^t i(\tau)\, d\tau + v_C(0)\, .$$

Differentiating this equation with respect to time and dividing by L, we arrive at the equation

$$\frac{d^2 i(t)}{dt^2} + \frac{R}{L}\frac{di(t)}{dt} + \frac{1}{CL}i(t) = \frac{1}{L}\frac{dv(t)}{dt}\, ,$$

which takes the form of Eq. (2.1) with input v and output i. The transfer function is then

$$G(s) = \frac{I(s)}{V(s)} = \frac{\frac{1}{L}s}{s^2 + \frac{R}{L}s + \frac{1}{CL}}\, . \qquad (2.12)$$

In the previous two examples we arrived at the transfer function of the circuit by first deriving a differential-equation model and then taking the Laplace transform. It is possible to arrive at the transfer function directly without writing a differential equation by working in the s-domain. For the R, L, and C components, take the Laplace transform of the v-i relationship, assuming zero initial conditions. This gives

$$V(s) = R\,I(s), \quad V(s) = Ls\,I(s), \quad I(s) = Cs\,V(s)\, . \qquad (2.13)$$

Using these relationships and applying Kirchhoff's laws in the s-domain, we can arrive at the transfer functions $G(s)$ given in Examples 2-3 and 2-4 without first writing the differential equation.

Example 2-5: Deriving a Transfer Function in the s-Domain

Reconsider the circuit of Fig. 2-3 and find the transfer function V/I by working in the s-domain.

Solution: Application of Kirchhoff's current law in the s-domain yields

$$I(s) = I_R(s) + I_L(s) + I_C(s)$$
$$= \frac{1}{R} V(s) + \frac{1}{Ls} V(s) + Cs\, V(s) = \left(\frac{1}{R} + \frac{1}{Ls} + Cs \right) V(s).$$

The transfer function is given by

$$G(s) = \frac{V(s)}{I(s)} = \frac{1}{\frac{1}{R} + \frac{1}{Ls} + Cs} = \frac{\frac{1}{C}s}{s^2 + \frac{1}{CR}s + \frac{1}{CL}},$$

which is identical to the expression for $G(s)$ given by Eq. (2.11) in Example 2-3.

Deriving the transfer function of electric circuits is simplified by using the concept of impedance. The impedance of an electric circuit with input voltage $V(s)$ and input current $I(s)$ is defined as

$$Z(s) = \frac{V(s)}{I(s)}. \tag{2.14}$$

For the R, L, and C components,

$$Z(s) = R \quad \text{for Resistor,} \tag{2.15a}$$
$$Z(s) = sL \quad \text{for Inductor,} \tag{2.15b}$$
$$Z(s) = \frac{1}{sC} \quad \text{for Capacitor.} \tag{2.15c}$$

A companion definition of admittance is given by

$$Y(s) = \frac{I(s)}{V(s)} = \frac{1}{Z(s)}. \tag{2.16}$$

2-2. EXAMPLES

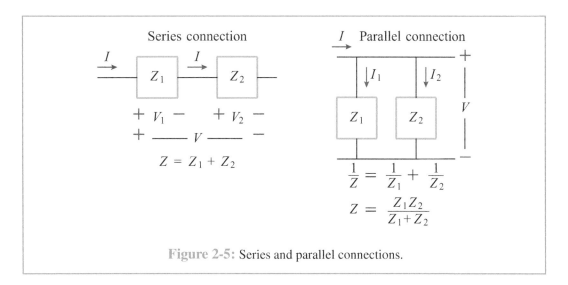

Figure 2-5: Series and parallel connections.

The use of impedance simplifies the calculation of transfer functions because we can replace certain configurations by their equivalent impedances. Two common configurations are shown in Fig. 2-5. For the series connection

$$V = V_1 + V_2 = Z_1 I + Z_2 I = (Z_1 + Z_2)I$$
$$\longrightarrow \quad Z = \frac{V}{I} = Z_1 + Z_2 \ . \tag{2.17a}$$

For the parallel connection

$$I = I_1 + I_2 = \frac{V}{Z_1} + \frac{V}{Z_2} = \left(\frac{1}{Z_1} + \frac{1}{Z_2}\right)V \quad \longrightarrow \quad \frac{I}{V} = \frac{1}{Z} = \frac{1}{Z_1} + \frac{1}{Z_2}$$
$$\longrightarrow \quad Z = \frac{Z_1 Z_2}{Z_1 + Z_2} \ . \tag{2.17b}$$

Another common configuration is the potential divider of Fig. 2-6.

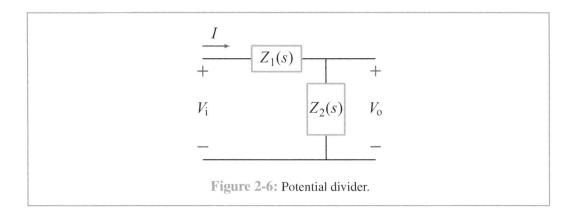

Figure 2-6: Potential divider.

The transfer function V_o/V_i of the potential divider is calculated as follows:

$$V_o = Z_2 I = Z_2 \frac{V_i}{Z_1 + Z_2}$$

$$\rightarrow \quad \frac{V_o}{V_i} = \frac{Z_2}{Z_1 + Z_2} \,. \tag{2.18}$$

Example 2-6: Potential Divider

Find V_o/V_i of the circuit shown in Fig. 2-7(a).

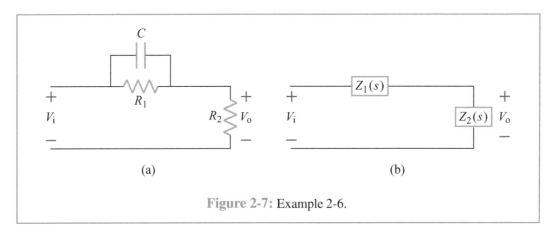

Figure 2-7: Example 2-6.

Solution: The circuit is redrawn in Fig. 2-7(b) as a potential divider with Z_1 representing the impedance of the parallel connection of R_1 and C while Z_2 represents the impedance of R_2. Hence,

$$Z_1 = \frac{R_1 \dfrac{1}{Cs}}{R_1 + \dfrac{1}{Cs}} = \frac{R_1}{CR_1 s + 1}\,,$$

$$Z_2 = R_2\,,$$

and

$$\frac{V_o}{V_i} = \frac{Z_2}{Z_1 + Z_2} = \frac{R_2}{\dfrac{R_1}{CR_1 s + 1} + R_2} = \frac{R_2(CR_1 s + 1)}{R_1 + R_2(CR_1 s + 1)} = \frac{s + \dfrac{1}{CR_1}}{s + \dfrac{R_1 + R_2}{CR_1 R_2}}\,.$$

2-2. EXAMPLES

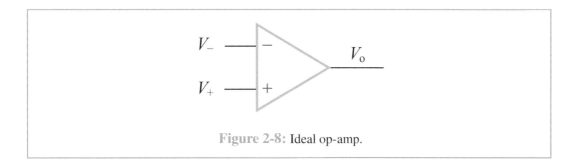

Figure 2-8: Ideal op-amp.

We now introduce the ideal *op-amp* (operational amplifier) as a new component in RLC circuits. The circuit symbol of the op-amp is shown in Fig. 2-8. It has two inputs, an inverting input with voltage V_- and a non-inverting input with voltage V_+. The output voltage is V_o. In an ideal op-amp, it is assumed that the input currents are zero and $V_- = V_+$. The output voltage V_o is determined solely by the connected circuit. In a practical op-amp, the input impedance is very high and the output is given by $V_o = K(V_+ - V_-)$, with very high gain K. By idealizing the high input impedance to infinity we have zero input currents and by idealizing the high gain K to infinity we end up with $V_- = V_+$. We shall deal with ideal op-amps only.

Example 2-7: Op-Amp Circuits

Find V_o/V_i of the circuits shown in Fig. 2-9.

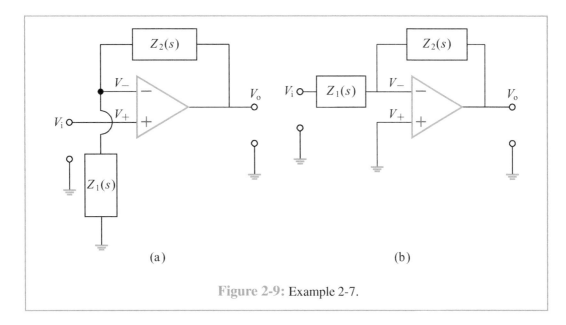

Figure 2-9: Example 2-7.

Solution: For circuit (a), $V_i(s) = V_+(s) = V_-(s)$. The current through Z_2 is the same as the current through Z_1 because the input current in the inverting input is zero.

$$\frac{V_i}{Z_1} = \frac{V_o - V_i}{Z_2} \quad \rightarrow \quad \left(\frac{1}{Z_1} + \frac{1}{Z_2}\right) V_i = \frac{V_o}{Z_2}$$

$$\rightarrow \quad \left(\frac{Z_2}{Z_1} + 1\right) V_i = V_o \quad \rightarrow \quad \frac{V_o}{V_i} = \frac{Z_1 + Z_2}{Z_1}. \tag{2.19a}$$

For circuit (b), $V_- = V_+ = 0$. The current through Z_1 is the same as the current through Z_2.

$$\frac{V_i}{Z_1} = \frac{-V_o}{Z_2} \quad \rightarrow \quad \frac{V_o}{V_i} = -\frac{Z_2}{Z_1}. \tag{2.19b}$$

Example 2-8: Op-Amp Circuits with R-C Components

Find V_o/V_i of the circuit shown in **Fig. 2-10**.

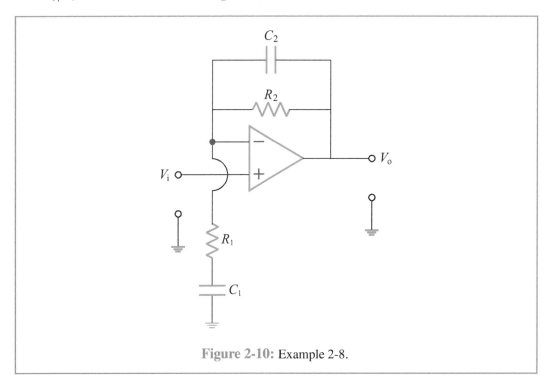

Figure 2-10: Example 2-8.

Solution: The circuit is in the form of the circuit of **Fig. 2-9(a)** with

$$Z_1 = R_1 + \frac{1}{C_1 s} = \frac{C_1 R_1 s + 1}{C_1 s}, \quad Z_2 = \frac{R_2 \dfrac{1}{C_2 s}}{R_2 + \dfrac{1}{C_2 s}} = \frac{R_2}{C_2 R_2 s + 1},$$

2-2. EXAMPLES

and

$$\frac{V_o}{V_i} = \frac{Z_1 + Z_2}{Z_1} = \frac{\dfrac{C_1 R_1 s + 1}{C_1 s} + \dfrac{R_2}{C_2 R_2 s + 1}}{\dfrac{C_1 R_1 s + 1}{C_1 s}} = \frac{(C_1 R_1 s + 1)(C_2 R_2 s + 1) + C_1 R_2 s}{(C_1 R_1 s + 1)(C_2 R_2 s + 1)}$$

$$= \frac{C_1 C_2 R_1 R_2 s^2 + (C_1 R_1 + C_1 R_2 + C_2 R_2) s + 1}{C_1 C_2 R_1 R_2 s^2 + (C_1 R_1 + C_2 R_2) s + 1}$$

$$= \frac{s^2 + \left(\dfrac{1}{C_1 R_1} + \dfrac{1}{C_2 R_1} + \dfrac{1}{C_2 R_2}\right) s + \dfrac{1}{C_1 C_2 R_1 R_2}}{s^2 + \left(\dfrac{1}{C_1 R_1} + \dfrac{1}{C_2 R_2}\right) s + \dfrac{1}{C_1 C_2 R_1 R_2}}.$$

Example 2-9: A Circuit with Two Op-Amps

Find V_o/V_i of the circuit shown in Fig. 2-11.

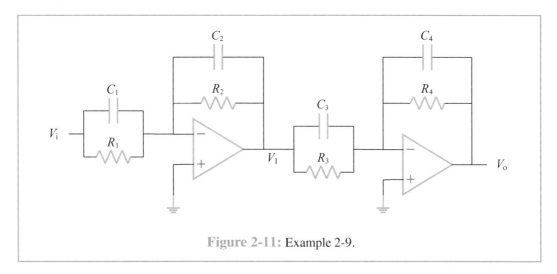

Figure 2-11: Example 2-9.

Solution: From V_i to V_1, the circuit takes the form of Fig. 2-9(b) with

$$Z_1 = \frac{R_1 \dfrac{1}{C_1 s}}{R_1 + \dfrac{1}{C_1 s}} = \frac{R_1}{C_1 R_1 s + 1} \quad \text{and} \quad Z_2 = \frac{R_2 \dfrac{1}{C_2 s}}{R_2 + \dfrac{1}{C_2 s}} = \frac{R_2}{C_2 R_2 s + 1}.$$

Hence,

$$\frac{V_1}{V_i} = -\frac{Z_2}{Z_1} = -\frac{R_2(C_1 R_1 s + 1)}{R_1(C_2 R_2 s + 1)}.$$

From V_1 to V_o, the circuit takes the form of Fig. 2-9(b) with

$$Z_1 = \frac{R_3 \frac{1}{C_3 s}}{R_3 + \frac{1}{C_3 s}} = \frac{R_3}{C_3 R_3 s + 1} \quad \text{and} \quad Z_2 = \frac{R_4 \frac{1}{C_4 s}}{R_4 + \frac{1}{C_4 s}} = \frac{R_4}{C_4 R_4 s + 1}.$$

Hence,

$$\frac{V_o}{V_1} = -\frac{Z_2}{Z_1} = -\frac{R_4(C_3 R_3 s + 1)}{R_3(C_4 R_4 s + 1)}.$$

Thus

$$\frac{V_o}{V_i} = \frac{V_o}{V_1}\frac{V_1}{V_i} = \left(\frac{R_2(C_1 R_1 s + 1)}{R_1(C_2 R_2 s + 1)}\right)\left(\frac{R_4(C_3 R_3 s + 1)}{R_3(C_4 R_4 s + 1)}\right)$$

$$= \frac{C_1 C_3}{C_2 C_4}\left(\frac{s + \frac{1}{C_1 R_1}}{s + \frac{1}{C_2 R_2}}\right)\left(\frac{s + \frac{1}{C_3 R_3}}{s + \frac{1}{C_4 R_4}}\right). \quad (2.20)$$

2-2.2 Mechanical Systems

Translational Mechanical Systems

The three basic components of linear translational mechanical systems are the **spring**, **viscous damper**, and **mass**, whose graphical symbols are shown in Fig. 2-12. The mathematical model of the spring is determined by Hook's law, which states that the displacement (or elongation) of the spring $x(t)$ from its unstretched position and the force $f(t)$ applied to stretch the spring, are related by $f(t) = Kx(t)$, where K is the spring constant. When a force $f(t)$ is applied as shown in the figure, there is an equal and opposite restoring force in the spring. The viscous damper models friction or resistance to the motion of a body, such as moving a body on a surface with friction. There are different forms of friction, but the simplest is viscous friction where the resisting force $f(t)$ is proportional to the velocity of the moving body; that is, $f(t) = b\dot{x}(t)$, where b is the damping constant or friction coefficient. The velocity $\dot{x}(t)$ is in

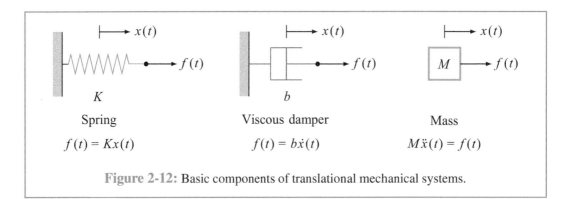

Figure 2-12: Basic components of translational mechanical systems.

2-2. EXAMPLES

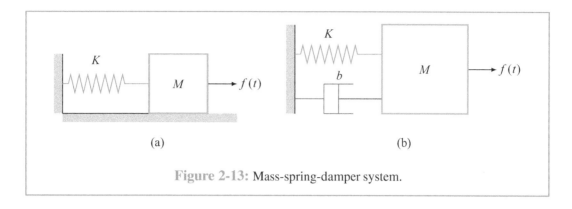

Figure 2-13: Mass-spring-damper system.

the same direction as the displacement $x(t)$. Such a resisting force is equal and opposite to the applied force. The motion of a mass M obeys Newton's second law of motion, which states that the rate of change of momentum equals the net force applied to the mass, or simply, the net force = mass × acceleration; $f(t) = M\ddot{x}(t)$. The acceleration $\ddot{x}(t)$ has the same direction as the displacement $x(t)$. In calculating the net force, all forces in the direction of x are added with a positive sign while those opposite to x are added with a negative sign.

Writing down the equation of motion for a mechanical system starts by drawing a free-body diagram for each mass and identifying all the forces acting on it. We illustrate this process by considering the system shown in Fig. 2-13(a), where a mass M is connected to a fixed surface by a spring with spring constant K. A force $f(t)$ is applied to the mass, which moves on a surface with friction. The system is redrawn in Fig. 2-13(b) where the surface friction is modeled by a viscous damper with damping constant b. The free-body diagram is shown in Fig. 2-14(a). The forces acting on the mass due to the spring and damper are opposite to the direction of motion. By *Newton's law*,

$$M\ddot{x}(t) = f(t) - b\dot{x}(t) - Kx(t) . \tag{2.21a}$$

The transfer function X/F can be found by taking the Laplace transform assuming zero initial conditions. That is,

$$Ms^2 X(s) = F(s) - bsX(s) - KX(s) , \tag{2.21b}$$

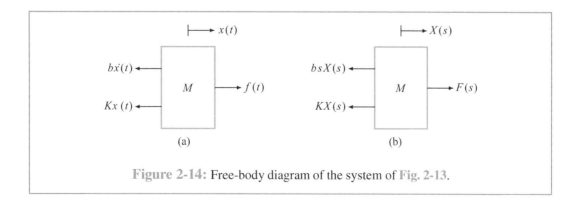

Figure 2-14: Free-body diagram of the system of Fig. 2-13.

which leads to

$$\frac{X}{F} = \frac{1}{Ms^2 + bs + K} = \frac{1/M}{s^2 + (b/M)s + K/M} \ . \quad (2.22)$$

Following the same logic we employed earlier with electrical circuits, we can arrive at the transfer function of the mechanical system without writing the differential equation. To do so we draw the free-body diagram with all variables in the s-domain, as in **Fig. 2-14(b)**. Newton's second law is applied in the s-domain by taking the Laplace transform of $M\ddot{x}(t) = f_{net}(t)$, assuming zero initial conditions, to obtain $Ms^2X(s) = F_{net}(s)$, where $F_{net}(s)$ is the net force. Applying this procedure to the free-body diagram of **Fig. 2-14(b)** we arrive at

$$Ms^2X(s) = F(s) - bsX(s) - KX(s) \ . \quad (2.23)$$

The motion of the mass in **Fig. 2-13** is in the horizontal direction, so its weight does not appear in Newton's law. What about vertical motion? Do we have to include the forces due to gravity in calculating the net force? To answer this question, consider the system shown in **Fig. 2-15(a)**, where the mass M is in vertical motion, connected to a fixed surface by a spring and a damper and subject to an applied force $f(t)$. The free-body diagram is shown in **Fig. 2-15(b)**. The force due to the spring, $Kx(t)$, and the force due to the damper, $b\dot{x}(t)$, are opposite to the direction of the displacement $x(t)$, which is measured from the fixed surface, while the force $f(t)$ and the weight Mg are along $x(t)$. Application of Newton's law yields

$$M\ddot{x}(t) = f(t) + Mg - b\dot{x}(t) - Kx(t) \ . \quad (2.24)$$

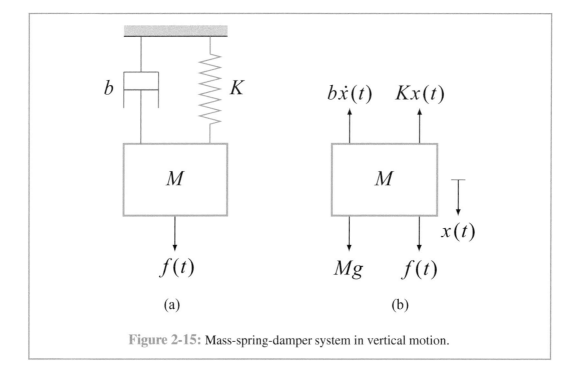

Figure 2-15: Mass-spring-damper system in vertical motion.

When $f(t) = 0$, the system is in equilibrium where x is a constant. To find the equilibrium position, set $f = \dot{x} = \ddot{x} = 0$ in the foregoing equation to obtain

$$0 = Mg - K\bar{x} . \tag{2.25}$$

where $\bar{x}$ is the equilibrium position. Subtracting Eq. (2.25) from Eq. (2.24), and setting $x_\delta = x - \bar{x}$, we obtain

$$M\ddot{x}_\delta(t) = f(t) - b\dot{x}_\delta(t) - Kx_\delta(t) . \tag{2.26}$$

In arriving at this equation we used the fact that the derivatives of the constant $\bar{x}$ are zero. Thus, when the displacement is measured from the equilibrium position, the weight Mg does not appear in the equation of motion. In dealing with vertical motion we will not include forces due to gravity, with the understanding that displacements are defined from equilibrium positions.

Example 2-10: A Mass-Spring-Damper System

Find X_2/F of the mechanical system shown in Fig. 2-16.

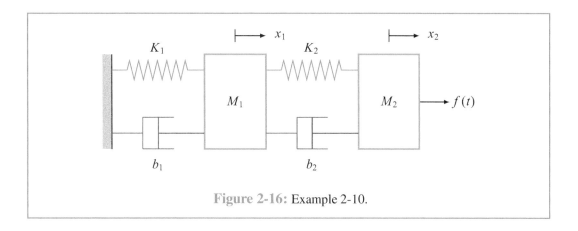

Figure 2-16: Example 2-10.

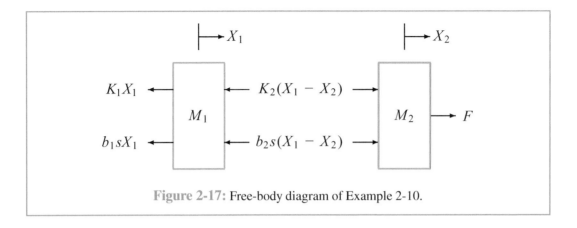

Figure 2-17: Free-body diagram of Example 2-10.

Solution: The equivalent free-body diagram of the system in Fig. 2-16 is shown in Fig. 2-17, where all variables are expressed in the *s*-domain. Note that the forces due to the spring and damper are always opposite to the direction of the displacement. For a spring connected between two masses, the force at one end is equal and opposite to the force at the other end and proportional to the difference between the displacements of the two masses. In Fig. 2-17 this force is written as $K_2(X_1 - X_2)$. Because the force acting on M_1 is opposing motion, the direction of the force is opposite to the direction of X_1. Immediately the force acting on M_2 is in the opposite direction. The arrowhead of the force acting on M_2 is in the same direction as X_2, yet this force is still opposing the motion because the force depends on $-X_2$. Had we written the force as $K_2(X_2 - X_1)$, the arrowheads on both ends would have been reversed. A similar discussion applies to a damper connected between two masses. Writing Newton's second law for the two masses results in

$$M_1 s^2 X_1(s) = -K_1 X_1(s) - b_1 s X_1(s) - K_2[X_1(s) - X_2(s)] - b_2 s[X_1(s) - X_2(s)] ,$$

$$M_2 s^2 X_2(s) = K_2[X_1(s) - X_2(s)] + b_2 s[X_1(s) - X_2(s)] + F(s) .$$

These two equations can be written as a matrix equation.[2]

$$\begin{bmatrix} M_1 s^2 + (b_1 + b_2)s + (K_1 + K_2) & -(b_2 s + K_2) \\ -(b_2 s + K_2) & M_2 s^2 + b_2 s + K_2 \end{bmatrix} \begin{bmatrix} X_1(s) \\ X_2(s) \end{bmatrix} = \begin{bmatrix} 0 \\ F(s) \end{bmatrix} ,$$

which yields

$$\begin{bmatrix} X_1(s) \\ X_2(s) \end{bmatrix} = \frac{1}{\Delta(s)} \begin{bmatrix} M_2 s^2 + b_2 s + K_2 & b_2 s + K_2 \\ b_2 s + K_2 & M_1 s^2 + (b_1 + b_2)s + (K_1 + K_2) \end{bmatrix} \begin{bmatrix} 0 \\ F(s) \end{bmatrix} ,$$

with

$$\Delta(s) = [M_1 s^2 + (b_1 + b_2)s + (K_1 + K_2)][M_2 s^2 + b_2 s + K_2] - (b_2 s + K_2)^2 .$$

---

[2] The inverse of a 2×2 matrix is given by

$$\begin{bmatrix} a & b \\ c & d \end{bmatrix}^{-1} = \frac{1}{ad - bc} \begin{bmatrix} d & -b \\ -c & a \end{bmatrix} ,$$

2-2. EXAMPLES

Hence,

$$\frac{X_2}{F} = \frac{M_1 s^2 + (b_1 + b_2)s + (K_1 + K_2)}{\Delta(s)}.$$

Example 2-11: Automobile Suspension

Figure 2-18 shows an automobile suspension system. A quarter-car is modeled by the mass-spring-damper system shown in the figure. The mass of the body (called sprung mass) is M_s, the mass of the wheel (called unsprung mass) is M_{us}, the suspension spring has spring constant k_s, the shock absorber is modeled by a damper with damping constant C_s, the tire is modeled by a spring with spring constant k_t, and the road surface displacement from a fixed level is S. The displacements of the sprung and unsprung masses are y_s and y_{us}, respectively. The purpose of the suspension system is to reduce the effect of the road-surface deviations on the passengers. Studies have shown that passengers are most affected by the acceleration of the sprung mass $a_s = \ddot{y}_s$. Find the transfer function A_s/S, where A_s is the Laplace transform of a_s.

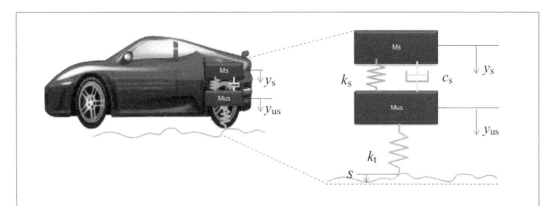

Figure 2-18: Automobile suspension (photo downloaded with permission from http://www.sharetechnote.com).

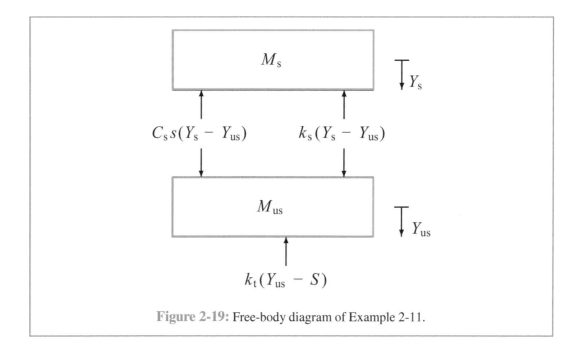

Figure 2-19: Free-body diagram of Example 2-11.

Solution: The free-body diagram of the system is shown in Fig. 2-19, in which variables are expressed in the *s*-domain. Application of Newton's second law to the two masses results in

$$M_s s^2 Y_s = -k_s(Y_s - Y_{us}) - C_s s(Y_s - Y_{us})$$
$$M_{us} s^2 Y_{us} = k_s(Y_s - Y_{us}) + C_s s(Y_s - Y_{us}) - k_t(Y_{us} - S),$$

$$\begin{bmatrix} M_s s^2 + C_s s + k_s & -(C_s s + k_s) \\ -(C_s s + k_s) & M_{us} s^2 + C_s s + (k_s + k_t) \end{bmatrix} \begin{bmatrix} Y_s \\ Y_{us} \end{bmatrix} = \begin{bmatrix} 0 \\ k_t S \end{bmatrix},$$

which leads to

$$\begin{bmatrix} Y_s \\ Y_{us} \end{bmatrix} = \frac{1}{\Delta(s)} \begin{bmatrix} M_{us} s^2 + C_s s + (k_s + k_t) & C_s s + k_s \\ C_s s + k_s & M_s s^2 + C_s s + k_s \end{bmatrix} \begin{bmatrix} 0 \\ k_t S \end{bmatrix},$$

with

$$\Delta(s) = [M_s s^2 + C_s s + k_s][M_{us} s^2 + C_s s + (k_s + k_t)] - (C_s s + k_s)^2.$$

Hence,

$$\frac{Y_s}{S} = \frac{(C_s s + k_s)k_t}{\Delta(s)} \quad \longrightarrow \quad \frac{A_s}{S} = \frac{s^2 Y_s}{S} = \frac{s^2(C_s s + k_s)k_t}{\Delta(s)}.$$

2-2. EXAMPLES

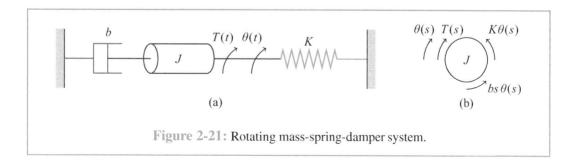

Figure 2-20: Basic components of rotational mechanical systems.

Rotational Mechanical Systems

The three basic components of linear rotational mechanical systems are the ***torsional spring***, ***viscous damper***, and ***rotational mass***, whose graphical symbols are shown in Fig. 2-20. When a torque $T(t)$ is applied to a torsional spring it rotates with an angle $\theta(t)$ according to the relation $T(t) = K\theta(t)$, where K is the spring constant. When the torque $T(t)$ is applied as shown in the figure, there is an equal and opposite restoring torque in the spring. The viscous damper models friction or resistance to rotational motion. A viscous damper is modeled by $T(t) = b\dot{\theta}(t)$, where $T(t)$ is the applied torque, which is equal and opposite to the resisting torque of the damper, and b is the damping constant. The motion of a rotating mass obeys Newton's second law for rotational motion: $T(t) = J\ddot{\theta}(t)$, where $T(t)$ is the net torque and J is the moment of inertia of the mass. The angular velocity $\dot{\theta}$ and angular acceleration $\ddot{\theta}$ have the same direction as θ. In calculating the net torque, all torques in the direction of θ are added with a positive sign while those whose directions are opposite to θ are added with a negative sign.

The first step in modeling a rotational mechanical system is to draw a free-body diagram for each rotating mass. We illustrate this step by considering the system shown in Fig. 2-21(a), whose free-body diagram is shown in Fig. 2-21(b) with all variables in the s-domain. The torques due to the spring and the damper are opposite to the direction of θ because they oppose the rotation. Newton's law yields

$$Js^2\,\theta(s) = -bs\,\theta(s) - K\,\theta(s) + T(s),$$

Figure 2-21: Rotating mass-spring-damper system.

which yields

$$\frac{\theta}{T} = \frac{1}{Js^2+bs+K} = \frac{1/J}{s^2+\frac{b}{J}s+\frac{K}{J}}. \qquad (2.27)$$

Example 2-12: Torsional Shaft

Figure 2-22(a) shows a torque applied at the left end of a shaft with torsion. The shaft is mounted on bearings at both ends. The torsional shaft is modeled in Fig. 2-22(b) as two rotating masses connected by a spring. The friction in the bearings is modeled by dampers. Find the transfer function θ_2/T.

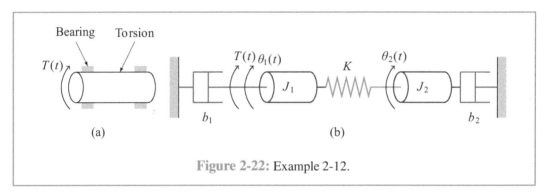

Figure 2-22: Example 2-12.

Solution: The free-body diagrams of the two rotating masses are shown in Fig. 2-23, where the variables are expressed in the s-domain. Two torques are opposing the motion of J_1. The torque due to the damper b_1 is $b_1 s\theta_1$ and the torque due to the spring K, which is connected between J_1 and J_2, is $K(\theta_1 - \theta_2)$. Two torques are acting on J_2. The torque $K(\theta_1 - \theta_2)$, due to the spring, is equal and opposite to the torque acting on J_1, while the torque $b_2 s\theta_2$ due to the damper b_2 is opposite to the direction of θ_2. The equations of motions are

$$J_1 s^2 \theta_1(s) = -b_1 s\theta_1(s) - K[\theta_1(s) - \theta_2(s)] + T(s),$$

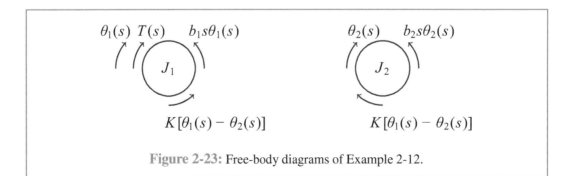

Figure 2-23: Free-body diagrams of Example 2-12.

2-2. EXAMPLES

$$J_2 s^2 \theta_2(s) = K[\theta_1(s) - \theta_2(s)] - b_2 s \theta_2(s) ,$$

which can be combined into the matrix equation

$$\begin{bmatrix} J_1 s^2 + b_1 s + K & -K \\ -K & J_2 s^2 + b_2 s + K \end{bmatrix} \begin{bmatrix} \theta_1(s) \\ \theta_2(s) \end{bmatrix} = \begin{bmatrix} T(s) \\ 0 \end{bmatrix}$$

to yield the solution:

$$\begin{bmatrix} \theta_1(s) \\ \theta_2(s) \end{bmatrix} = \frac{1}{\Delta(s)} \begin{bmatrix} J_2 s^2 + b_2 s + K & K \\ K & J_1 s^2 + b_1 s + K \end{bmatrix} \begin{bmatrix} T(s) \\ 0 \end{bmatrix} ,$$

where

$$\Delta(s) = (J_1 s^2 + b_1 s + K)(J_2 s^2 + b_2 s + K) - K^2 .$$

The solution for θ_2/T is

$$\frac{\theta_2}{T} = \frac{K}{\Delta(s)} . \tag{2.28}$$

Example 2-13: Rotating Mass-Torsional Spring-Damper System

Find the transfer function θ_3/T of the system shown in Fig. 2-24.

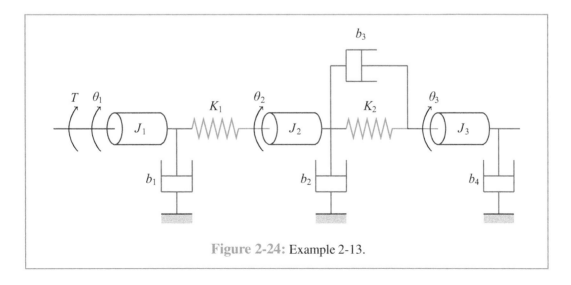

Figure 2-24: Example 2-13.

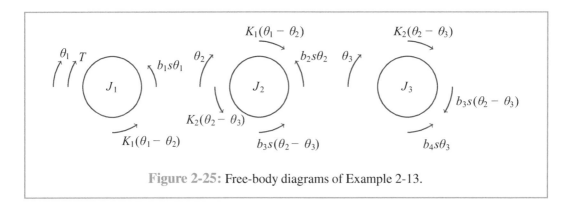

Figure 2-25: Free-body diagrams of Example 2-13.

Solution: The free-body diagrams of the three rotating masses are shown in Fig. 2-25, where all variables are in the s-domain. Three torques are acting on J_1, the applied torque T, the torque due to the damper $b_1 s \theta_1$ and the torque due to the spring $K_1(\theta_1 - \theta_2)$. The latter two torques are opposite to the direction of θ_1. Four torques are acting on J_2. The torque $K_1(\theta_1 - \theta_2)$ due to the spring K_1 is equal and opposite to its direction in the diagram of J_1. The torques due to the damper b_2, the damper b_3 and the spring K_2 are all opposite to the direction of θ_2 and are given by $b_2 s \theta_2$, $b_3 s (\theta_2 - \theta_3)$, and $K_2(\theta_2 - \theta_3)$, respectively. In the free-body diagram of J_3, the torques due to the spring and damper connected between J_2 and J_3 are equal and opposite to their directions in the diagram of J_2. The torque due to the damper b_4 is $b_4 s \theta_3$ and is opposite to the direction of θ_3. The equations of motion are given by

$$J_1 s^2 \theta_1 = -b_1 s \theta_1 - K_1(\theta_1 - \theta_2) + T ,$$

$$J_2 s^2 \theta_2 = -b_2 s \theta_2 - b_3 s (\theta_2 - \theta_3) + K_1(\theta_1 - \theta_2) - K_2(\theta_2 - \theta_3) ,$$

$$J_3 s^2 \theta_3 = b_3 s (\theta_2 - \theta_3) - b_4 s \theta_3 + K_2(\theta_2 - \theta_3) ,$$

which together form the matrix equation

$$\begin{bmatrix} J_1 s^2 + b_1 s + K_1 & -K_1 & 0 \\ -K_1 & J_2 s^2 + (b_2 + b_3)s + K_1 + K_2 & -(b_3 s + K_2) \\ 0 & -(b_3 s + K_2) & J_3 s^2 + (b_3 + b_4)s + K_2 \end{bmatrix} \begin{bmatrix} \theta_1 \\ \theta_2 \\ \theta_3 \end{bmatrix} = \begin{bmatrix} T \\ 0 \\ 0 \end{bmatrix} .$$

Hence,

$$\begin{bmatrix} \theta_1 \\ \theta_2 \\ \theta_3 \end{bmatrix} = \frac{1}{\Delta(s)} \begin{bmatrix} \star & \star & \star \\ \star & \star & \star \\ K_1(b_3 s + K_2) & \star & \star \end{bmatrix} \begin{bmatrix} T \\ 0 \\ 0 \end{bmatrix} ,$$

where $\Delta(s)$ is the determinant of the 3×3 matrix and $\star$ denotes a value that is not needed in calculating the transfer function. The solution for θ_3/T is

$$\frac{\theta_3}{T} = \frac{K_1(b_3 s + K_2)}{\Delta(s)} . \tag{2.29}$$

Gears

An important component of rotational mechanical systems is the *gear*, which translates motion between two rotating shafts. A schematic diagram representation of the gear assembly is shown in Fig. 2-26.

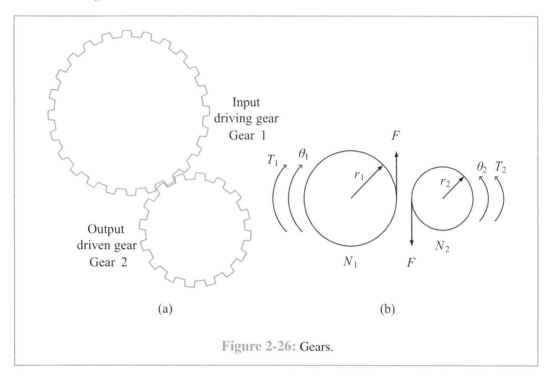

Figure 2-26: Gears.

The input gear, Gear 1, has radius r_1, N_1 teeth, and rotates an angle θ_1 clockwise. The output gear, Gear 2, has radius r_2, N_2 teeth, and rotates an angle θ_2 counterclockwise. Gear 1 applies torque T_2 to Gear 2, which reacts with torque T_1. We assume ideal gears.[3] At the point of contact, the arc lengths at both gears must be the same:

$$r_1 \theta_1 = r_2 \theta_2 .$$

Since the angular speed $\omega = \dot{\theta}$,

$$r_1 \omega_1 = r_2 \omega_2 .$$

The number of teeth on each gear is proportional to its radius. Hence

$$\frac{\theta_2}{\theta_1} = \frac{\omega_2}{\omega_1} = \frac{r_1}{r_2} = \frac{N_1}{N_2} \stackrel{\text{def}}{=} n \text{ (gear ratio)} .$$

At the point of contact, the force applied from Gear 1 to Gear 2 is equal to the reaction force from Gear 2 to Gear 1. Since Torque = Force × Radius,

$$\frac{T_1}{r_1} = \frac{T_2}{r_2} .$$

---

[3] Ideal gear analysis ignores backlash and deadzone nonlinearities.

Therefore,
$$\frac{T_2}{T_1} = \frac{r_2}{r_1} = \frac{N_2}{N_1} = \frac{1}{n}.$$

Hence,
$$\begin{cases} n < 1 & \rightarrow \quad \text{speed reduction and torque increase,} \\ n > 1 & \rightarrow \quad \text{speed increase and torque reduction.} \end{cases}$$

Example 2-14: A Motor Connected to a Load through Gears

Figure 2-27 shows a schematic diagram of a motor connected to a load through gears. Both the motor and the load are modeled as rotating masses with viscous damping. Find the transfer functions θ_m/T_m and θ_L/T_m.

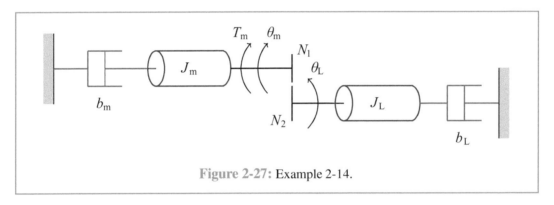

Figure 2-27: Example 2-14.

Solution: The free-body diagrams of the motor and load are shown in Fig. 2-28. Let us start on the load side. The torque applied on the load from the motor is T_2, which causes the load to rotate an angle θ_L in the same direction as T_2. This motion is opposed by the torque due to the load damper, $b_L s \theta_L$. On the motor side, the torque induced by the motor T_m rotates the motor

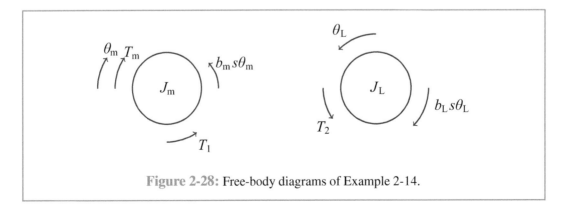

Figure 2-28: Free-body diagrams of Example 2-14.

2-2. EXAMPLES

shaft with angle θ_m in the same direction of T_m. This rotation is opposed by the torque due to the motor damper, $b_m s \theta_m$, and the reaction torque of the gear T_1, where $T_2/T_1 = N_2/N_1$. Recall also that $\theta_L/\theta_m = N_1/N_2$. The equation of motion of the load is

$$J_L s^2 \theta_L = T_2 - b_L s \theta_L ,$$

and the equation of motion of the motor is

$$J_m s^2 \theta_m = T_m - b_m s \theta_m - T_1 .$$

Substituting

$$T_2 = (J_L s^2 + b_L s)\theta_L = (J_L s^2 + b_L s)\frac{N_1}{N_2}\theta_m$$

from the load equation into the motor equation and using $T_2/T_1 = N_2/N_1$ yields

$$J_m s^2 \theta_m = T_m - b_m s \theta_m - \frac{N_1}{N_2}(J_L s^2 + b_L s)\frac{N_1}{N_2}\theta_m ,$$

which can be rearranged as

$$(J_m + n^2 J_L)s^2 \theta_m + (b_m + n^2 b_L)s \theta_m = T_m$$

where $n = N_1/N_2$. By defining the equivalent moment of inertia $J_e = J_m + n^2 J_L$ and the equivalent damping constant $b_e = b_m + n^2 b_L$, we have

$$J_e s^2 \theta_m + b_e s \theta_m = T_m$$

and

$$\frac{\theta_m}{T_m} = \frac{1}{s(J_e s + b_e)} , \quad (2.30a)$$

$$\frac{\theta_L}{T_m} = \frac{n}{s(J_e s + b_e)} . \quad (2.30b)$$

2-2.3 Electromechanical Systems

Electromechanical systems have electrical and mechanical components, which are modeled using the techniques discussed in the previous two subsections. We illustrate such systems by the dc (direct current) motor.

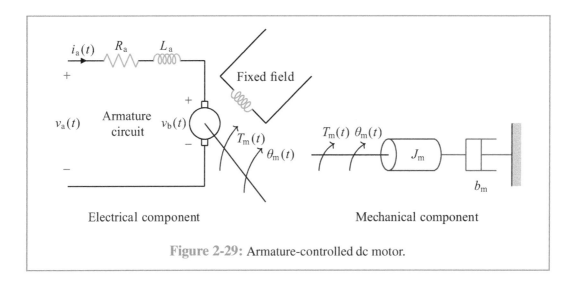

Figure 2-29: Armature-controlled dc motor.

Armature-Controlled dc Motor

A schematic diagram of the dc motor is shown in Fig. 2-29. The motor has two circuits, *field* and *armature*. The magnetic flux in the field circuit is kept constant, while the motor is controlled through the armature circuit, which has inductance L_a and resistance R_a. The input voltage is v_a and the current in the armature circuit is i_a. The torque induced by the motor is given by $T_m = K_t i_a$, where K_t is the torque constant. When the motor's shaft rotates, a back e.m.f. (electromotive force) $v_b = K_b \omega_m = K_b \dot{\theta}_m$ is induced in the armature circuit, where K_b is the back e.m.f. constant. The constants K_t and K_b have different units but in an MKS unit system they are numerically equal. The motor's shaft is modeled by moment of inertia J_m and damping constant b_m. Application of Kirchhoff's voltage law to the armature circuit yields

$$V_a(s) = R_a I_a(s) + L_a s I_a(s) + V_b(s) = R_a I_a(s) + L_a s I_a(s) + K_b \omega_m(s).$$

Therefore,

$$I_a(s) = \frac{V_a(s) - K_b \omega_m(s)}{L_a s + R_a}, \qquad T_m(s) = K_t I_a(s) = \frac{K_t [V_a(s) - K_b \omega_m(s)]}{L_a s + R_a}.$$

Application of Newton's second law to the mechanical part yields

$$T_m(s) = J_m s^2\, \theta_m(s) + b_m s\, \theta_m(s) = J_m s\, \omega_m(s) + b_m\, \omega_m(s).$$

Hence,

$$(J_m s + b_m)\omega_m(s) = \frac{K_t[V_a(s) - K_b\, \omega_m(s)]}{L_a s + R_a},$$

$$[(J_m s + b_m)(L_a s + R_a) + K_t K_b]\omega_m(s) = K_t\, V_a(s),$$

which leads to

$$\frac{\omega_m}{V_a} = \frac{K_t}{(J_m s + b_m)(L_a s + R_a) + K_t K_b}. \qquad (2.31a)$$

2-2. EXAMPLES

The numerical values of L_a and b_m are typically small relative to the other parameters. A simplified model of the motor is obtained by setting $L_a = b_m = 0$.

$$\frac{\omega_m}{V_a} = \frac{K_t}{J_m R_a s + K_t K_b} = \frac{1/K_b}{\frac{J_m R_a}{K_t K_b} s + 1} \, . \tag{2.31b}$$

It is common practice to write this simplified transfer function as

$$\frac{\omega_m}{V_a} = \frac{K_m}{\tau_m s + 1}, \tag{2.31c}$$

where

$$K_m = \frac{1}{K_b} \quad \text{and} \quad \tau_m = \frac{J_m R_a}{K_t K_b}.$$

Given the moment of inertia J_m, the motor constant K_m and its time constant τ_m can be calculated from the steady-state torque-speed characteristics, given by the equation

$$T_m = \frac{K_t}{R_a} (v_a - K_b \omega_m), \tag{2.32}$$

which is sketched in Fig. 2-30 for each of two values of the input voltage v_a. The intersections of the line with the axes define the no-load speed and stall torque as

$$\omega_{\text{no-load}} = \frac{v_a}{K_b}, \tag{2.33a}$$

$$T_{\text{stall}} = \frac{K_t v_a}{R_a} \, . \tag{2.33b}$$

The back e.m.f. constant K_b is determined using Eq. (2.33a), and since K_t and K_b are numerically equal, Eq. (2.33b) is used to determine R_a. Then, using J_m, τ_m is calculated.

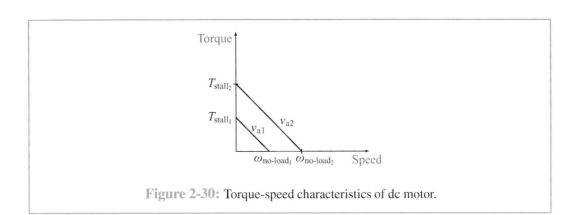

Figure 2-30: Torque-speed characteristics of dc motor.

2-3 Block Diagrams

The modeling techniques described in the previous section allow us to derive the transfer function of a component of a system. Practical systems are composed of several components and there is need to develop a systematic way to establish the transfer function of such systems. The block-diagram representation is one such technique.[4] In block diagrams, a transfer function $Y(s) = G(s)\,U(s)$ is represented by a block with input signal U and output signal Y, as shown in Fig. 2-31. Connecting transfer functions uses the elements of summing junctions and pick-off point, examples of which are shown in Fig. 2-32.

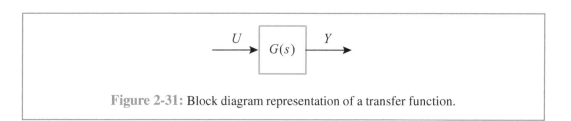

Figure 2-31: Block diagram representation of a transfer function.

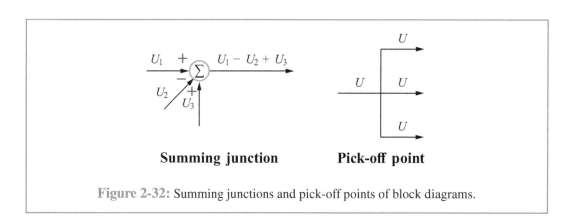

Figure 2-32: Summing junctions and pick-off points of block diagrams.

To illustrate the construction of a block diagram, consider the dc motor from the previous section. It is modeled by the equations

$$I_a = \frac{V_a - K_b \omega_m}{L_a s + R_a}, \qquad T_m = K_t I_a, \qquad \omega_m = \frac{T_m}{J_m s + b_m}, \qquad \theta_m = \frac{\omega_m}{s}.$$

---

[4]Other techniques are graphical methods such as signal flow graphs and bond graphs.

2-3. BLOCK DIAGRAMS

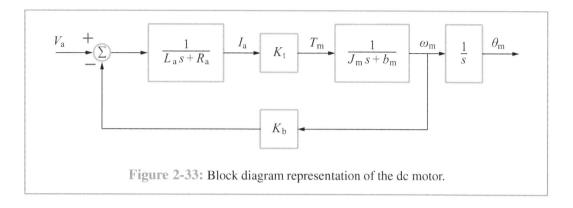

Figure 2-33: Block diagram representation of the dc motor.

These equations can be represented by the block diagram in Fig. 2-33. We will come back to this block diagram to see how to find the transfer function connecting the input to the output.

Reducing a block diagram to a single transfer function uses reduction rules for three basic types of connections, as shown in Fig. 2-34. For the series connection, $Y = G_3(s) X_2$, $X_2 = G_2(s) X_1$, and $X_1 = G_1(s) U$. Hence

$$Y = G_3(s) X_2 = G_3(s) G_2(s) X_1 = G_3(s) G_2(s) G_1(s) U . \qquad (2.34)$$

For the parallel connection, $X_1 = G_1(s) U$, $X_2 = G_2(s) U$, and $X_3 = G_3(s) U$. Hence

$$Y = X_1 + X_2 + X_3 = G_1(s) U + G_2(s) U + G_3(s) U = [G_1(s) + G_2(s) + G_3(s)] U . \qquad (2.35)$$

For the negative feedback connection, $Y = G_1(s) E$, $E = U - F$, and $F = G_2(s) Y$. Hence

$$Y = G_1(s) E = G_1(s) (U - F) = G_1(s) [U - G_2(s) Y] ,$$

$$[1 + G_1(s) G_2(s)] Y = G_1(s) U ,$$

which leads to

$$Y = \frac{G_1(s)}{1 + G_1(s) G_2(s)} U . \qquad (2.36)$$

Similarly, for the positive feedback connection,

$$Y = \frac{G_1(s)}{1 - G_1(s) G_2(s)} U . \qquad (2.37)$$

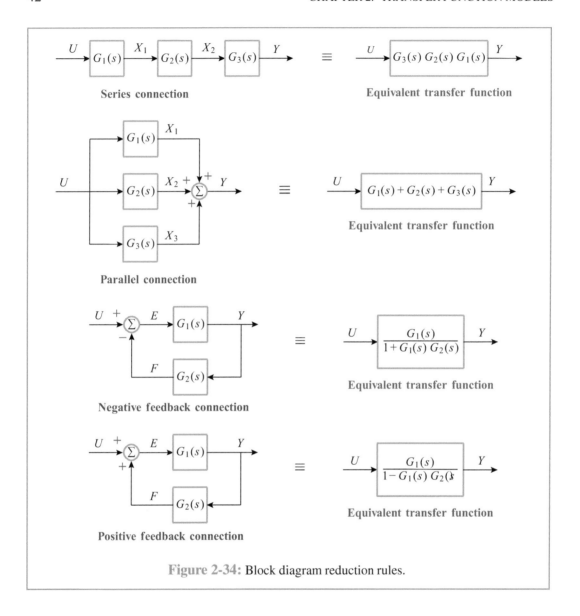

Figure 2-34: Block diagram reduction rules.

Example 2-15: dc Motor

Find the transfer functions ω_m/V_a and θ_m/V_a from the block diagram of the dc motor in Fig. 2-33.

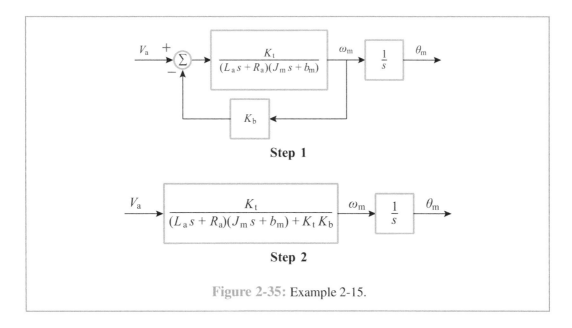

Figure 2-35: Example 2-15.

Solution: The block diagram is reduced in two steps, as shown in Fig. 2-35. In the first step, the series connection of the three transfer functions in the forward path of the loop are replaced by their equivalent transfer function. In the second step, the feedback connection is replaced by its equivalent transfer function:

$$\frac{\left[\dfrac{K_t}{(L_a s + R_a)(J_m s + b_m)}\right]}{\left[1 + \dfrac{K_b K_t}{(L_a s + R_a)(J_m s + b_m)}\right]} = \frac{K_t}{(L_a s + R_a)(J_m s + b_m) + K_t K_b} \; .$$

Hence,

$$\frac{\omega_m}{V_a} = \frac{K_t}{(L_a s + R_a)(J_m s + b_m) + K_t K_b} \; . \qquad (2.38)$$

The series connection of this transfer function with $\frac{1}{s}$ results in

$$\frac{\theta_m}{V_a} = \frac{K_t}{s[(L_a s + R_a)(J_m s + b_m) + K_t K_b]} \; . \qquad (2.39)$$

To reduce more complicated block diagrams we need additional rules, which are shown in **Figs. 2-36** to **2-38**.

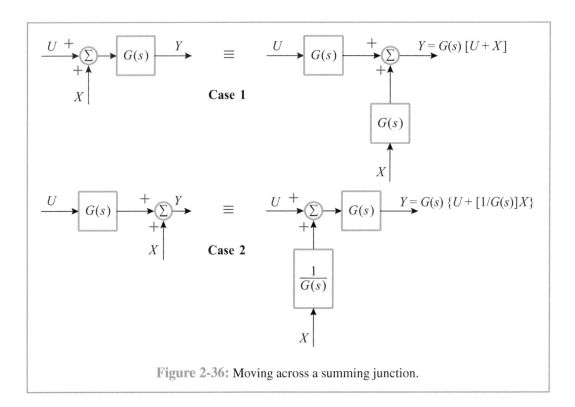

Figure 2-36: Moving across a summing junction.

In Case 1 of **Fig. 2-36**,

$$Y = G(s)\,(U+X) = G(s)\,U + G(s)\,X\ .$$

In Case 2 of the same figure,

$$Y = G(s)\,U + X = G(s)\left[U + \frac{1}{G(s)}\,X\right]\ .$$

2-3. BLOCK DIAGRAMS

In Case 1 of Fig. 2-37,

$$Y_2 = U = \frac{1}{G(s)} G(s) U ,$$

and the same for Y_3. In Case 2 of the same figure, $Y_i = G(s) U$ in the two equivalent representations.

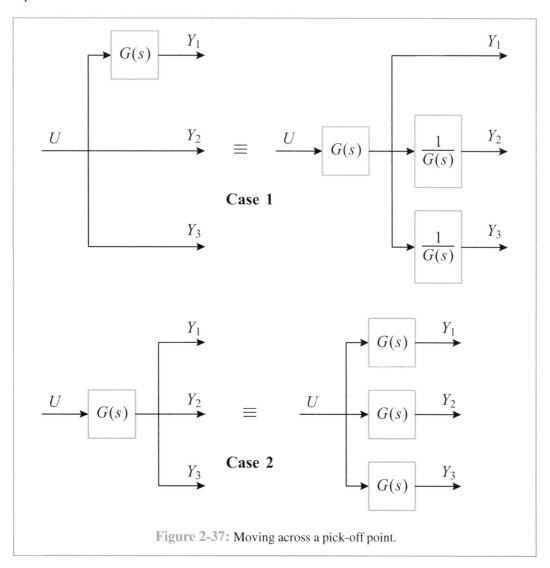

Figure 2-37: Moving across a pick-off point.

Finally, Fig. 2-38 shows equivalent ways of drawing block diagrams. In Case 1,

$$Y = X_1 + X_2 - X_3 = (X_1 + X_2) - X_3 \ .$$

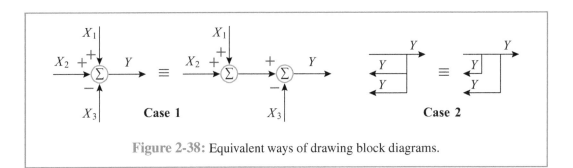

Figure 2-38: Equivalent ways of drawing block diagrams.

Example 2-16: Block Diagram Reduction

Find the transfer function Y/U of the block diagram in Fig. 2-39.

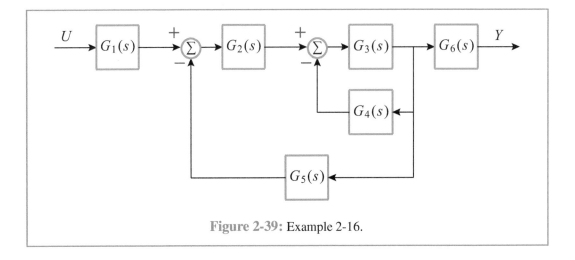

Figure 2-39: Example 2-16.

Solution: The block diagram reduction is realized in four steps in Fig. 2-40. In Step 1, a part of the block diagram is redrawn to separate the two feedback signals from the output of G_3. In Step 2, the feedback connection of G_3 and G_4 is replaced by its equivalent. In Step 3, the

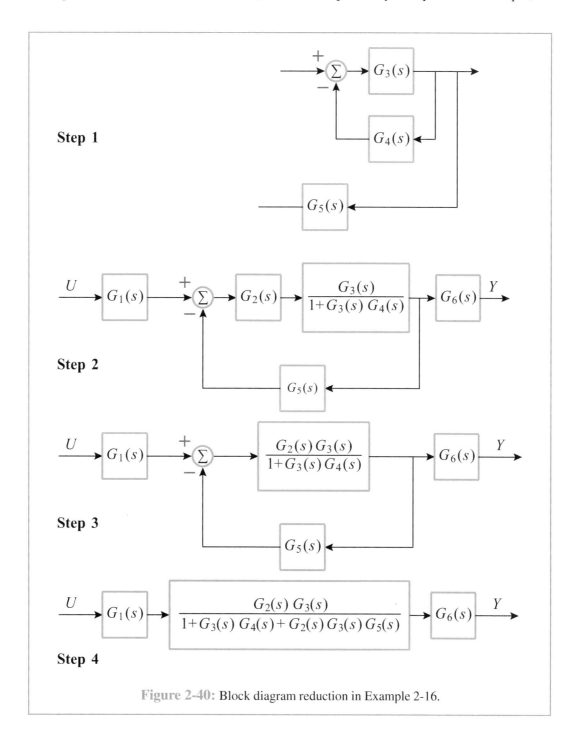

Figure 2-40: Block diagram reduction in Example 2-16.

series connection of the two transfer functions in the forward path of the loop is replaced by its equivalent. In Step 4, the feedback loop is replaced by its equivalent. At this point we have a series connection of three transfer functions. Hence,

$$\frac{Y}{U} = \frac{G_1(s)\,G_2(s)\,G_3(s)\,G_6(s)}{1 + G_3(s)\,G_4(s) + G_2(s)\,G_3(s)\,G_5(s)}.$$

Example 2-17: Block Diagram Reduction

Find the transfer function Y/U of the block diagram in Fig. 2-41.

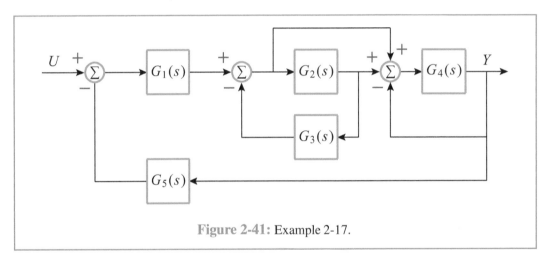

Figure 2-41: Example 2-17.

Solution: The block diagram reduction is accomplished in three steps in Fig. 2-42. In Step 1, the forward signal from the input of G_2 is moved to its output, the summation junction at the input of G_4 is split into two summation junctions, and the feedback from Y is redrawn as two feedback signals. In Step 2, two feedback connections are replaced by their equivalents and the parallel connection of $1/G_2$ and one is repalced by its equivalent. In Step 3, the series connection of the four transfer functions in the forward path of the loop is replaced by its equivalent. In Step 4, the feedback loop is replaced by its equivalent. Thus,

$$\frac{Y}{U} = \frac{G_1(s)[1+G_2(s)]G_4(s)}{[1+G_2(s)\,G_3(s)][1+G_4(s)] + G_1(s)[1+G_2(s)]\,G_4(s)\,G_5(s)}.$$

2-3. BLOCK DIAGRAMS

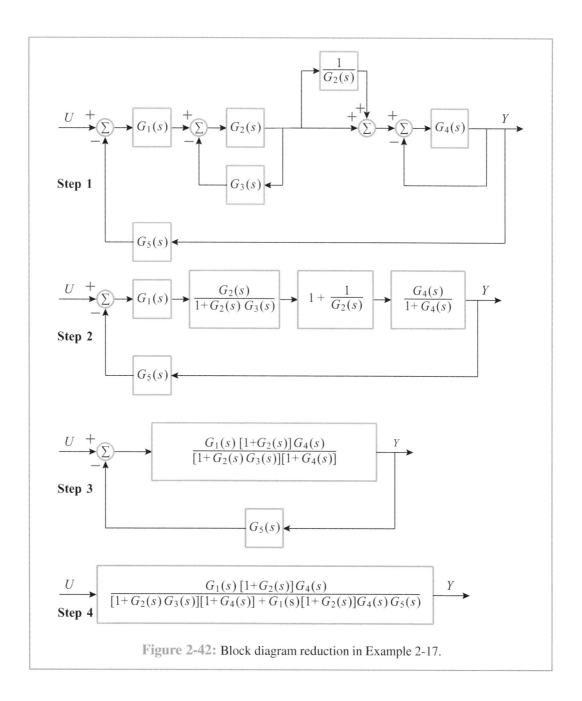

Figure 2-42: Block diagram reduction in Example 2-17.

2-4 Linearization

Throughout the book we deal with the control of linear systems. However, many physical systems are nonlinear. How can we then focus our attention on the control of only linear systems? The answer lies in the fact that many nonlinear systems can be modeled by linear equations in the neighborhood of an operating point or an equilibrium point. Figure 2-43 shows the v–i characteristics of a nonlinear resistor. In the neighborhood of the operating point (v_0, i_0), the nonlinear curve can be approximated by its slope at the operating point. Similarly, many physical components that might have nonlinear characteristics over a wide operation range can be approximated by a linear relationship in the neighborhood of an operating point. This process is known as linearization.

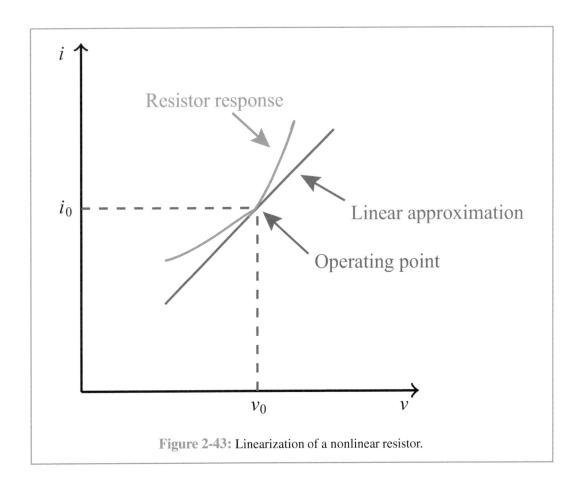

Figure 2-43: Linearization of a nonlinear resistor.

Another form of linearization takes place when a system operates near an equilibrium point. We illustrate this by the following example.

2-4. LINEARIZATION

Example 2-18: Linearization of the Pendulum Equation

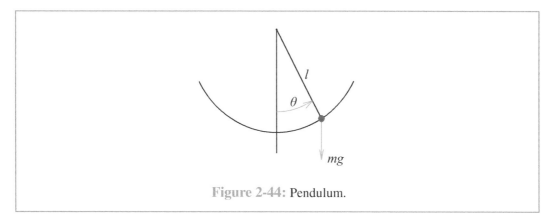

Figure 2-44: Pendulum.

Consider the simple pendulum shown in Fig. 2-44, where l denotes the length of the rod and m denotes the mass of the bob. Assume the rod is rigid and has zero mass. Let θ denote the angle subtended by the rod and the vertical axis through the pivot point. The pendulum is free to swing in the vertical plane. The bob of the pendulum moves in a circle of radius l. To write the equation of motion of the pendulum, let us identify the forces acting on the bob. There is a downward gravitational force equal to mg, where g is the acceleration due to gravity. There is also a frictional force resisting the motion, which we assume to be proportional to the speed of the bob with a coefficient of friction k. Using Newton's second law of motion, we can write the equation of motion in the tangential direction as

$$ml\ddot{\theta}(t) = -mg\sin\theta(t) - kl\dot{\theta}(t) \;. \tag{2.40}$$

Writing the equation of motion in the tangential direction has the advantage that the rod tension, which is in the normal direction, does not appear in the equation. The system is in equilibrium when θ is constant. To find the equilibrium points, set $\dot{\theta} = \ddot{\theta} = 0$ in the foregoing equation to obtain $\sin\theta = 0$. Hence the pendulum has equilibrium points at $\theta = 0, \pm\pi, \pm 2\pi, \ldots$. Of the infinitely many equilibrium points, two have physical meaning: the downward position where $\theta = 0$ and the upward position where $\theta = \pi$, while all the others are repetitions of them. If we are interested in the motion near $\theta = 0$, we can use the approximation $\sin\theta \approx \theta$ to arrive at the linear differential equation

$$ml\ddot{\theta}(t) = -mg\theta(t) - kl\dot{\theta}(t) \;.$$

If we are interested in the motion near $\theta = \pi$, we write $\theta = \pi + \phi$ and use the approximation

$$\sin\theta = \sin(\pi + \phi) = -\sin\phi \approx -\phi$$

to arrive at the linear equation

$$ml\ddot{\phi}(t) = mg\phi(t) - kl\dot{\phi}(t) \;,$$

where we used the fact that $\dot{\theta} = \dot{\phi}$ and $\ddot{\theta} = \ddot{\phi}$.

Summary

Concepts

- Differential-equation and transfer-function models are equivalent.
- Transfer-function models of electrical, mechanical, and electromechanical systems can be derived by application of physical laws in the Laplace-domain.
- Transfer-function models of electric circuits can be derived using impedances.
- Block diagrams can be used to represent systems that are formed from the interconnection of several components.
- Block-diagram reduction rules can be used to simplify a block diagram to a single transfer function.
- A nonlinear system can be approximated by a linear system in the neighborhood of an operating point.

Important Terms Provide definitions or explain the meaning of the following terms:

dynamic component	pick-off point	strictly proper transfer function
free-body diagram	poles	summing junction
gear ratio	potential divider	time delay
linearization	proper transfer function	torque-speed characteristics
op-amp	rational transfer function	transfer function
Padé approximation	static component	zeros

Mathematical Models

Transfer functions:

Rational $\quad G(s) = \dfrac{N(s)}{D(s)}$

$\qquad\qquad N(s)$ and $D(s)$ are polynomials

Time-delay $\quad G(s) = e^{-Ts}$

s-domain relations of electric circuits:

Resistor	$V(s) = R\,I(s)$
Inductor	$V(s) = Ls\,I(s)$
Capacitor	$I(s) = Cs\,V(s)$
Kirchhoff's voltage law	$V_{ab} + V_{bc} + V_{cd} + V_{da} = 0$
Kirchhoff's current law	$I_a + I_b + I_c + I_d = 0$

s-domain relations of translational mechanical systems:

Mass	$F(s) = Ms^2\,X(s)$
Spring	$F(s) = K\,X(s)$
Viscous damper	$F(s) = bs\,X(s)$
Newton's second law	$Ms^2\,X(s) = $ net force

s-domain relations of rotational mechanical systems:

Rotating mass	$F(s) = Js^2\,\theta(s)$
Torsional spring	$T(s) = K\,\theta(s)$
Viscous damper	$T(s) = bs\,\theta(s)$
Newton's second law	$Js^2\,\theta(s) = $ net torque
Gear	$\dfrac{\omega_2}{\omega_1} = n,\quad \dfrac{T_2}{T_1} = \dfrac{1}{n}$

Block diagram reduction rules:

Series connection	$G = G_1 G_2 G_3$
Parallel connection	$G = G_1 + G_2 + G_3$
Negative feedback	$\dfrac{G_1}{1 + G_1 G_2}$
Positive feedback	$\dfrac{G_1}{1 - G_1 G_2}$

PROBLEMS

2.1 For each of the following differential equations, find the transfer function Y/U. Determine if the transfer function is proper or strictly proper. If it is not strictly proper, find the strictly proper part.

(a) $y^{(3)} = -3y^{(2)} - 3y^{(1)} - 2y + u^{(2)} - u$

(b) $y^{(3)} = -3.5y^{(2)} - 3.5y^{(1)} - y + u^{(3)} - 3.5u^{(2)} + 3.5u^{(1)} + 3u$

(c) $y^{(4)} = -8y - 14y^{(1)} - 15y^{(2)} - 7y^{(3)} + 4u + 4u^{(1)} + 3u^{(2)} + u^{(3)}$

(d) $\dot{y} - \dot{u} = -y - u$

2.2 For each of the following transfer functions, $G(s) = Y(s)/U(s)$, find the differential equation that relates the input $u(t)$ to the output $y(t)$.

(a) $G(s) = \dfrac{(s+2)(s+3)}{(s+1)(s+4)}$

(b) $G(s) = \dfrac{(s^2+0.4s+1.04)(s+3)}{(s^2+0.2s+1)(s+2)(s+4)}$

(c) $G(s) = 1 + \dfrac{s-1}{(s+1)(s+2)}$

(d) $G(s) = \dfrac{a}{s^2 - a}$

2.3 Find the poles and zeros of each of the transfer functions of the previous problem.

2.4 Find the transfer function V/I of the circuits in Fig. P2.4.

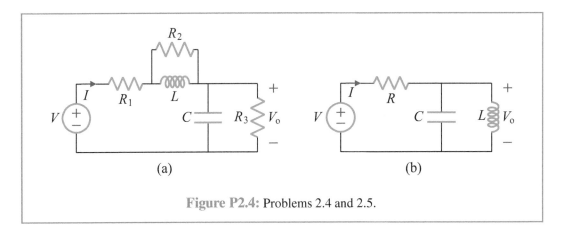

Figure P2.4: Problems 2.4 and 2.5.

2.5 Find the transfer function V_o/V of the circuits in Fig. P2.4.

PROBLEMS

2.6 Consider the circuit of Fig. P2.6. Show that

$$I_1 = I \frac{Z_2(s)}{Z_1(s)+Z_2(s)} \quad \text{and} \quad I_2 = I \frac{Z_1(s)}{Z_1(s)+Z_2(s)}$$

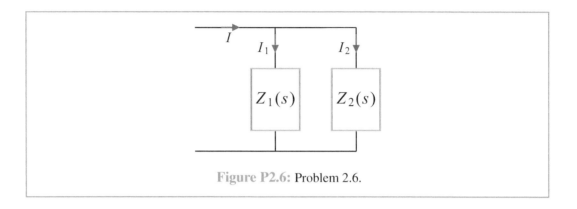

Figure P2.6: Problem 2.6.

2.7 Find the transfer function I/V of the circuit in Fig. P2.7.

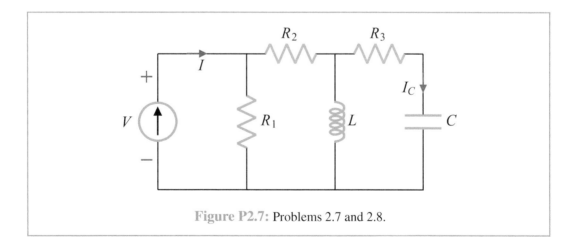

Figure P2.7: Problems 2.7 and 2.8.

2.8 Find the transfer function I_c/I of the circuit in Fig. P2.7.

2.9 Find the transfer function V_o/V_i of the circuits in Fig. P2.9.

2.10 Find the transfer function V_o/V_i of the circuits in Fig. P2.10.

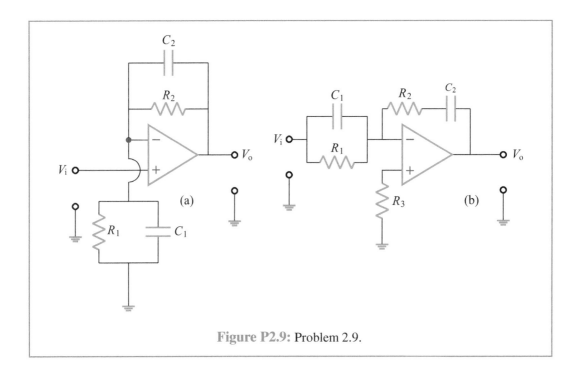

Figure P2.9: Problem 2.9.

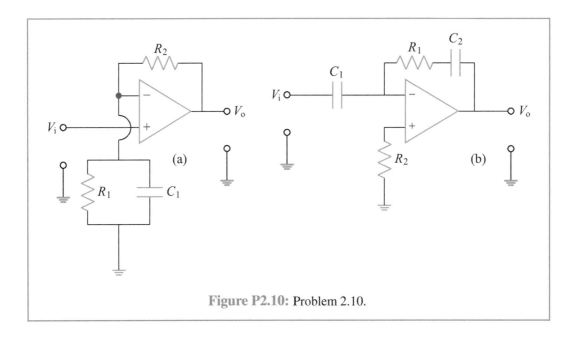

Figure P2.10: Problem 2.10.

2.11 Find the transfer function X_2/F of the mechanical system of Fig. P2.11.

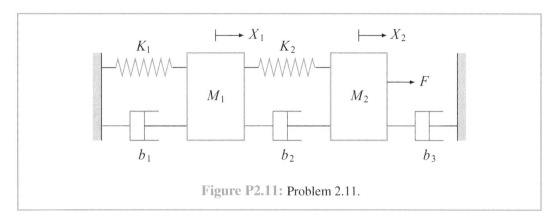

Figure P2.11: Problem 2.11.

2.12 Find the transfer function X_2/F of the mechanical system of Fig. P2.12.

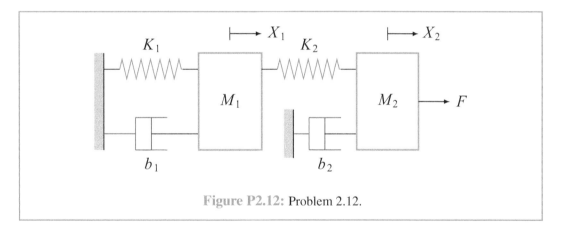

Figure P2.12: Problem 2.12.

2.13 Find the transfer function X_2/F of the mechanical system of Fig. P2.13.

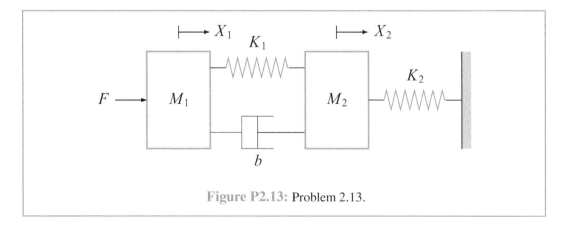

Figure P2.13: Problem 2.13.

2.14 Find the transfer function X_2/F of the mechanical system of Fig. P2.14.

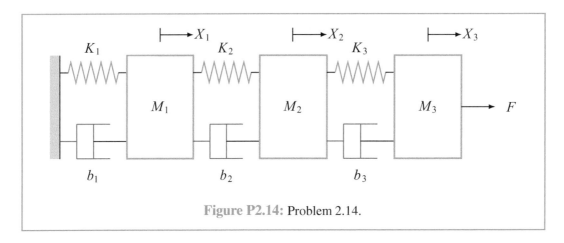

Figure P2.14: Problem 2.14.

2.15 Find the transfer function X_2/F of the mechanical system of Fig. P2.15.

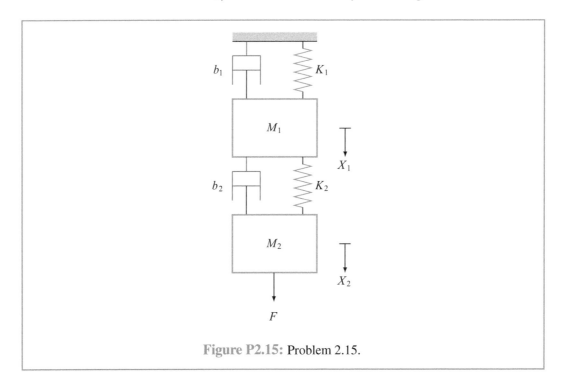

Figure P2.15: Problem 2.15.

2.16 In the automobile suspension model of Example 2-11, the tire is modeled by a spring with spring constant k_t. In other models the damping of the tire is modeled by a viscous damper in parallel with the spring, with damping constant C_t. Find the transfer function A_s/S of this model.

PROBLEMS

2.17 Find the transfer function θ_2/T of the mechanical system of Fig. P2.17.

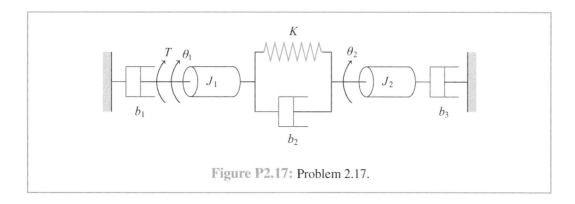

Figure P2.17: Problem 2.17.

2.18 Find the transfer function θ_2/T of the mechanical system of Fig. P2.18.

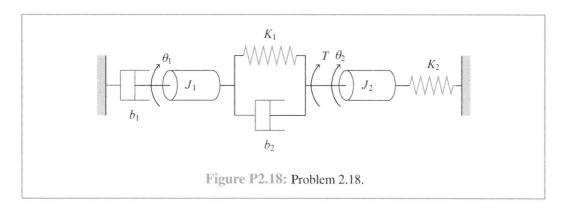

Figure P2.18: Problem 2.18.

2.19 Find the transfer function θ_2/T of the mechanical system of Fig. P2.19.

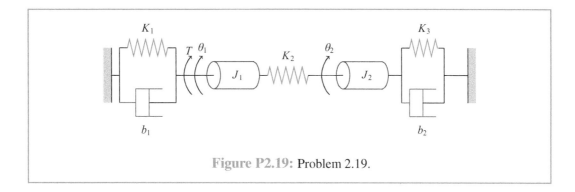

Figure P2.19: Problem 2.19.

2.20 Find the transfer function θ_3/T of the mechanical system of Fig. P2.20.

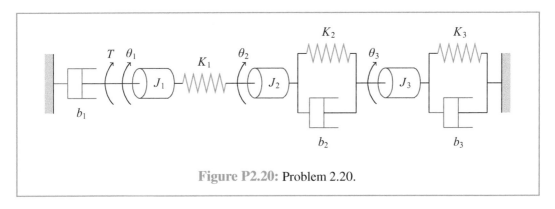

Figure P2.20: Problem 2.20.

2.21 Figure P2.21 shows a motor connected to a load through two gears. Find the transfer function θ_L/T_m in terms of $n_1 = N_1/N_2$ and $n_2 = N_3/N_4$.

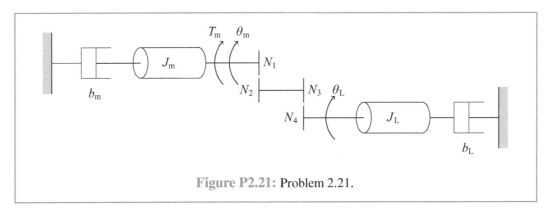

Figure P2.21: Problem 2.21.

2.22 Figure P2.22 shows a motor connected to a flexible load through gears. The flexible load is modeled as two rotating masses connected by a spring. Find the transfer function θ_2/T_m.

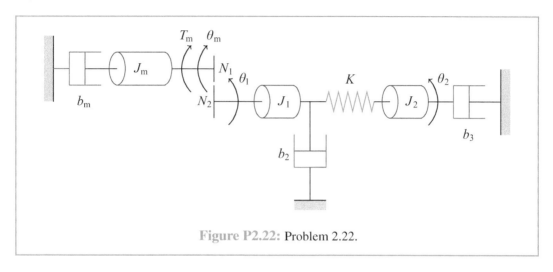

Figure P2.22: Problem 2.22.

2.23 The torque-speed characteristics of a dc motor have no-load speed = 9760 rpm and stall torque = 222 mN-m at nominal voltage = 15 V. The motor's moment of inertia is 10 g-cm^2. Find the constants K_m (in rad/sec/volt) and τ_m (in sec).

2.24 Repeat Problem 2.23 for no-load speed = 9970 rpm, stall torque = 246 mN-m, nominal voltage = 30 V, and moment of inertia = 10.1 g-cm^2.

2.25 Repeat Problem 2.23 for no-load speed = 8320 rpm, stall torque = 207 mN-m, nominal voltage = 48 V, and moment of inertia = 9.96 g-cm^2.

2.26 For each of the systems in Fig. P2.26, find the transfer function Y/U.

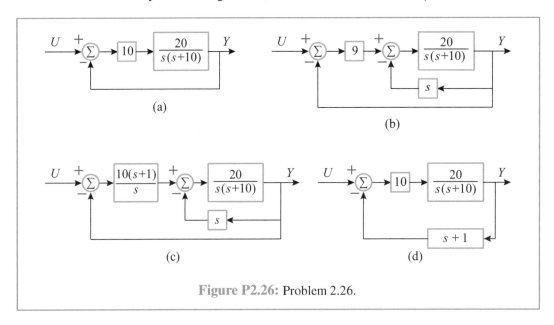

Figure P2.26: Problem 2.26.

2.27 Find the transfer function Y/U for the system in Fig. P2.27.

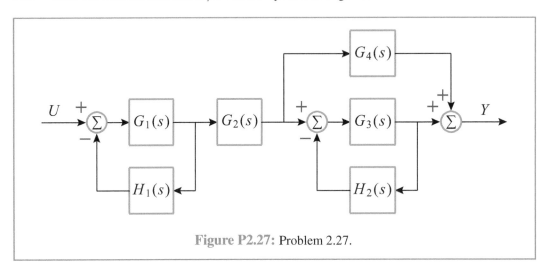

Figure P2.27: Problem 2.27.

2.28 Find the transfer function Y/U for the system in Fig. P2.28.

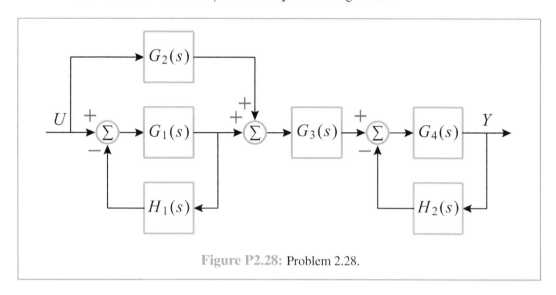

Figure P2.28: Problem 2.28.

2.29 Use MATLAB to find the equivalent transfer function of the block diagram of Fig. 2-39 when

$$G_1 = \frac{1}{s+1}, \quad G_2 = \frac{s+0.2}{s}, \quad G_3 = \frac{5}{s+2}, \quad G_4 = 2, \quad G_5 = \frac{10}{s+10}, \text{ and } G_6 = 3.$$

2.30 Use MATLAB to find the equivalent transfer function of the block diagram of Fig. 2-41 when

$$G_1 = \frac{s+1}{s}, \quad G_2 = \frac{s+2}{s+1}, \quad G_3 = 1, \quad G_4 = \frac{1}{s^2+s+1}, \text{ and } G_5 = \frac{10}{s+10}.$$

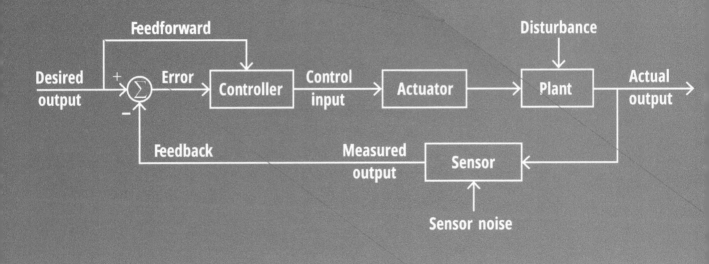

Chapter 3
Stability and Dynamic Response

Chapter Contents

    Overview, 65
3-1 BIBO Stability, 65
3-2 Routh-Hurwitz Criterion, 70
3-3 Dynamic Response, 78
3-4 Step Response, 81
    Chapter Summary, 98
    Problems, 100

Objectives

Upon learning the material presented in this chapter, you should be able to:

1. Define stability of a linear system.
2. Determine stability in terms of the impulse response and the transfer function.
3. Determine stability without calculating the poles of the transfer function.
4. Divide the dynamic response into natural and forced responses.
5. Divide the dynamic response into transient and steady-state responses.
6. Determine the steady-state response to step, sinusoidal, and periodic inputs.
7. Study the step response of first-order and second-order transfer functions.
8. Approximate the step response of higher-order transfer functions.

Overview

It is important in the analysis of dynamical systems to determine whether or not the system is stable and to study its response to various inputs. In Section 3-1 we define input-output–stability for linear time-invariant systems and characterize it in terms of the impulse response and the poles of the transfer function. Section 3-2 presents the Routh-Hurwitz criterion, which allows us to determine the stability of a system without calculating its poles. Section 3-3 examines the response of the system to various inputs, while Section 3-4 focuses on the step response.

3-1 BIBO Stability

The response of a linear time-invariant system can be characterized by its transfer function $G(s)$ using the equation

$$Y(s) = G(s)\, U(s), \qquad (3.1)$$

where $U(s)$ and $Y(s)$ are the Laplace transforms of the input and output signals $u(t)$ and $y(t)$, respectively. Equivalently, the system can be characterized by its *impulse response* $g(t) = \mathcal{L}^{-1}\{G(s)\}$ using the convolution integral

$$y(t) = \int_0^t g(\tau)\, u(t-\tau)\, d\tau. \qquad (3.2)$$

Equation (3.1) is the Laplace transform of Eq. (3.2). The function $g(t)$ is called the impulse response because an impulse input $u(t) = \delta(t)$ results in the output $y(t) = g(t)$.

In the preceding chapter we described systems whose transfer functions are rational functions of s. It is worthwhile to note that there are systems of interest in which the transfer function is not a rational function of s. An important example is the delay system defined by

$$y(t) = u(t-T),$$

where the output signal is a replica of the input signal but delayed by the time delay T. Taking the Laplace transform of this equation shows that

$$Y(s) = e^{-sT}\, U(s) \quad \longrightarrow \quad G(s) = e^{-sT}.$$

The function e^{-sT} is not a ratio of two polynomials, which shows that a delay element cannot be represented by a differential equation of the form of Eq. (2.1).

Signals can be classified as *bounded signals* versus unbounded signals. A signal $x(t)$ is bounded if there is a positive constant k such that $|x(t)| \leq k$ for all $t \geq 0$; otherwise it is unbounded. Examples of bounded signals are the unit step function

$$1(t) = \begin{cases} 1 & \text{if } t \geq 0, \\ 0 & \text{if } t < 0, \end{cases}$$

the sinusoidal $\sin(\omega t + \phi)$, and the decaying exponential $e^{-at}\, 1(t)$, $a > 0$. Examples of unbounded signals are the ramp $t\, 1(t)$ and the growing exponential $e^{at}\, 1(t)$ for $a > 0$. The system is *Bounded-Input–Bounded-Output* (BIBO) stable, or simply stable, if for *every bounded input*, the output is bounded. We emphasize that this property has to hold for every bounded input. If there is even one bounded input for which the output is unbounded, the system is considered unstable. In the rest of this section we characterize stability in terms of the impulse response and the transfer function.

Theorem 3.1 *A system represented by the convolution integral Eq. (3.2) is stable if and only if*

$$\int_0^\infty |g(t)|\, dt < \infty. \tag{3.3}$$

Proof: The statement "if and only if" means that the condition defined by Eq. (3.3) is both necessary and sufficient.

Proof of sufficiency: Suppose $\int_0^\infty |g(t)|\, dt \leq k_g$ and consider a bounded input with $|u(t)| \leq k_u$ for all $t \geq 0$.

$$\begin{aligned}
|y(t)| &= \left| \int_0^t g(\tau)\, u(t-\tau)\, d\tau \right| \\
&\leq \int_0^t |g(\tau)\, u(t-\tau)|\, d\tau \\
&= \int_0^t |g(\tau)|\, |u(t-\tau)|\, d\tau \\
&\leq \int_0^t |g(\tau)|\, k_u\, d\tau \\
&= \int_0^t |g(\tau)|\, d\tau\, k_u \leq k_g k_u,
\end{aligned}$$

where we used the properties that the absolute value of an integral is less than or equal to the integral of the absolute value of the integrand, and the absolute value of a product is equal to the product of absolute values.

Proof of necessity: We use a contradiction argument to show that BIBO stability cannot hold if condition Eq. (3.3) is not satisfied. Given $k_u > 0$, suppose there is a $k_y > 0$ such that

$$|u(t)| \leq k_u, \forall\, t \geq 0 \quad \rightarrow \quad |y(t)| \leq k_y, \forall\, t \geq 0,$$

but $\int_0^\infty |g(t)|\, dt$ is not finite. Then there is a time t_1 such that

$$\int_0^{t_1} |g(t)|\, dt > \frac{k_y}{k_u}.$$

Construct a signal $u(t)$ such that

$$u(t_1 - t) = \begin{cases} k_u & \text{when } g(t) > 0, \\ 0 & \text{when } g(t) = 0, \\ -k_u & \text{when } g(t) < 0. \end{cases}$$

3-1 BIBO STABILITY

The signal $u(t)$ has the properties

$$|u(t_1 - t)| \leq k_u, \quad \text{and} \quad g(t)\,u(t_1 - t) = k_u |g(t)|, \quad \text{for } 0 \leq t \leq t_1 \,.$$

Then,

$$y(t_1) = \int_0^{t_1} g(t)\,u(t_1 - t)\,dt = \int_0^{t_1} k_u |g(t)|\,dt > k_u \frac{k_y}{k_u} = k_y\,,$$

which contradicts the claim that $|y(t)| \leq k_y$ for all $t \geq 0$.

Example 3-1: Stability of the Delay Element

Is the delay element $y(t) = u(t - T)$ BIBO stable?

Solution: Yes, $y(t) = u(t - T)$ is BIBO stable because the output signal is a delayed version of the input signal. But we can also see that the condition of Theorem 3.1 is satisfied because $g(t) = \mathcal{L}^{-1}\{e^{-sT}\} = \delta(t - T)$ and $\int_0^\infty \delta(t - T)\,dt = 1$.

Theorem 3.2 *A system represented by a rational proper transfer function $G(s)$ is stable if and only if all poles of $G(s)$ have negative real parts.*

Proof: $G(s)$ can be represented as $G(s) = G(\infty) + G_{\text{sp}}(s)$ and, using partial-fraction expansion, the **strictly proper** transfer function $G_{\text{sp}}(s)$ can be expanded as

$$G_{\text{sp}}(s) = \sum_{i=1}^{r} G_i(s)\,,$$

where $G_i(s)$ takes the general form

$$G_i(s) = \frac{K}{(s - p)^\alpha}\,,$$

where p is a pole of $G(s)$, $\alpha \geq 1$ is an integer, and K is a constant. If all poles of $G(s)$ are simple, $\alpha = 1$ in the foregoing form for all the $G_i(s)$ terms; otherwise for a pole with multiplicity $m > 1$, there is a $G_i(s)$ term with $\alpha = m$. For example,

$$G(s) = \frac{8}{s(s^2 + 2s + 2)(s + 2)^2}$$

$$= \frac{1}{s} + \frac{1 - j}{s + 1 + j} + \frac{1 + j}{s + 1 - j} - \frac{3}{s + 2} - \frac{2}{(s + 2)^2}\,,$$

where $j = \sqrt{-1}$. For the system to be stable, each $G_i(s)$ term has to be stable. With

$$g_i(t) = \mathcal{L}^{-1}\left\{\frac{K}{(s - p)^\alpha}\right\} = \frac{K\,t^{\alpha - 1} e^{pt}}{(\alpha - 1)!}\,.$$

$G_i(s)$ is stable if and only if $g_i(t)$ satisfies the condition of Theorem 3.1; that is,

$$\int_0^\infty |g_i(t)|\, dt = \int_0^\infty \left| \frac{K\, t^{\alpha-1} e^{pt}}{(\alpha-1)!} \right| dt < \infty.$$

The pole p can be complex, so we write it as $p = \sigma + j\omega$. Hence,

$$\left| \frac{K\, t^{\alpha-1} e^{(\sigma+j\omega)t}}{(\alpha-1)!} \right| = \frac{|K|\, t^{\alpha-1} e^{\sigma t}}{(\alpha-1)!}.$$

Consider first the case $\sigma = 0$:

$$\lim_{T\to\infty} \int_0^T |g_i(t)|\, dt = \lim_{T\to\infty} \int_0^T \frac{|K|\, t^{\alpha-1}}{(\alpha-1)!}\, dt = \frac{|K|}{\alpha!} \lim_{T\to\infty} T^\alpha = \infty.$$

Hence, the system is unstable if $G(s)$ has a pole on the imaginary axis. Consider now the case $\sigma \neq 0$. If $\alpha = 1$,

$$\lim_{T\to\infty} \int_0^T |g_i(t)|\, dt = \lim_{T\to\infty} \int_0^T |K| e^{\sigma t}\, dt = \frac{|K|}{\sigma} \lim_{T\to\infty} (e^{\sigma T} - 1).$$

The limit is finite if and only if $\sigma < 0$. If $\alpha = 2$,

$$\lim_{T\to\infty} \int_0^T |g_i(t)|\, dt = \lim_{T\to\infty} \int_0^T |K|\, t e^{\sigma t}\, dt = |K| \lim_{T\to\infty} \left[\left(\frac{T}{\sigma} - \frac{1}{\sigma^2}\right) e^{\sigma T} + \frac{1}{\sigma^2} \right].$$

The limit is finite if and only if $\sigma < 0$. In taking the limit, recall that

$$\lim_{T\to\infty} T^a e^{\sigma T} = 0,$$

when $\sigma < 0$ for any integer $a > 0$. Since

$$\int x^n e^{ax}\, dx = \frac{x^n e^{ax}}{a} - \frac{n}{a} \int x^{n-1} e^{ax}\, dx,$$

repeating the foregoing calculation shows that for any $\alpha \geq 1$, $G(s)$ is stable if and only if $\sigma < 0$ for all poles of $G(s)$.

▶ A transfer function is unstable if it has one or more poles with nonnegative real part. ◀

A special case of unstable transfer functions is the case when it has simple poles on the imaginary axis (zero real part) with the remaining poles, if any, in the left-half plane (negative real part). In this special case the system is called *marginally stable*. A marginally stable transfer function is still unstable but its impulse response is bounded. For example, for the transfer function

$$G(s) = \frac{s+1}{s(s^2+\omega^2)} = \frac{1}{\omega^2 s} + \frac{1}{s^2+\omega^2} - \frac{s}{\omega^2(s^2+\omega^2)},$$

3-1 BIBO STABILITY

which has poles at $s = 0$ and $s = \pm j\omega$, the impulse response

$$g(t) = \left(\frac{1}{\omega^2} + \frac{1}{\omega}\sin\omega t - \frac{1}{\omega^2}\cos\omega t\right)1(t)$$

is bounded. Moreover, the output of a marginally stable transfer function will be bounded for almost all bounded inputs except those whose Laplace transforms have the same poles as the simple poles of the transfer function on the imaginary axis. For example, the transfer function $G(s) = 1/s$ will produce an unbounded ramp when driven by a constant input, but will produce a bounded output when driven by a periodic input with zero dc component. The transfer function $1/(s^2 + \omega^2)$ will produce unbounded output if driven by the sinusoidal signal $\sin(\omega t + \phi)$ but will produce a bounded output if driven by the sinusoidal signal $\sin(\Omega t + \phi)$ with $\Omega \neq \omega$.

Example 3-2: Stability of Transfer Functions

For each of the following transfer functions, determine if the system is stable or unstable. If it is unstable, determine if it is marginally stable.

(a) $\quad G(s) = \dfrac{(s+2)(s+3)}{(s+1)(s+4)}$

(b) $\quad G(s) = \dfrac{(s+2)(s+3)}{(s-1)(s+4)}$

(c) $\quad G(s) = \dfrac{(s-2)(s+3)}{(s+1)(s+4)}$

(d) $\quad G(s) = \dfrac{(s^2 + 0.4s + 1.04)(s+3)}{(s^2 + 0.2s + 1)(s+2)(s+4)}$

(e) $\quad G(s) = \dfrac{(s^2 + 0.4s + 1.04)(s+3)}{(s^2 - 0.2s + 1)(s+2)(s+4)}$

(f) $\quad G(s) = 1 + \dfrac{s-1}{(s+1)(s+2)}$

(g) $\quad G(s) = \dfrac{a}{s^2 - a}, \quad a > 0$

(h) $\quad G(s) = \dfrac{1}{(s+2)(s^2+1)}$

(i) $\quad G(s) = \dfrac{1}{(s-2)(s^2+1)}$

(j) $\quad G(s) = \dfrac{1}{(s^2+1)^2}$

Solution:

(a) The poles of $G(s) = (s+2)(s+3)/[(s+1)(s+4)]$ are -1 and -4. It is stable.

(b) The poles of $G(s) = (s+2)(s+3)/[(s-1)(s+4)]$ are 1 and -4. It is unstable because of the pole at 1. It is not marginally stable because the unstable pole is in the right-half plane.

(c) The poles of $G(s) = (s-2)(s+3)/[(s+1)(s+4)]$ are -1 and -4. It is stable. There is a zero in the right-half plane, but zeros do not affect stability.

(d) The poles of $G(s) = (s^2 + 0.4s + 1.04)(s+3)/[(s^2 + 0.2s + 1)(s+2)(s+4)]$ are $-0.1 \pm j\sqrt{0.99}$, -2, and -4. It is stable.

(e) The poles of $G(s) = (s^2 + 0.4s + 1.04)(s+3)/[(s^2 - 0.2s + 1)(s+2)(s+4)]$ are $0.1 \pm j\sqrt{0.99}$, -2, and -4. It is unstable because of the poles at $0.1 \pm j\sqrt{0.99}$. It is not marginally stable because the unstable poles are in the right-half plane.

(f) The poles of $G(s) = 1 + (s-1)/[(s+1)(s+2)]$ are -1 and -2. It is stable.

(g) The poles of $G(s) = a/(s^2 - a)$ $(a > 0)$ are $\pm\sqrt{a}$. It is unstable because of the pole at $\sqrt{a}$. It is not marginally stable because the unstable pole is in the right-half plane.

(h) The poles of $G(s) = 1/[(s+2)(s^2 + 1)]$ are -2 and $\pm j$. It is unstable because of the poles at $\pm j$. It is marginally stable because the unstable poles are simple poles on the imaginary axis.

(i) The poles of $G(s) = 1/[(s-2)(s^2 + 1)]$ are 2 and $\pm j$. It is unstable because the poles are not in the left-half plane. Even though there are simple poles on the imaginary axis, it is not marginally stable because of the pole at 2.

(j) There are multiple poles of $G(s) = 1/(s^2 + 1)^2$ at $\pm j$. It is unstable because of these poles. Even though the unstable poles are on the imaginary axis, it is not marginally stable because the poles are not simple.

3-2 Routh-Hurwitz Criterion

Investigating the stability of a rational proper transfer function requires us to calculate its poles; that is, to calculate the roots of its denominator polynomial. Calculating the roots of a polynomial is easy if its degree is one or two. For higher degrees it is hard to calculate the roots manually. Decades ago engineers used the Routh-Hurwitz criterion to determine whether the roots of a polynomial are in the left-half plane without calculating such roots. Even though we now have computer programs, such as MATLAB, that can easily calculate the roots of a polynomial of a fairly high degree, it turns out that the Routh-Hurwitz criterion is still useful in the analysis of feedback control systems because it can determine the range of system parameters for the which the system is stable.

Starting with the 4th degree polynomial

$$s^4 + a_3 s^3 + a_2 s^2 + a_1 s + a_0,$$

we construct a table, known as a *Routh table*, as shown below.

$$
\begin{array}{c|ccc}
s^4 & 1 & a_2 & a_0 \\
s^3 & a_3 & a_1 \\
s^2 & b_1 & b_2 \\
s & c_1 \\
1 & d_1
\end{array}
$$

The table has five rows, designated by the powers of s from s^4 to $s^0 = 1$. The first two rows are constructed from the coefficients of the polynomial in a zig-zag manner. In the first row and first column we write the coefficient of s^4. Moving down in the first column to the second row, we write the coefficient of s^3. Back to the first row, we write the coefficient of s^2 in the second column. Moving down in the second column to the second row, we write the coefficient of s. Back to the first row, we write the absolute term in the third column. At this point all the coefficients of the polynomial are covered, so the remaining elements in the first and second rows are zeros, which we leave as blank. The coefficients of the third row are calculated from the coefficients of the previous two rows as follows:

$$b_1 = \frac{-1}{a_3}\begin{vmatrix} 1 & a_2 \\ a_3 & a_1 \end{vmatrix} = \frac{a_2 a_3 - a_1}{a_3}, \qquad b_2 = \frac{-1}{a_3}\begin{vmatrix} 1 & a_0 \\ a_3 & 0 \end{vmatrix} = \frac{a_0 a_3}{a_3} = a_0.$$

b_1 equals the negative of the determinant of the 2×2 matrix formed by the first and second columns of the previous two rows, divided by the first coefficient of the previous row; b_2 equals the negative of the determinant of the 2×2 matrix formed by the first and third columns of the previous two rows, divided by the first coefficient of the previous row. This pattern continues for all the elements of the third row. The 2×2 matrix is formed by the first column and the column next to the column we are calculating and the division is always by the first element of the previous row. The next element of the third row would have been

$$b_3 = \frac{-1}{a_3}\begin{vmatrix} 1 & 0 \\ a_3 & 0 \end{vmatrix} = 0,$$

and that is why this element is left blank. The next row is constructed using the same pattern from the previous two rows.

$$c_1 = \frac{-1}{b_1}\begin{vmatrix} a_3 & a_1 \\ b_1 & b_2 \end{vmatrix} = \frac{a_1 b_1 - a_3 b_2}{b_1}, \qquad c_2 = \frac{-1}{b_1}\begin{vmatrix} a_3 & 0 \\ b_1 & 0 \end{vmatrix} = 0.$$

Finally,

$$d_1 = \frac{-1}{c_1}\begin{vmatrix} b_1 & b_2 \\ c_1 & 0 \end{vmatrix} = b_2 = a_0.$$

For the nth degree polynomial,

$$s^n + a_{n-1} s^{n-1} + a_{n-2} s^{n-2} \cdots + a_1 s + a_0.$$

The Routh table is constructed in the same way:

$$
\begin{array}{c|cccc}
s^n & 1 & a_{n-2} & a_{n-4} & \cdots \\
s^{n-1} & a_{n-1} & a_{n-3} & a_{n-5} & \cdots \\
s^{n-2} & b_1 & b_2 & b_3 & \cdots \\
s^{n-3} & c_1 & c_2 & c_3 & \cdots \\
\vdots & \vdots & & & \\
1 & h_1 & & &
\end{array}
$$
(3.4)

with

$$b_1 = \frac{-1}{a_{n-1}} \begin{vmatrix} 1 & a_{n-2} \\ a_{n-1} & a_{n-3} \end{vmatrix}, \quad b_2 = \frac{-1}{a_{n-1}} \begin{vmatrix} 1 & a_{n-4} \\ a_{n-1} & a_{n-5} \end{vmatrix}, \quad \cdots$$

$$c_1 = \frac{-1}{b_1} \begin{vmatrix} a_{n-1} & a_{n-3} \\ b_1 & b_2 \end{vmatrix}, \quad c_2 = \frac{-1}{b_1} \begin{vmatrix} a_{n-1} & a_{n-5} \\ b_1 & b_3 \end{vmatrix}, \quad \cdots$$

Routh-Hurwitz Criterion[a]

- The roots of the polynomial have negative real parts if and only if all the elements of the first column are positive.

- If the elements of the first column are nonzero and change signs between positive and negative, then the number of roots with positive real parts is equal to the number of sign changes.

[a]The proof of the Routh-Hurwiz criterion can be found in [27] or [10].

Example 3-3: Stability by the Routh-Hurwitz Criterion

Investigate the stability of the system

$$G(s) = \frac{s+2}{s^3 + 9s^2 + 10s + 20}.$$

Solution: Construct the Routh table of the denominator polynomial

$$s^3 + 9s^2 + 10s + 20:$$

$$
\begin{array}{c|cc}
s^3 & 1 & 10 \\
s^2 & 9 & 20 \\
s & 70/9 & \\
1 & 20 &
\end{array}
\qquad
b_1 = \frac{-1}{9} \begin{vmatrix} 1 & 10 \\ 9 & 20 \end{vmatrix} = \frac{70}{9},
$$

$$c_1 = \frac{-1}{70/9} \begin{vmatrix} 9 & 20 \\ 70/9 & 0 \end{vmatrix} = 20.$$

All elements in the first column are positive. Hence, the system is stable.

3-2 ROUTH-HURWITZ CRITERION

Example 3-4: Stability by the Routh-Hurwitz Criterion

Investigate the stability of the system

$$G(s) = \frac{s+2}{s^4 + 4s^3 + 5s^2 + 12s + 10}.$$

Solution: Construct the Routh table of the denominator polynomial

$$s^4 + 4s^3 + 5s^2 + 12s + 10.$$

s^4	1	5	10
s^3	4	12	
s^2	2	10	
s	-8		
1	10		

$$b_1 = \frac{-1}{4}\begin{vmatrix} 1 & 5 \\ 4 & 12 \end{vmatrix} = 2,$$

$$c_1 = \frac{-1}{2}\begin{vmatrix} 4 & 12 \\ 2 & 10 \end{vmatrix} = -8.$$

There are two sign changes in the first column, from row 3 to row 4 and from row 4 to row 5. The system is unstable. There are two poles in the right-half plane.

Example 3-5: Determination of the Range of a Gain for Closed-Loop Stability

Consider the feedback control system of Fig. 3-1, where the plant transfer function is

$$G_p(s) = \frac{6}{s^3 + 6s^2 + 11s + 6},$$

and the transfer function of a proportional controller is $G_c(s) = K$. Find the range of K for which the closed-loop system is stable.

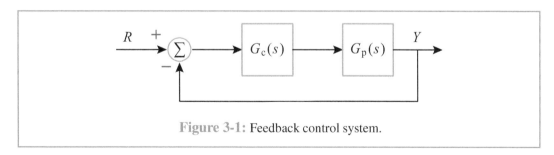

Figure 3-1: Feedback control system.

Solution: The closed-loop transfer function is

$$\frac{Y}{R} = \frac{6K}{s^3 + 6s^2 + 11s + 6(1+K)}.$$

The Routh table of the denominator polynomial is

s^3	1	11
s^2	6	$6(1+K)$
s	$(10-K)$	
1	$6(1+K)$	

The system is stable when all elements in the first column are positive. This is the case only when $-1 < K < 10$.

Example 3-6: Determination of the Ranges of Two Gains for Closed-Loop Stability

---

Reconsider the feedback control system of the previous example, but change the controller to a Proportional-Integral (PI) controller whose transfer function is $G_c(s) = K_P + K_I/s$. Find the ranges of K_P and K_I for which the closed-loop system is stable.

Solution: The closed-loop transfer function is

$$\frac{Y}{R} = \frac{6(K_P s + K_I)}{s^4 + 6s^3 + 11s^2 + 6(1+K_P)s + 6K_I}.$$

The Routh table of the denominator polynomial is

s^4	1	11	$6K_I$
s^3	6	$6(1+K_P)$	
s^2	$(10-K_P)$	$6K_I$	
s	$\dfrac{6[(10-K_P)(1+K_P) - 6K_I]}{(10-K_P)}$		
1	$6K_I$		

The system is stable when all elements in the first column are positive. This is the case when

$$K_P < 10 \quad \text{and} \quad 0 < K_I < \frac{(10-K_P)(1+K_P)}{6}.$$

For example, if we choose $K_P = 4$, then the system is stable only if $0 < K_I < 5$.

3-2 ROUTH-HURWITZ CRITERION

> Example 3-7: A Necessary Condition for Stability

Show that for the roots of a polynomial to have negative real parts, it is necessary that all the coefficients of the polynomial be positive.

Solution: We show it for a polynomial of degree five. The general case is done similarly.

$$s^5 + a_4 s^4 + a_3 s^3 + a_2 s^2 + a_1 s + a_0.$$

s^5	1	a_3	a_1
s^4	a_4	a_2	a_0
s^3	b_1	b_2	
s^2	c_1	a_0	
s	d_1		
1	a_0		

with

$$b_1 = \frac{a_3 a_4 - a_2}{a_4}, \quad b_2 = \frac{a_1 a_4 - a_0}{a_4}, \quad c_1 = \frac{a_2 b_1 - a_4 b_2}{b_1}, \quad \text{and} \quad d_1 = \frac{b_2 c_1 - a_0 b_1}{c_1}.$$

For the roots of the polynomial to have negative real parts it must be true that all elements in the first column are positive. Thus, a_4 and a_0 are positive. Now,

$$c_1 > 0 \ \& \ d_1 > 0 \longrightarrow b_2 > 0 \longrightarrow a_1 > 0; \quad b_1 > 0 \ \& \ c_1 > 0 \longrightarrow a_2 > 0; \quad b_1 > 0 \longrightarrow a_3 > 0.$$

The two cases mentioned under the Routh-Hurwitz criterion require all elements in the first column to be nonzero. What if an element of the first column is zero? In that case the system is unstable, but it will be useful to find more information about the roots to determine if the system is marginally stable. Two cases are discussed below.

Case 1: An entire row is zero

In this case an ***even polynomial*** is a factor of the original polynomial. A polynomial is even when all powers of s are even. The roots of an even polynomial, of degree 2 or higher, are symmetric with respect to the origin, as in the cases

$$s^2 - \sigma^2 = (s - \sigma)(s + \sigma),$$

$$s^2 + \omega^2 = (s + j\omega)(s - j\omega),$$

$$s^4 + 2(\omega^2 - \sigma^2)s^2 + (\sigma^2 + \omega^2)^2 = (s^2 + 2\sigma s + \sigma^2 + \omega^2)(s^2 - 2\sigma s + \sigma^2 + \omega^2)$$
$$= (s + \sigma + j\omega)(s + \sigma - j\omega)(s - \sigma + j\omega)(s - \sigma - j\omega).$$

When we encounter a row of zeros, we form an ***auxiliary*** polynomial using the elements of the row immediately above the row of zeros. The polynomial starts with the power of s in the label column and continues by decreasing the power by two. We then divide the auxiliary polynomial by the coefficient of the highest power of s. The resultant polynomial is a factor of the original polynomial.

Example 3-8: Routh Table with a Row of Zeros

Construct the Routh table for the polynomial

$$s^5 + 2s^4 + 5s^3 + 10s^2 + 4s + 8.$$

Solution:

s^5	1	5	4
s^4	2	10	8
s^3	0	0	0

The third row is zero. We construct the auxiliary polynomial from the second row as $2s^4 + 10s^2 + 8$. Division by 2 yields $s^4 + 5s^2 + 4$ as a factor of the original polynomial. Factoring the auxiliary polynomial as

$$s^4 + 5s^2 + 4 = (s^2 + 1)(s^2 + 4)$$

shows that the original polynomial has roots at $\pm j$ and $\pm 2j$. We can also determine the remaining factor of the original polynomial by polynomial long division:

$$
\begin{array}{r}
s + 2 \\
s^4 + 5s^2 + 4 \enclose{longdiv}{s^5 + 2s^4 + 5s^3 + 10s^2 + 4s + 8} \\
\underline{s^5 + 5s^3 + 4s } \\
2s^4 + 10s^2 + 8 \\
\underline{2s^4 + 10s^2 + 8} \\
0   0   0
\end{array}
$$

Thus, the original polynomial can be factored as $(s+2)(s^4 + 5s^2 + 4)$ and its roots are -2, $\pm j$, and $\pm 2j$.

Example 3-9: A Marginally Stable Closed-Loop System

Reconsider the system of Example 3-5. The Routh table of the denominator polynomial is

s^3	1	11
s^2	6	$6(1+K)$
s	$(10-K)$	
1	$6(1+K)$	

3-2 ROUTH-HURWITZ CRITERION

There is a row of zeros at $K = 10$ or $K = -1$. Consider first $K = 10$. The Routh table becomes

s^3	1	11
s^2	6	66
s	0	

The zero is in the third row, so the auxiliary polynomial obtained from the row above it is $(6s^2 + 66)$, which simplifies to $s^2 + 11$. By polynomial long division, the remaining factor is $s + 6$. Thus, the system has poles at -6 and $\pm\sqrt{11}\,j$; it is marginally stable. Next, consider $K = -1$.

s^3	1	11
s^2	6	0
s	11	
1	0	

The auxiliary polynomial is s. The remaining factor is $s^2 + 6s + 11$. Thus, the system has poles at 0 and $-3 \pm \sqrt{2}\,j$; it is marginally stable.

Case 2: An element in the first column is zero but there are nonzero elements in the same row

In this case the zero element is replaced with a positive constant ε and we continue to construct the table. When done, we determine the signs of the elements of the first column as $\varepsilon \to 0$.

Example 3-10: Routh Table with a Zero in the First Column

Construct the Routh table for the polynomial

$$s^5 + 2s^4 + 2s^3 + 4s^2 + 11s + 10\,.$$

Solution:

s^5	1	2	11
s^4	2	4	10
s^3	0	6	

Replace 0 in the third row with ε and continue:

s^5	1	2	11
s^4	2	4	10
s^3	ε	6	
s^2	c_1	10	
s	d_1		
1	10		

$$c_1 = \frac{4\varepsilon - 12}{\varepsilon}, \qquad d_1 = 6 - \frac{10\varepsilon^2}{4\varepsilon - 12}\,.$$

As $\varepsilon \to 0$, $c_1 \to -12/\varepsilon < 0$ and $d_1 \to 6 > 0$. There are two sign changes. Hence, the polynomial has two roots in the right-half plane.

> Example 3-11: Routh Table for a PI Controller Case with a Zero in the First Column

Reconsider the system of Example 3-6. The Routh table of the denominator polynomial is

s^4	1	11	$6K_I$
s^3	6	$6(1+K_P)$	
s^2	$(10-K_P)$	$6K_I$	
s	$\dfrac{6[(10-K_P)(1+K_P)-6K_I])}{(10-K_P)}$		
1	$6K_I$		

At $K_P = 10$ while $K_I \neq 0$, the first element in the third row is zero, while the second element is nonzero. Replace $10 - K_P$ with $\varepsilon > 0$ and continue:

s^4	1	11	$6K_I$
s^3	6	$6(11-\varepsilon)$	
s^2	ε	$6K_I$	
s	c_1		
1	$6K_I$		

with

$$c_1 = \frac{6\varepsilon(11-\varepsilon) - 36K_I}{\varepsilon}.$$

As $\varepsilon \to 0$, $c_1 \to -36K_I/\varepsilon$; $\text{sign}(c_1) = -\text{sign}(K_I)$. Thus there is a sign change and the system is unstable. The number of poles in the right-half plane depends on the sign of K_I. If $K_I > 0$, there are two sign changes; hence, there are two poles in the right-half plane. If $K_I < 0$, there is one sign change; hence, there is one pole in the right-half plane. For example, if $K_I = 1$, the polynomial is

$$s^4 + 6s^3 + 11s^2 + 66s + 6.$$

and its roots are -5.9785, -0.0923, and $0.0354 \pm 3.298j$, while if $K_I = -1$, the polynomial is

$$s^4 + 6s^3 + 11s^2 + 66s - 6$$

and its roots are -6.0211, $-0.0342 \pm 3.3365j$, and 0.0895.

3-3 Dynamic Response

The response of a linear system to a wide class of inputs, including step, ramp, sinusoidal, etc., is divided into a *natural response* and a *forced response*. The natural response is determined by the poles of the transfer function, while the forced response has the same form as the input. For a stable transfer function, all poles have negative real parts, which causes the natural response to converge to zero as time tends to infinity. In this case, the natural and forced

3-3 DYNAMIC RESPONSE

responses are called *transient* and *steady-state*, respectively. We shall look at the response of a stable linear system to step, sinusoidal, and periodic inputs.

Step input:

Let $u(t) = A\, 1(t)$, where $1(t)$ is the unit step function and A is the amplitude of the step. To start with, let us consider the case where the transfer function has simple poles:

$$G(s) = \frac{N(s)}{(s-p_1)(s-p_2)\cdots(s-p_n)}, \quad p_i \neq p_j. \tag{3.5}$$

The Laplace transform of the output is given by

$$Y(s) = G(s)\, U(s) = G(s)\,\frac{A}{s} = \frac{N(s)\, A}{s(s-p_1)(s-p_2)\cdots(s-p_n)}.$$

Expanding the expression of $Y(s)$ using partial-fraction expansion results in

$$Y(s) = \frac{G(0)\, A}{s} + \sum_{i=1}^{i=n} \frac{K_i}{s-p_i},$$

where

$$K_i = \left.\frac{A\, G(s)\, (s-p_i)}{s}\right|_{s=p_i}.$$

Taking the inverse Laplace transform yields

$$y(t) = \underbrace{G(0)\, A\, 1(t)}_{\text{forced (steady-state) response}} + \underbrace{\sum_{i=1}^{i=n} K_i e^{p_i t}\, 1(t)}_{\text{natural (transient) response}}$$

and

$$\text{Re}\{p_i\} < 0,\ \forall i \quad \rightarrow \quad \lim_{t\to\infty} \sum_{i=1}^{i=n} K_i e^{p_i t}\, 1(t) = 0 \quad \rightarrow \quad \lim_{t\to\infty} y(t) = G(0)\, A.$$

Thus, the steady-state response of a stable transfer function to a step input is a constant equal to the amplitude of the step multiplied by $G(0)$, which is called the dc gain of the transfer function. Even if $G(s)$ has multiple poles, the output takes the form

$$y(t) = G(0)\, A\, 1(t) + f(t), \tag{3.6}$$

where $f(t)$ is a summation of terms of the form $K t^\alpha e^{pt}/\alpha!$, for a nonnegative integer α. Once again, because the poles have negative real parts, these terms converge to zero as time tends to infinity and the steady-state response is $G(0)\, A$.

▶ The steady-state response of a stable transfer function $G(s)$ to the step input $A\, 1(t)$ is $G(0)\, A$, where $G(0)$ is the dc gain of $G(s)$. ◀

Sinusoidal input:

Let $u(t) = A\sin\omega t$, for $t \geq 0$:

$$u(t) = A\sin\omega t = \frac{A}{2j}\left(e^{j\omega t} - e^{-j\omega t}\right), \quad (3.7)$$

and therefore

$$U(s) = \frac{A}{2j}\left(\frac{1}{(s-j\omega)} - \frac{1}{(s+j\omega)}\right)$$

and

$$Y(s) = G(s)\,U(s) = G(s)\frac{A}{2j}\left(\frac{1}{(s-j\omega)} - \frac{1}{(s+j\omega)}\right).$$

Expanding the expression of $Y(s)$ by partial-fraction expansion results in

$$Y(s) = \frac{A}{2j}G(j\omega)\frac{1}{(s-j\omega)} - \frac{A}{2j}G(-j\omega)\frac{1}{(s+j\omega)} + \text{terms dependent on the poles of } G(s).$$

Let $M = |G(j\omega)|$ and $\phi = \angle G(j\omega)$ so that $G(j\omega) = Me^{j\phi}$. It can be verified that

$$G(-j\omega) = Me^{-j\phi}.$$

Hence,

$$\frac{A}{2j}G(j\omega)\frac{1}{(s-j\omega)} - \frac{A}{2j}G(-j\omega)\frac{1}{(s+j\omega)} = \frac{AM}{2j}\left(\frac{e^{j\phi}}{(s-j\omega)} - \frac{e^{-j\phi}}{(s+j\omega)}\right),$$

and

$$\mathcal{L}^{-1}\left\{\frac{AM}{2j}\left(\frac{e^{j\phi}}{(s-j\omega)} - \frac{e^{-j\phi}}{(s+j\omega)}\right)\right\} = \frac{AM}{2j}\left[e^{j(\omega t+\phi)} - e^{-j(\omega t+\phi)}\right] = AM\sin(\omega t+\phi).$$

Thus,

$$y(t) = AM\sin(\omega t + \phi) + f(t), \quad \text{for } t \geq 0, \quad (3.8)$$

where $f(t)$ is a summation of terms of the form $Kt^{\alpha}e^{pt}/\alpha!$ with a nonnegative integer α. Note that

$$\operatorname{Re}\{p_i\} < 0, \forall i \;\;\rightarrow\;\; \lim_{t\to\infty} f(t) = 0 \;\;\rightarrow\;\; \lim_{t\to\infty}[y(t) - AM\sin(\omega t+\phi)] = 0.$$

Hence, $y(t)$ approaches the sinusoidal signal $A|G(j\omega)|\sin(\omega t + \angle G(j\omega))$; $G(j\omega)$ is called the *frequency response* of $G(s)$. This result can be extended to the sinusoidal input $A\sin(\omega t + \theta)$ for $\theta \neq 0$, where the steady-state response is $A|G(j\omega)|\sin(\omega t + \theta + \angle G(j\omega))$.

▶ The steady-state response of a stable transfer function $G(s)$ to the sinusoidal input $A\sin(\omega t + \theta)$ is a sinusoidal signal of the same frequency with the amplitude and phase modified by the frequency response $G(j\omega)$. ◀

Periodic input:

A periodic signal can be expanded using Fourier series as a summation of a dc component, a fundamental-frequency component and components of higher harmonics. A linear system obeys the *superposition principle*; that is, if y_1, y_2, and y_3 are the outputs to the inputs u_1, u_2, and u_3, respectively, then $a_1 y_1 + a_2 y_2 + a_3 y_3$ is the output to $a_1 u_1 + a_2 u_2 + a_3 u_3$, for any constants a_1, a_2, and a_3. Using this principle, the steady-state output of a stable linear system driven by a periodic input will be a summation of a dc component, a fundamental-frequency component, and components of higher harmonics, with the same fundamental frequency of the input; that is, the steady-state output is a periodic signal of the same period as the input.

> ▶ The steady-state response of a stable transfer function to a periodic input is a periodic signal of the same period as that of the input. ◀

3-4 Step Response

In the analysis and design of control systems we pay special attention to the step response. The shape of the transient part of the step response is indicative of the transient response to other inputs. Characteristics like how fast the response reaches steady state or whether the natural response is monotonic or oscillatory can be seen from the step response. In this section we look at the response of first and second-order transfer functions, how the zero of a second-order transfer function affects its response, and when high-order transfer functions can be approximated by lower-order transfer functions.

3-4.1 First-Order Transfer Functions

When the step input $u(t) = A\, 1(t)$ is applied to the first-order stable transfer function

$$G(s) = \frac{K}{s+a}, \qquad K > 0, \quad a > 0, \tag{3.9}$$

the Laplace transform of the output is given by

$$Y(s) = G(s)\, U(s) = \frac{KA}{s(s+a)}.$$

Using partial-fraction expansion, we obtain

$$Y(s) = \frac{K_1}{s} + \frac{K_2}{s+a},$$

where

$$K_1 = sY(s)\big|_{s=0} = \frac{KA}{a}, \qquad K_2 = (s+a)Y(s)\big|_{s=-a} = -\frac{KA}{a}.$$

The inverse Laplace transform of $Y(s)$ is given by

$$y(t) = \frac{KA}{a}(1 - e^{-at})\, 1(t). \tag{3.10}$$

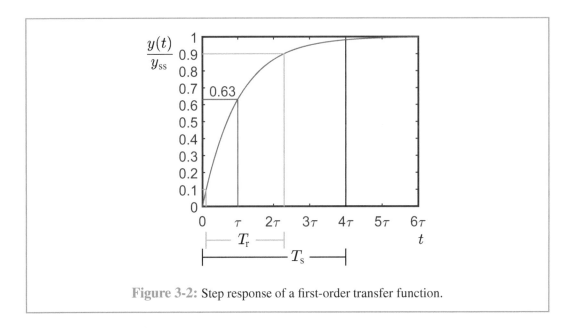

Figure 3-2: Step response of a first-order transfer function.

The steady-state output is

$$y_{ss} = \lim_{t \to \infty} y(t) = \frac{KA}{a} = G(0)\,A\ .$$

Upon normalizing the output by its steady-state value and defining the time constant $\tau = 1/a$, we have

$$\frac{y(t)}{y_{ss}} = \left(1 - e^{-t/\tau}\right) 1(t)\ . \tag{3.11}$$

This function is plotted in **Fig. 3-2**. The time response is characterized by two parameters: the rise time T_r, which is the time it take $y(t)$ to go from 10% to 90% of its steady-state value, and the settling time T_s, which is the time it takes $y(t)$ to go from zero to 98% of its steady-state value (that is, within 2% of its final value):

$$T_r = t_2 - t_1, \quad \text{where } 1 - e^{-t_1/\tau} = 0.1 \quad \text{and} \quad 1 - e^{-t_2/\tau} = 0.9\ .$$

It can be verified that $t_1 \approx 0.1\tau$ and $t_2 \approx 2.3\tau$. Hence, $T_r \approx 2.2\tau$. Similarly,

$$1 - e^{-T_s/\tau} = 0.98 \quad \longrightarrow \quad T_s \approx 4\tau\ .$$

Another point of interest is that at $t = \tau$, $y/y_{ss} = 1 - e^{-1} \approx 0.63$. This feature is used to find τ experimentally. After a step input is applied to the system, if the step response is close to the plot of **Fig. 3-2**, the system is modeled by a first-order transfer function and τ is the time at which the response is 63% of the steady-state value. This determines $a = 1/\tau$. The constant K is determined from the steady-state output, since $y_{ss} = K/a$.

In summary,

$$G(s) = \frac{K}{s+a}, \quad K > 0, \quad a > 0, \quad \frac{y(t)}{y_{ss}} = (1 - e^{-t/\tau})\, 1(t),$$

$$\tau = \frac{1}{a}, \quad T_r = 2.2\tau, \quad T_s = 4\tau, \quad y = 0.63\, y_{ss} \text{ at } t = \tau.$$

3-4.2 Second-Order Transfer Functions

In this section we examine the step response of a second-order stable transfer function with no zeros. The examination is divided into three cases depending on the location of the poles:

> *Overdamped*: Two real distinct poles
> *Critically damped*: Two real identical poles
> *Underdamped*: A pair of complex poles

Figure 3-3 shows typical responses for the three cases. The overdamped response is monotonic; that is, the output approaches the steady state monotonically. It does not overshoot the steady state. The critically damped response is also monotonic and does not overshoot the steady state, but it is faster than the overdamped response. The underdamped response is oscillatory and overshoots the steady state. We now show the calculation of the output in each case.

Overdamped:

For

$$G(s) = \frac{K}{(s+a)(s+b)}, \quad K > 0, \quad b > a > 0, \quad (3.12)$$

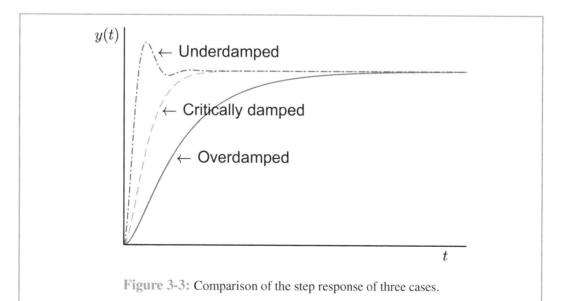

Figure 3-3: Comparison of the step response of three cases.

the Laplace transform of the output to the step input $u(t) = A\,1(t)$ is

$$Y(s) = \frac{KA}{s(s+a)(s+b)}\ .$$

Using partial-fraction expansion, it can be shown that

$$Y(s) = \frac{K_1}{s} + \frac{K_2}{s+a} + \frac{K_3}{s+b}\ ,$$

with

$$K_1 = \frac{KA}{ab}\ ,\quad K_2 = \frac{-KA}{a(b-a)}\ ,\quad K_3 = \frac{KA}{b(b-a)}\ .$$

The inverse Laplace transform of $Y(s)$ is

$$y(t) = (K_1 + K_2 e^{-at} + K_3 e^{-bt})\,1(t)\ . \tag{3.13}$$

The steady-state output is

$$y_{ss} = \lim_{t \to \infty} y(t) = K_1 = \frac{KA}{ab} = G(0)\,A$$

and

$$\frac{y(t)}{y_{ss}} = \left(1 - \frac{b}{b-a}e^{-at} + \frac{a}{b-a}e^{-bt}\right)1(t)\ . \tag{3.14}$$

The fact that the response is monotonic and does not overshoot the steady state can be seen by calculating the derivative of the foregoing expression:

$$\frac{d}{dt}\left[\frac{y(t)}{y_{ss}}\right] = \frac{ab}{b-a}(e^{-at} - e^{-bt})\ ,$$

$$b > a \;\rightarrow\; e^{-at} > e^{-bt},\quad \text{for } t > 0 \;\rightarrow\; \frac{d}{dt}\left[\frac{y(t)}{y_{ss}}\right] > 0,\quad \text{for } t > 0\ .$$

Hence, the output increases monotonically towards the steady state. It does not overshoot the steady state, because if it did so it would have to move toward the steady state with negative slope. Similar to a first-order transfer function, the monotonic step response of an overdamped transfer functions is characterized by the rise time T_r (to go from 10% to 90%) and the settling time T_s (to go from 0 to 98%). However, in this case there are no simple formulas to calculate T_r and T_s in terms of a and b. They are determined from the simulation of the response.

Critically damped:

For

$$G(s) = \frac{K}{(s+a)^2}\ ,\quad K > 0,\quad a > 0\ , \tag{3.15}$$

the Laplace transform of the output to the step input $u(t) = A\,1(t)$ is

$$Y(s) = \frac{KA}{s(s+a)^2}\ .$$

3-4 STEP RESPONSE

Using partial-fraction expansion, it can be shown that

$$Y(s) = \frac{K_1}{s} + \frac{K_2}{(s+a)} + \frac{K_3}{(s+a)^2},$$

with

$$K_1 = \frac{KA}{a^2}, \quad K_2 = \frac{-KA}{a^2}, \quad K_3 = \frac{-KA}{a}.$$

Then,

$$y(t) = \mathcal{L}^{-1}\{Y(s)\} = \frac{KA}{a^2}\left[1 - (1+at)e^{-at}\right] 1(t), \qquad (3.16)$$

$$y_{ss} = \lim_{t \to \infty} y(t) = \frac{KA}{a^2} = G(0)\,A,$$

and

$$\frac{y(t)}{y_{ss}} = \left[1 - (1+at)e^{-at}\right] 1(t). \qquad (3.17)$$

Once again, calculating the derivative of the foregoing expression shows that

$$\frac{d}{dt}\left[\frac{y(t)}{y_{ss}}\right] = a^2 t e^{-at} > 0, \quad \text{for } t > 0.$$

Hence, the output increases monotonically towards the steady state and does not overshoot it. As in the previous case, the monotonic step response is characterized by the rise time T_r and the settling time T_s, which are determined from the simulation of the response.

Underdamped:

For

$$G(s) = \frac{K\omega_n^2}{s^2 + 2\zeta\omega_n s + \omega_n^2}, \quad \omega_n > 0, \quad 0 < \zeta < 1, \qquad (3.18)$$

$$s^2 + 2\zeta\omega_n s + \omega_n^2 = (s + \zeta\omega_n)^2 + \omega_n^2(1 - \zeta^2) = (s + \zeta\omega_n)^2 + \omega_d^2,$$

where

$$\omega_d \stackrel{\text{def}}{=} \omega_n \sqrt{1 - \zeta^2}.$$

The transfer function has a pair of complex poles at $-\zeta\omega_n \pm j\omega_d$; ω_n is the undamped natural frequency, ω_d is the damped natural frequency, and ζ is the damping ratio. The natural response of the system oscillates with frequency ω_d. If there is no damping; that is, if $\zeta = 0$, $G(s)$ will have a pair of poles on the imaginary axis at $\pm j\omega_n$ and the natural response will have sustained oscillation with frequency ω_n.

The Laplace transform of the output to the step input $u(t) = A\,1(t)$ is

$$Y(s) = \frac{K\omega_n^2 A}{s(s^2 + 2\zeta\omega_n s + \omega_n^2)}.$$

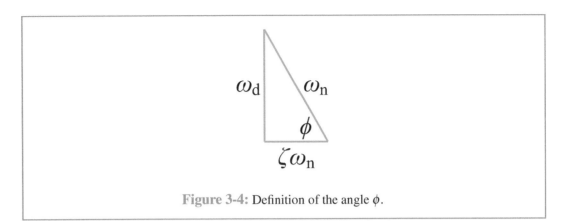

Figure 3-4: Definition of the angle ϕ.

By partial-fraction expansion,

$$Y(s) = \frac{KA}{s} + \frac{-KA(s+\zeta\omega_n) - \zeta\omega_n KA}{(s+\zeta\omega_n)^2 + \omega_d^2},$$

$$y(t) = \mathcal{L}^{-1}\{Y(s)\} = KA\left(1 - e^{-\zeta\omega_n t}\cos\omega_d t - e^{-\zeta\omega_n t}\frac{\zeta\omega_n}{\omega_d}\sin\omega_d t\right) 1(t), \quad (3.19)$$

$$y_{ss} = \lim_{t\to\infty} y(t) = KA = G(0)A,$$

$$\frac{y(t)}{y_{ss}} = \left(1 - e^{-\zeta\omega_n t}\cos\omega_d t - e^{-\zeta\omega_n t}\frac{\zeta\omega_n}{\omega_d}\sin\omega_d t\right) 1(t). \quad (3.20)$$

Consider the right-angled triangle of Fig. 3-4 and define the angle ϕ as shown in the figure. Then

$$\cos\phi = \frac{\zeta\omega_n}{\omega_n} = \zeta \quad \text{and} \quad \sin\phi = \frac{\omega_d}{\omega_n} = \sqrt{1-\zeta^2},$$

$$\cos\omega_d t + \frac{\zeta\omega_n}{\omega_d}\sin\omega_d t = \frac{\omega_n}{\omega_d}\left(\frac{\omega_d}{\omega_n}\cos\omega_d t + \zeta\sin\omega_d t\right)$$

$$= \frac{1}{\sqrt{1-\zeta^2}}(\sin\phi\cos\omega_d t + \cos\phi\sin\omega_d t)$$

$$= \frac{1}{\sqrt{1-\zeta^2}}\sin(\omega_d t + \phi).$$

Hence, the step response can be expressed as

$$\frac{y(t)}{y_{ss}} = \left(1 - \frac{1}{\sqrt{1-\zeta^2}}e^{-\zeta\omega_n t}\sin(\omega_d t + \phi)\right) 1(t). \quad (3.21)$$

3-4 STEP RESPONSE

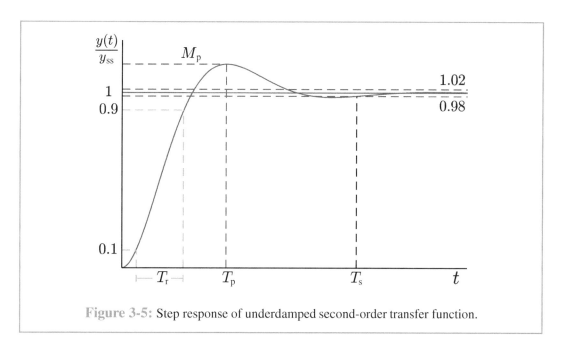

Figure 3-5: Step response of underdamped second-order transfer function.

A sketch of this response is shown in Fig. 3-5. The step response is characterized by the following parameters:

Rise time T_r:	Time to go from 10% to 90%
Settling time T_s:	Time to go from zero to within 2% of steady state
Peak time T_p:	Time of the first peak $y_{\max}$
Overshoot M_p:	$(y_{\max} - y_{ss})/y_{ss}$
Percent overshoot (PO):	$100 \times M_p$

Next, we determine how each of these response parameters depends on the parameters of the transfer function. Let us start with T_p. The times of maximum and minimum points of the function $y(t)/y_{ss}$ are determined by differentiating the function with respect to t and equating the derivative to zero.

$$\begin{aligned}
\frac{d}{dt}\left[\frac{y(t)}{y_{ss}}\right] &= \frac{-1}{\sqrt{1-\zeta^2}}\left[-\zeta\omega_n e^{-\zeta\omega_n t}\sin(\omega_d t + \phi) + \omega_d e^{-\zeta\omega_n t}\cos(\omega_d t + \phi)\right] \\
&= \frac{-\omega_n}{\sqrt{1-\zeta^2}}e^{-\zeta\omega_n t}\left[-\zeta\sin(\omega_d t + \phi) + \sqrt{1-\zeta^2}\cos(\omega_d t + \phi)\right] \\
&= \frac{-\omega_n}{\sqrt{1-\zeta^2}}e^{-\zeta\omega_n t}\left[-\cos\phi\sin(\omega_d t + \phi) + \sin\phi\cos(\omega_d t + \phi)\right] \\
&= \frac{-\omega_n}{\sqrt{1-\zeta^2}}e^{-\zeta\omega_n t}\sin(\omega_d t + \phi - \phi) = \frac{-\omega_n}{\sqrt{1-\zeta^2}}e^{-\zeta\omega_n t}\sin\omega_d t.
\end{aligned}$$

For

$$\frac{d}{dt}\left[\frac{y(t)}{y_{ss}}\right] = 0 \quad \rightarrow \quad \sin\omega_d t = 0 \quad \rightarrow \quad t = \frac{k\pi}{\omega_d}, \quad \text{for } k = 0, 1, 2, 3, \ldots$$

$k = 1, 2, 3, \ldots$ correspond to points where the response is a maximum or a minimum; $k = 1$ for the first maximum, $k = 2$ for the first minimum, $k = 3$ for the second maximum, and so on. Thus, for the first maximum $T_p = \pi/\omega_d$, and the maximum is

$$\frac{y_{\max}}{y_{ss}} = 1 - \frac{1}{\sqrt{1-\zeta^2}} e^{-\zeta \omega_n T_p} \sin(\omega_d T_p + \phi) \,.$$

Since

$$\omega_d T_p = \pi \quad \text{and} \quad \zeta \omega_n T_p = \zeta \omega_n \frac{\pi}{\omega_d} = \frac{\zeta \pi}{\sqrt{1-\zeta^2}} \,,$$

the foregoing expression simplifies to

$$\frac{y_{\max}}{y_{ss}} = 1 - \frac{1}{\sqrt{1-\zeta^2}} e^{-\zeta\pi/\sqrt{1-\zeta^2}} \sin(\pi + \phi)$$

$$= 1 + \frac{1}{\sqrt{1-\zeta^2}} e^{-\zeta\pi/\sqrt{1-\zeta^2}} \sin\phi$$

$$= 1 + e^{-\zeta\pi/\sqrt{1-\zeta^2}} \,.$$

Hence

$$M_p = \frac{y_{\max} - y_{ss}}{y_{ss}} = \frac{y_{\max}}{y_{ss}} - 1 = e^{-\zeta\pi/\sqrt{1-\zeta^2}} \,. \tag{3.22}$$

The overshoot depends only on ζ. As ζ increases towards 1, the overshoot decreases. Two values of ζ that are useful to remember are

ζ	PO (% overshoot)
0.5	16.3% < 17%
0.7	4.3% < 5%

The settling time T_s is the first time at which

$$\left| \frac{y(t) - y_{ss}}{y_{ss}} \right| \leq 0.02, \quad \text{for all } t \geq T_s \,,$$

$$\left| \frac{y(t) - y_{ss}}{y_{ss}} \right| = \left| \frac{1}{\sqrt{1-\zeta^2}} e^{-\zeta \omega_n t} \sin(\omega_d t + \phi) \right| \leq \frac{1}{\sqrt{1-\zeta^2}} e^{-\zeta \omega_n t} \,.$$

This upper bound is conservative because we replace the absolute value of a sine by its maximum value of 1. Equating the upper bound at $t = T_s$ to 0.02 yields

$$\zeta \omega_n T_s = \ln\left(\frac{1}{0.02\sqrt{1-\zeta^2}}\right) \,.$$

The right-hand side varies from 3.917 at $\zeta = 0.1$ to 4.742 at $\zeta = 0.9$, so we approximate it by 4 to obtain $T_s = 4/(\zeta \omega_n)$.

3-4 STEP RESPONSE

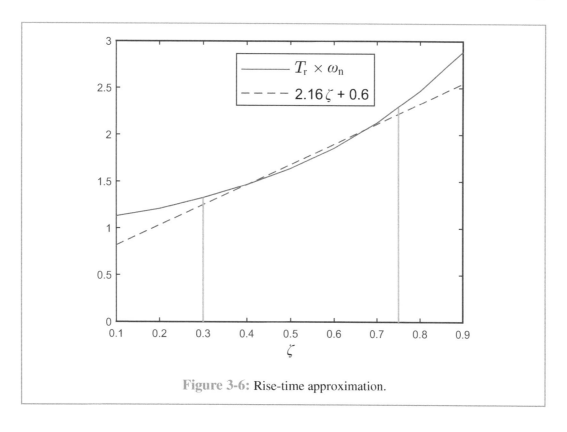

Figure 3-6: Rise-time approximation.

The rise time T_r depends on both ζ and ω_n, and there is no simple formula for it. However, for $0.3 \leq \zeta \leq 0.75$, it can be approximated as

$$T_r \approx \frac{2.16\zeta + 0.6}{\omega_n}.$$

Figure 3-6 compares $T_r \times \omega_n$ with $2.16\zeta + 0.6$ for different values of ζ. For $0.3 \leq \zeta \leq 0.75$, the relative approximation error ($100 \times \Delta T_r / T_r$) is less than 6%.

In summary,

$$G(s) = \frac{K\omega_n^2}{s^2 + 2\zeta\omega_n s + \omega_n^2}, \quad \omega_n > 0, \quad 0 < \zeta < 1,$$

$$\frac{y(t)}{y_{ss}} = \left[1 - \frac{1}{\sqrt{1-\zeta^2}} e^{-\zeta\omega_n t} \sin(\omega_d t + \phi)\right] 1(t), \quad \omega_d = \omega_n\sqrt{1-\zeta^2}, \quad \cos\phi = \zeta,$$

$$T_p = \frac{\pi}{\omega_d}, \quad PO = 100 \times e^{-\zeta\pi/\sqrt{1-\zeta^2}}, \quad T_s = \frac{4}{\zeta\omega_n},$$

$$T_r \approx \frac{2.16\zeta + 0.6}{\omega_n}, \quad \text{for } 0.3 \leq \zeta \leq 0.75.$$

Example 3-12: The Step-Response Parameters of a Second-Order Transfer Function

Compute T_r, T_p, T_s, PO, and y_{ss} to a unit step input, for each of the following transfer functions.

$$\text{(a)} \quad G(s) = \frac{100}{s^2 + 8\sqrt{2}s + 64}$$

$$\text{(b)} \quad G(s) = \frac{9}{s^2 + 3s + 9}$$

Solution:
(a) The form of $G(s)$ matches the underdamped transfer function defined by Eq. (3.18). Hence,

$$\omega_n^2 = 64 \text{ and } 2\zeta\omega_n = 8\sqrt{2} \quad \rightarrow \quad \omega_n = 8, \quad \zeta = \frac{1}{\sqrt{2}}, \quad \omega_d = \omega_n\sqrt{1-\zeta^2} = \frac{8}{\sqrt{2}},$$

$$T_r \approx \frac{2.16\zeta + 0.6}{\omega_n} = \frac{2.16 \times 0.707 + 0.6}{8} = 0.2659 \text{ s},$$

$$T_p = \frac{\pi}{\omega_d} = \frac{\pi}{8/\sqrt{2}} = 0.5554 \text{ s}, \quad T_s = \frac{4}{\zeta\omega_n} = \frac{4}{4\sqrt{2}} = 0.707 \text{ s},$$

$$PO = 100 \times e^{-\zeta\pi/\sqrt{1-\zeta^2}} = 100 \times e^{-\pi} = 4.3\%,$$

$$y_{ss} = G(0) = \frac{100}{64}.$$

(b) This is also an underdamped transfer function with

$$\omega_n^2 = 9 \text{ and } 2\zeta\omega_n = 3 \quad \rightarrow \quad \omega_n = 3, \quad \zeta = \frac{1}{2}, \quad \omega_d = \omega_n\sqrt{1-\zeta^2} = \frac{3\sqrt{3}}{2},$$

$$T_r \approx \frac{2.16\zeta + 0.6}{\omega_n} = \frac{2.16 \times 0.5 + 0.6}{3} = 0.56 \text{ s},$$

$$T_p = \frac{\pi}{\omega_d} = \frac{\pi}{3\sqrt{3}/2} = 1.21 \text{ s}, \quad T_s = \frac{4}{\zeta\omega_n} = \frac{4}{1.5} = 2.67 \text{ s},$$

$$PO = 100 \times e^{-\zeta\pi/\sqrt{1-\zeta^2}} = 100 \times e^{-\pi/\sqrt{3}} = 16.3\%,$$

$$y_{ss} = G(0) = 1.$$

3-4.3 Effect of Zeros

The step response for the underdamped transfer function calculated in the previous examples, together with T_r, T_p, T_s, and PO, are valid when the transfer function has no zero. The presence of a zero can affect the response in a significant way. The zero could be in the left-half plane or the right-half plane. We consider these two cases separately.

3-4 STEP RESPONSE

Effect of zero in the left-half plane (minimum-phase system)

The transfer function
$$G(s) = \frac{K\omega_n^2 \left(1 + \frac{s}{a}\right)}{s^2 + 2\zeta\omega_n s + \omega_n^2}, \quad a > 0 \tag{3.23}$$

has a zero at $s = -a < 0$. With a step input $u(t) = A\,1(t)$, we have

$$Y(s) = \frac{KA\omega_n^2 \left(1 + \frac{s}{a}\right)}{s(s^2 + 2\zeta\omega_n s + \omega_n^2)} = \underbrace{\frac{KA\omega_n^2}{s(s^2 + 2\zeta\omega_n s + \omega_n^2)}}_{Y_1(s)} + \underbrace{\frac{KA\omega_n^2 s/a}{s(s^2 + 2\zeta\omega_n s + \omega_n^2)}}_{\frac{s}{a}Y_1(s)}, \tag{3.24}$$

where $y_1(t)$ is the step response of the transfer function with no zero. The inverse Laplace transform yields

$$y(t) = y_1(t) + \frac{1}{a}\frac{dy_1}{dt}.$$

Since $y_1(t)$ is increasing for small t, its derivative dy_1/dt is positive, and $(1/a)\,dy_1/dt$ contributes to faster rise and larger overshoot. When the zero is far to the left; that is, a is large, $y(t) \approx y_1(t)$. As the zero approaches the origin, the overshoot increases.

> **Example 3-13: Effect of Zero in the Left-Half Plane**

Figure 3-7 shows the step response of the transfer function $(1 + s/a)/(s^2 + s + 1)$ for three different values of a, compared with the step response of $1/(s^2 + s + 1)$ (no zero). As the zero approaches the origin, the rise time is smaller and the overshoot is larger. Table 3-1 compares the percent overshoot for $a = 10, 5, 2, 1, 0.5$, and 0.25 with the percent overshoot when there is no zero.

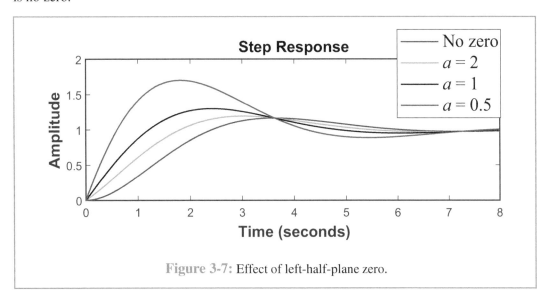

Figure 3-7: Effect of left-half-plane zero.

Table 3-1: Percent overshoot for different values of a.

a	No zero	10	5	2	1	0.5	0.25
PO	16.29	16.38	16.68	19.1	29.8	69.9	171.2

Effect of zero in the right-half plane (non-minimum-phase system)

The transfer function

$$G(s) = \frac{K\omega_n^2\left(1 - \frac{s}{a}\right)}{s^2 + 2\zeta\omega_n s + \omega_n^2}, \quad a > 0 \tag{3.25}$$

has a zero at $s = a > 0$. Repeating the steps in the previous case, the step response can be expressed as

$$y(t) = y_1(t) - \frac{1}{a}\frac{dy_1}{dt},$$

where y_1 is the step response when there is no zero. Once again, dy_1/dt is positive for small t, so that the effect of the zero is to push the response in the negative direction. In particular, at the starting point

$$y(0^+) = \underbrace{y_1(0^+)}_{=0} - \frac{1}{a}\underbrace{\frac{dy_1}{dt}(0^+)}_{>0} < 0.$$

The response undershoots the commanded input because it moves in the opposite direction before it turns around to approach the steady state. When the zero is far to the right, that is, when a is large, $y(t) \approx y_1(t)$. As the zero approaches the origin, the undershoot increases.

Example 3-14: Effect of Zero in the Right-Half Plane

Figure 3-8 shows the step response of the transfer function $(1 - s/a)/(s^2 + s + 1)$ for three different values of a, compared with the step response of $1/(s^2 + s + 1)$ (no zero). As the zero approaches the origin, the undershoot is larger.

In summary,

▶ When the zero is far to the left or far to the right, its effect on the step response is negligible. As the zero approaches the origin, its effect increases. For minimum-phase systems the rise time decreases and the overshoot increases, while for non-minimum-phase systems the undershoot increases. ◀

3-4 STEP RESPONSE 93

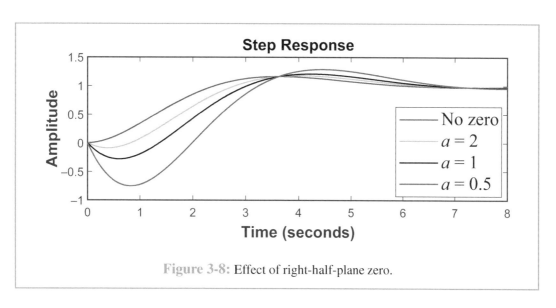

Figure 3-8: Effect of right-half-plane zero.

3-4.4 Approximation of Higher-Order Transfer Functions

We have derived the step response for a first-order transfer function of the form $b_0/(s+a_0)$ and a second-order transfer function of the form $b_0/(s^2+a_1s+a_0)$. What about higher-order transfer functions? In general, calculating the step response would be very complicated and we need to use computer simulation to find it. However, there are special cases where the step response of a higher-order transfer function can be approximated by a first-order or second-order transfer function. Two such cases are discussed next for approximation by a second-order transfer function; similar discussion can be carried out for approximation by a first-order transfer function.

Case 1: Additional poles are farther to the left than a pair of dominant poles

We illustrate this case by considering the transfer function

$$G(s) = \frac{b_0}{(s^2+a_1s+a_0)(s+\alpha_1)(s+\alpha_2)}, \qquad (3.26)$$

where all the coefficients are positive. Suppose the poles $-\alpha_1$ and $-\alpha_2$ are farther to the left than the dominant poles, which are the roots of $s^2+a_1s+a_0$. For convenience suppose $\alpha_1 \neq \alpha_2$, and let us represent them as $\alpha_1 = \gamma_1\alpha$ and $\alpha_2 = \gamma_2\alpha$, where $\gamma_2 > \gamma_1 > 0$. As α increases, these two poles move farther to the left. The dc gain of $G(s)$ is $G(0) = b_0/(a_0\alpha_1\alpha_2)$. Writing $G(s)$ in terms of $G(0)$, we obtain

$$G(s) = \frac{G(0)a_0\alpha_1\alpha_2}{(s^2+a_1s+a_0)(s+\alpha_1)(s+\alpha_2)} = \frac{G(0)a_0\gamma_1\gamma_2\alpha^2}{(s^2+a_1s+a_0)(s+\gamma_1\alpha)(s+\gamma_2\alpha)}.$$

The Laplace transform of the output to a step input of amplitude A is

$$Y(s) = \frac{G(0)a_0\gamma_1\gamma_2\alpha^2 A}{s(s^2+a_1s+a_0)(s+\gamma_1\alpha)(s+\gamma_2\alpha)}.$$

By partial-fraction expansion,

$$Y(s) = \frac{K_1}{s+\gamma_1\alpha} + \frac{K_2}{s+\gamma_2\alpha} + \cdots$$

The dots stand for the additional terms of the expansion. The residues K_1 and K_2 are given by

$$K_1 = \frac{G(0)\,a_0\gamma_1\gamma_2\alpha^2 A}{-\gamma_1\alpha(\gamma_1^2\alpha^2 - a_1\gamma_1\alpha + a_0)(-\gamma_1\alpha+\gamma_2\alpha)} = \frac{G(0)\,a_0\gamma_1\gamma_2 A}{-\gamma_1(\gamma_1^2\alpha^2 - a_1\gamma_1\alpha + a_0)(\gamma_2-\gamma_1)},$$

$$K_2 = \frac{G(0)\,a_0\gamma_1\gamma_2\alpha^2 A}{-\gamma_2\alpha(\gamma_2^2\alpha^2 - a_1\gamma_2\alpha + a_0)(-\gamma_2\alpha+\gamma_1\alpha)} = \frac{G(0)\,a_0\gamma_1\gamma_2 A}{\gamma_2(\gamma_2^2\alpha^2 - a_1\gamma_2\alpha + a_0)(\gamma_2-\gamma_1)}.$$

The inverse Laplace transform yields

$$y(t) = \left(K_1 e^{-\gamma_1\alpha t} + K_2 e^{-\gamma_2\alpha t} + \cdots\right) 1(t). \tag{3.27}$$

For large α, the exponential terms $e^{-\gamma_1\alpha t}$ and $e^{-\gamma_2\alpha t}$ decay to zero faster than the terms due to the dominant poles. Moreover, the residues K_1 and K_2 are small for large α; in particular, $\lim_{\alpha\to\infty} K_1 = 0$ and $\lim_{\alpha\to\infty} K_2 = 0$; Therefore, for large α, the contribution of the additional poles to the step response is negligible. How much farther to the left should the additional poles be for the approximation to be acceptable? The rule of thumb is:

> ▶ The additional poles should be five times farther to the left than the dominant poles for acceptable approximation. A very good approximation is typically achieved if they are ten times farther to the left. ◀

In neglecting the additional poles, care should be taken not to change the dc gain of the transfer function. This can be done by expressing the term in the denominator due to the pole in the form $(\tau s + 1)$ and then setting $\tau = 0$. For the foregoing example, let $\tau_1 = 1/\alpha_1$ and $\tau_2 = 1/\alpha_2$, and rewrite the transfer function as

$$G(s) = \frac{G(0)\,a_0}{(s^2 + a_1 s + a_0)(\tau_1 s + 1)(\tau_2 s + 1)}, \tag{3.28}$$

with

$$\tau_1 \approx 0 \text{ and } \tau_2 \approx 0 \quad \longrightarrow \quad G(s) \approx \frac{G(0)a_0}{(s^2 + a_1 s + a_0)}.$$

Example 3-15: Approximation of a Third-Order Transfer Function

The transfer function

$$\frac{1}{(s^2+s+1)(\tau s+1)}$$

has a pair of complex poles at $-0.5 \pm 0.5\sqrt{3}j$ and a third pole at $-1/\tau$. According to the rule of thumb, it can be approximated by

$$\frac{1}{s^2+s+1}$$

if

$$\frac{1/\tau}{0.5} \geq 5, \quad \text{i.e., } \tau < 0.4.$$

Figure 3-9 shows the step responses of the second-order and third-order transfer functions for two different value of τ:

$$\tau = 1 \quad \rightarrow \quad \frac{1/\tau}{0.5} = \frac{1}{0.5} = 2,$$

$$\tau = 0.25 \quad \rightarrow \quad \frac{1/\tau}{0.5} = \frac{4}{0.5} = 8.$$

The case $\tau = 0.25$ is within the range of the rule of thumb, while $\tau = 1$ is not. The figure shows good approximation in the first case, and not so good approximation in the second case.

Case 2: Additional poles are very close to zeros (almost pole-zero cancellation)

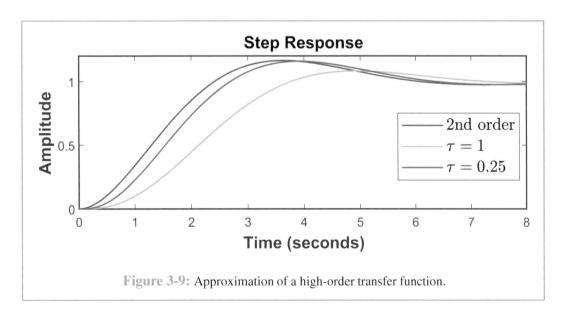

Figure 3-9: Approximation of a high-order transfer function.

We illustrate this case by the transfer function

$$G(s) = \frac{b_0(s+\beta)}{(s^2+a_1 s+a_0)(s+\alpha)},$$

where all coefficients are positive and $\alpha \approx \beta$. For the step input $u(t) = A\,1(t)$,

$$Y(s) = \frac{b_0(s+\beta)A}{s(s^2+a_1 s+a_0)(s+\alpha)}.$$

Partial-fraction expansion yields

$$Y(s) = \frac{K}{s+\alpha} + \cdots,$$

where

$$K = \frac{b_0(-\alpha+\beta)A}{-\alpha(\alpha^2 - a_1\alpha + a_0)},$$

$$\alpha \approx \beta \;\;\longrightarrow\;\; \beta - \alpha \approx 0 \;\;\longrightarrow\;\; K \approx 0.$$

Thus, the effect of the pole at $-\alpha$ on the step response is negligible. The pole-zero cancellation is performed while preserving the dc gain by writing $G(s)$ as

$$G(s) = \frac{b_0(\beta/\alpha)(s/\beta+1)}{(s^2+a_1 s+a_0)(s/\alpha+1)} \approx \frac{b_0(\beta/\alpha)}{(s^2+a_1 s+a_0)}.$$

Example 3-16: Approximation of a Third-Order Transfer Function with Almost Pole-Zero Cancellation

---

The third-order transfer function

$$\frac{(s+z)}{(s^2+s+1)(s+1)}$$

can be approximated by the second-order transfer function

$$\frac{z}{(s^2+s+1)}$$

if z is close enough to 1.

Figure 3-10 shows the step responses of the second-order and third-order transfer functions for $z = 0.5$, 1.1, and 2. For $z = 0.5$ and 2, the zero is not close enough to the pole at -1 and the responses are not close to each other. For $z = 1.1$, the two responses are almost indistinguishable.

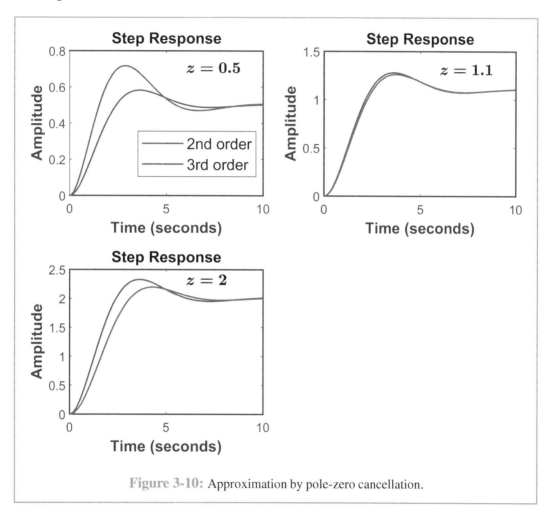

Figure 3-10: Approximation by pole-zero cancellation.

Summary

Concepts

- The system is stable when its output is bounded for every bounded input.
- The system is stable when
    - its impulse response function satisfies $\int_0^\infty |g(t)|\, dt < \infty$.
    - its poles have negative real parts.
- The system is marginally stable when all its poles have nonpositive real parts and poles with zero real parts are simple
- Using the Routh-Hurwitz criterion we can determine stability without calculating the poles of the transfer function
- The steady-state response of a stable transfer function to
    - a step input is a constant equal to the amplitude of the input multiplied by the dc gain.
    - a sinusoidal input is a sinusoidal signal having the same frequency as the input, with the amplitude and phase modified by the frequency response.
    - a periodic input is a periodic signal having the same period as the input.
- The step response is monotonic for first-order transfer functions and also for overdamped and critically damped second-order transfer functions.
- The step response is oscillatory for underdamped second-order transfer functions.
- A zero in the left-half complex plane increases the overshoot.
- A zero in the right-half complex plane causes undershoot.
- The effect of a zero is negligible if it is far to the left or far to the right in the complex plane relative to the dominant poles.
- A higher-order transfer function can be approximated by a second-order transfer function if
    - extra poles are far to the left in the complex plane relative to the dominant poles;
    - extra poles are very close to zeros (almost resulting in pole-zero cancellation).

Mathematical Models

Step input: $u(t) = A\,1(t)$

Steady-state response:
  dc gain $\quad\quad\quad\quad G(0)$
  Steady-state output $\quad y_{ss} = G(0)\,A$

Transient response:

$G(s) = K/(s+a)$:
  Output $\quad\quad\quad y(t) = y_{ss}(1 - e^{-t/\tau})\,1(t)$
  Time constant $\quad \tau = 1/a$
  Rise time $\quad\quad T_r = 2.2\tau$ (10% to 90% of y_{ss})
  Settling time $\quad T_s = 4\tau$ (0 to 98% of y_{ss})

$G(s) = K\omega_n^2/(s^2 + 2\zeta\omega_n s + \omega_n^2)$:
  Damping ratio $\quad\quad\quad\quad\quad\quad \zeta < 1$
  Undamped natural frequency $\quad \omega_n$
  Damped natural frequency $\quad\quad \omega_d = \omega_n\sqrt{1-\zeta^2}$
  Angle ϕ $\quad\quad\quad\quad\quad\quad\quad\quad \cos\phi = \zeta$

  Output $\quad\quad y(t) = y_{ss}\left[1 - \dfrac{1}{\sqrt{1-\zeta^2}}\,e^{-\zeta\omega_n t}\sin(\omega_d t + \phi)\right]1(t)$

  Rise time $\quad\quad T_r$ (10% to 90% of y_{ss})
  $\quad\quad\quad\quad\quad\quad \approx (2.16\zeta + 0.6)/\omega_n$, for $0.3 \le \zeta \le 0.75$
  Settling time $\quad T_s = 4/(\zeta\omega_n)$ (0 to within 2% of y_{ss})
  Peak time $\quad\quad T_p = \pi/\omega_d$ (time of first peak)
  Percent overshoot $\quad PO = 100\,(y_{max} - y_{ss})/y_{ss} = 100\,e^{-\zeta\pi/\sqrt{1-\zeta^2}}$

Important Terms Provide definitions or explain the meaning of the following terms:

BIBO stability
damped natural frequency
damping ratio
dc gain
forced response
frequency response
marginal stability

minimum-phase system
natural response
non-minimum-phase system
overshoot
peak time
rise time

Routh-Hurwitz criterion
settling time
steady-state response
transient response
undamped natural frequency
undershoot

PROBLEMS

3.1 For each of the following transfer functions, determine whether the system is stable or unstable. If it is unstable, determine whether it is marginally stable.

(a) $G(s) = \dfrac{s+3}{(s+1)(s+2)(s+5)}$

(b) $G(s) = \dfrac{s-3}{(s+1)(s-2)(s+5)}$

(c) $G(s) = \dfrac{1}{s^2}$

(d) $G(s) = \dfrac{s-2}{(s+1)(s^2+4)}$

(e) $G(s) = \dfrac{s-3}{(s+1)(s+2)(s+5)}$

(f) $G(s) = \dfrac{2(s^4+1)}{(s^2+s+1)(s+1)(s+2)}$

3.2 Repeat the previous problem for each of the following transfer functions.

(a) $G(s) = \dfrac{s+2}{(s+1)(s^2+4)}$

(b) $G(s) = \dfrac{4}{s^2-9}$

(c) $G(s) = \dfrac{s+2}{(s+1)(s^2+4)^2}$

(d) $G(s) = \dfrac{s+3}{(s+1)(s-2)(s+5)}$

(e) $G(s) = \dfrac{2}{(s^2-s+1)(s+1)(s+2)}$

(f) $G(s) = \dfrac{s+2}{(s-1)(s^2+4)(s+5)}$

3.3 For each of the following transfer functions, use the Routh-Hurwitz criterion to determine whether the system is stable or unstable. If it is unstable, determine whether it is marginally stable. If it is unstable but not marginally stable, determine how many poles are in the right-half plane.

(a) $G(s) = \dfrac{2s^2-8s-10}{s^4+8s^3+23s^2+28s+12}$

(b) $G(s) = \dfrac{5s+5}{s^4+3s^3+3s^2+3s+2}$

(c) $G(s) = \dfrac{s^2+3s+1}{s^4+2s^2+1}$

(d) $G(s) = \dfrac{1}{s^4 + 7s^3 + 17s^2 + 17s + 6}$

(e) $G(s) = \dfrac{8s + 1}{s^4 + 3s^3 + 3s^2 + 9s + 1}$

3.4 Repeat the previous problem for each of the following transfer functions.

(a) $G(s) = \dfrac{1}{s^4 + s^3 + s^2 + s + 1}$

(b) $G(s) = \dfrac{s^2 - 4s + 2}{s^4 + 5s^3 + 9s^2 + 7s + 2}$

(c) $G(s) = \dfrac{2}{s^5 + 5s^4 + 10s^3 + 10s^2 + 5s + 1}$

(d) $G(s) = \dfrac{2}{s^6 + 7s^5 + 20s^4 + 30s^3 + 25s^2 + 11s + 2}$

(e) $G(s) = \dfrac{4}{s^4 + 4s^3 + 3s^2 - 4s - 4}$

3.5 Consider the feedback control system of Fig. 3-1 where $G_c(s) = K \geq 0$. For each of the following cases of $G_p(s)$, find the range of K for which the system is stable.

(a) $G_p(s) = \dfrac{4(s+1)}{s^3 + 9s^2 + 26s + 24}$

(b) $G_p(s) = \dfrac{10}{s^4 + 10s^3 + 35s^2 + 50s + 24}$

(c) $G_p(s) = \dfrac{s+1}{s^3 + 4s^2 + s - 6}$

(d) $G_p(s) = \dfrac{4(s-1)}{s^3 + 9s^2 + 26s + 24}$

3.6 Estimate T_r, T_p, T_s, PO, and y_{ss} to a unit step input, for each of the following transfer functions. Then, use MATLAB to find these quantities and compare with your estimates.

(a) $G(s) = \dfrac{10}{s^2 + 2s + 5}$

(b) $G(s) = \dfrac{9}{s^2 + 4.2s + 9}$

(c) $G(s) = \dfrac{3}{s^2 + 0.6s + 1}$

(d) $G(s) = \dfrac{8}{(s^2 + 2s + 4)(s + 8)}$

(e) $G(s) = \dfrac{(s+8)}{s^2 + 1.73s + 3}$

(f) $G(s) = \dfrac{10}{(s^2+s+1)(s+10)}$

3.7 Repeat the previous problem for each of the following transfer functions.

(a) $G(s) = \dfrac{(8-s)}{s^2+2.45s+3}$

(b) $G(s) = \dfrac{(s+8)}{(s^2+2s+4)(s+10)}$

(c) $G(s) = \dfrac{(s+0.9)}{(s^2+2s+4)(s+1)}$

(d) $G(s) = \dfrac{(s+0.9)}{(s^2+s+1)(s+1)(s+10)}$

(e) $G(s) = \dfrac{80}{(s^2+s+1)(s+8)(s+10)}$

3.8 Use MATLAB to plot the response of the transfer function

$$G(s) = \dfrac{3}{(s^2+s+1)(s+1)(s+3)}$$

to each of the following inputs. For the first three cases determine the steady state and comment on it in view of the conclusions of Section 3-3.

(a) Unit step input $u(t) = 1(t)$

(b) Sinusoidal input $u(t) = \sin 3t$

(c) Periodic input $u(t) = \sin t + 0.3\sin(3t) + 0.1\sin(5t)$

(d) Ramp input $u(t) = t\, 1(t)$

3.9 Repeat Problem 3.8 for the transfer function

$$G(s) = \dfrac{s+2}{s^2+2s+4}.$$

3.10 Repeat Problem 3.8 for the transfer function

$$G(s) = \dfrac{s+2}{(s^2+2s+4)(s+3)}.$$

3.11 For each of the unit step responses shown in Fig. P3.11, model the system by a transfer function. Then, find the step response using MATLAB and compare with the one in Fig. P3.11.

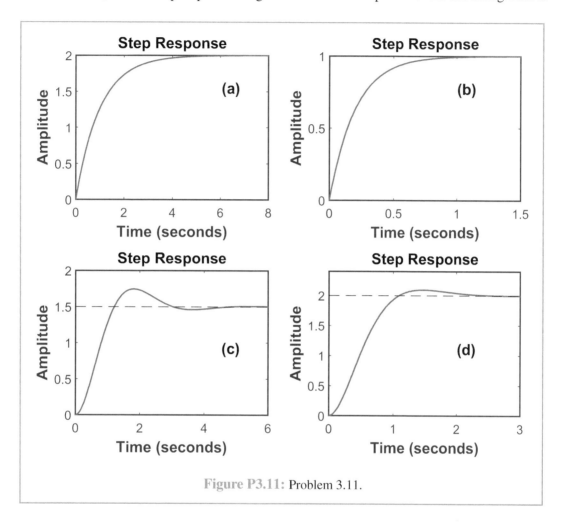

Figure P3.11: Problem 3.11.

3.12 Explain why the step responses of the following four transfer functions are close to each other:

(a) $\dfrac{4}{s^2+2s+4}$

(b) $\dfrac{4}{(s^2+2s+4)(0.1s+1)}$

(c) $\dfrac{4(0.1s+1)}{s^2+2s+4}$

(d) $\dfrac{4(2.1s+1)}{(s^2+2s+4)(2s+1)}$

3.13 Find the transfer function of the system whose unit step response is shown in Fig. P3.13.

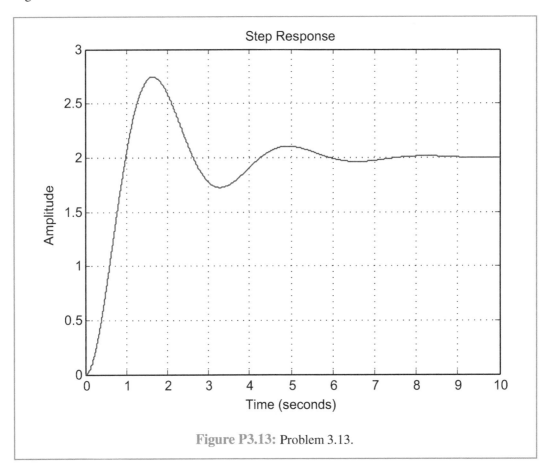

Figure P3.13: Problem 3.13.

3.14 Show that the automobile suspension system of Example 2-11 is stable.

3.15 Consider the automobile suspension system of Example 2-11 with the parameters $M_s = 973$ kg, $M_{us} = 114$ kg, $k_s = 42720$ N/m, $k_t = 101115$ N/m, and $C_s = 1095$ N·s/m [13]. We will simulate the response of the system to an isolated bump in an otherwise smooth surface. The road displacement is defined by

$$S(t) = \begin{cases} \frac{A}{2}\left[1 - \cos\left(\frac{2\pi v t}{L}\right)\right] & \text{if } 0 \leq t \leq \frac{L}{v}, \\ 0 & \text{if } t > \frac{L}{v}, \end{cases}$$

where A and L are the height and length of the bump and v is the forward speed of the automobile. $A = 0.06$ m and $L = 5$ m.

(a) Plot the body acceleration $\ddot{y}_s$ and the suspension stroke $y_s - y_{us}$ for $t \in [0, 10]$ s when $v = 15$ m/s and when $v = 30$ m/s.

(b) Repeat (a) when $C_s = 500$ N·s/m.

(c) Comment on the results.

3.16 Engine idle is frequently encountered in city driving. The objective of idle speed control is to maintain the engine speed at a desired set point in the presence of various load disturbances due to air conditioning, power windows, etc. Two control inputs are usually used in idle speed control: spark timing and air flow (regulated by the throttle or a bypass valve). For the purpose of this problem, we consider only air flow. Figure P3.16 shows a linearized model for a typical eight-cylinder engine [7]. In this figure, the control input Δu is the change of the bypass valve, the disturbance input ΔT_L is the change of the load torque, and the output ΔN is the change of the engine speed; all changes are from the corresponding nominal values. The time delay is due to the digital implementation of the controller. It is modeled in [7] to be 0.04 s. The transfer function of the delay should be $e^{-0.04s}$ but we will use the Padé approximation

$$H(s) = \frac{1 - (0.04/2)s + (0.04)^2 s^2/12}{1 + (0.04/2)s + (0.04)^2 s^2/12}.$$

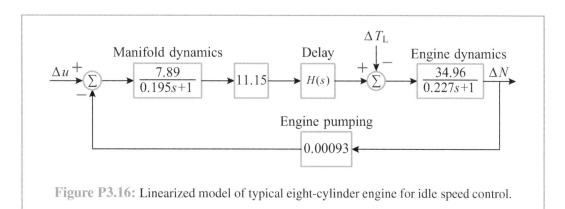

Figure P3.16: Linearized model of typical eight-cylinder engine for idle speed control.

(a) Find the transfer functions from Δu to ΔN and from ΔT_L to ΔN.

(b) Find the poles of the transfer function with and without delay. By "without delay" we mean ignoring the delay by taking $H(s) = 1$.

(c) Plot the unit step response from Δu and ΔT_L with and without delay.

(d) Comment on the effect of the delay.

3.17 A synchronous generator connected to an infinite bus is modeled by [30]

$$\ddot{\delta} = -b_1 E_\text{q} \sin \delta - b_2 \dot{\delta} + P,$$
$$\dot{E}_\text{q} = b_3 \cos \delta - b_4 E_\text{q} + E_\text{F},$$

where δ is the load angle in radians, E_q is the quadrature axis internal voltage, P is the mechanical input power, and E_F is the field voltage (input). The coefficients b_1 to b_4 are positive. When $P = P^\star$ and $E_\text{F} = E_\text{F}^\star$ are positive constants, the equilibrium points $(\delta^\star, E_\text{q}^\star)$ of the system satisfy the equations

$$0 = -b_1 E_\text{q}^\star \sin \delta^\star + P^\star, \qquad 0 = b_3 \cos \delta^\star - b_4 E_\text{q}^\star + E_\text{F}^\star.$$

It can be shown that when $P^\star < b_1 E_F^\star / b_4$, there is a unique equilibrium point in the range $0 < \delta < \pi/2$, which is the equilibrium point of interest. Assume that

$$b_4 E_q^\star \cos \delta^\star - b_3 \sin^2 \delta^\star > 0 \,.$$

Writing $\delta = \delta^\star + y$, $E_q = E_q^\star - x$, and $E_F = E_F^\star - u$, a linearized model of the system near the equilibrium point is shown in the block diagram of Fig. P3.17.

(a) Find the transfer function $G_p(s) = Y(s)/U(s)$.

(b) Using the Routh-Hurwitz criterion, verify that $G_p(s)$ is stable.

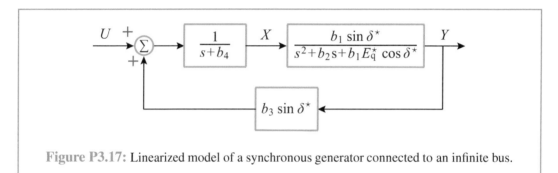

Figure P3.17: Linearized model of a synchronous generator connected to an infinite bus.

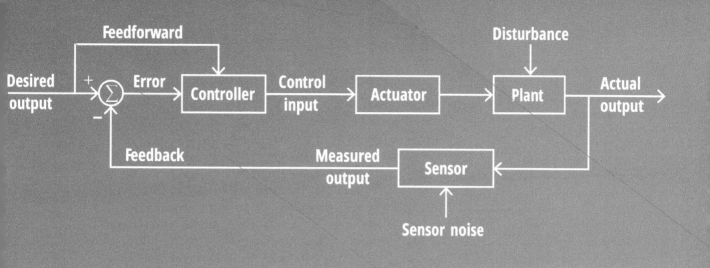

Chapter 4
Feedback Control Systems

Chapter Contents

	Overview, 108
4-1	Stability, 108
4-2	The Root Locus Method, 110
4-3	Steady-State Error, 126
4-4	Effect of Disturbance, 133
4-5	Plant Parameter Uncertainty, 136
4-6	Effect of Measurement Noise, 139
4-7	Control Constraints, 140
4-8	Why Use Closed-Loop Instead of Open-Loop?, 142
	Chapter Summary, 144
	Problems, 145

Objectives

Upon learning the material presented in this chapter, you should be able to:

1. Study the stability of feedback control systems.
2. Use the root locus method to study how the closed-loop poles change as the loop gain changes.
3. Compute the steady-state tracking error for step, ramp, and parabola reference inputs.
4. Study the effect of disturbance, plant parameter uncertainty, and measurement noise on the performance of the system.
5. Understand the effect of control constraints.
6. Compare open-loop and closed-loop control systems.

Overview

In this chapter we look at the main features of feedback control systems. Without doubt, the most important feature is stability. The feedback system must be designed to be stable, whether or not the plant is stable. Stability is characterized in Section 4-1 in terms of the roots of the characteristic equation. This characterization is carried out further in Section 4-2 by developing the root locus method that allows us to study how the roots of the characteristic equation change as some parameter of the system changes. Implicit in the stabilization of the system is the assignment of the closed-loop poles, which, together with the closed-loop zeros, shape the transient response of the system. Once we guarantee the stability of the system, the next most important feature is the ability of the system to track the desired reference signal. Section 4-3 characterizes the steady-state tracking error for three typical reference signals: the step, the ramp, and the parabola. The performance of the feedback system is affected by external factors beyond our control. Sections 4-4 to 4-6 describe three such factors: disturbance, plant parameter uncertainty, and measurement noise. The effects of the first two factors can be reduced by designing a tight feedback loop; that is, a feedback loop with high loop gain. Such a design does not reduce the affect of measurement noise, and design trade-offs, usually done in the frequency domain, need to be made to address all three effects. The performance of feedback control systems is limited by our inability to make the control signal arbitrarily large since the control input to any plant has to satisfy certain constraints. Section 4-7 discusses control constraints. Finally, Section 4-8 summaries the pros and cons of feedback control systems when compared with open-loop control.

4-1 Stability

A typical configuration of the feedback control system is shown in **Fig. 4-1(a)**, where $G_p(s)$ is the transfer function of the plant (that is, the system to be controlled), $G_c(s)$ is the transfer function of the controller or compensator, and $H(s)$ is the transfer function of the sensor that measures the output y. The input r is the reference signal to be tracked by the output y according to certain design specifications. Define $G(s) = G_p(s)\, G_c(s)$ and redraw the system

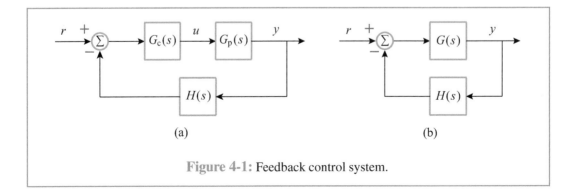

Figure 4-1: Feedback control system.

4-1. STABILITY

as in Fig. 4-1(b). The closed-loop transfer function is then given by

$$\frac{Y(s)}{R(s)} = \frac{G(s)}{1 + G(s)H(s)} \stackrel{\text{def}}{=} T(s) . \quad (4.1)$$

The closed-loop system is stable when the poles of $T(s)$ (called *closed-loop poles*) have negative real parts. The poles of $T(s)$ are the roots of the characteristic equation

$$1 + G(s)H(s) = 0 . \quad (4.2)$$

Investigating the stability of the closed-loop system is usually done by investigating the roots of the characteristic equation.

Example 4-1: PI Control of a dc Motor

We saw in Chapter 2 that an armature-controlled dc motor with the armature voltage v_a as the control input and the motor speed ω_m as the output is represented by the transfer function given by Eq. (2.31c):

$$G_p(s) = \frac{K_m}{\tau_m s + 1} .$$

In speed control problems, the motor speed ω_m is required to track a reference signal ω_r. A common controller in such problems is the PI (Proportional-Integral) controller, represented by

$$G_c(s) = K_P + \frac{K_I}{s} = \frac{K_P s + K_I}{s} ,$$

with positive gains K_P and K_I. Suppose that the sensor that measures the motor speed has transfer function $H(s) = 1$. Upon using $G(s) = G_p(s)G_c(s)$ in Eq. (4.2), the characteristic equation of the closed-loop system is given by

$$1 + \left(\frac{K_P s + K_I}{s}\right)\left(\frac{K_m}{\tau_m s + 1}\right) = 0 .$$

Multiplication by $s(\tau_m s + 1)$ and division by τ_m yields the polynomial equation

$$s^2 + \left(\frac{1 + K_P K_m}{\tau_m}\right) s + \frac{K_I K_m}{\tau_m} = 0 .$$

The roots of this equation are the closed-loop poles of the system. In the design of the controller gains K_P and K_I, suppose we fix the ratio K_I/K_P and then examine how changing the gain K_P changes the closed-loop poles. Then

$$G(s)H(s) = \frac{K_P K_m}{\tau_m} \frac{(s + K_I/K_P)}{s(s + 1/\tau_m)} .$$

Defining

$$K = \frac{K_P K_m}{\tau_m} \quad \text{and} \quad L(s) = \frac{(s + K_I/K_P)}{s(s + 1/\tau_m)} ,$$

the characteristic equation is then given by

$$1 + K L(s) = 0,$$

where $L(s)$ is given and the roots of the equation change with K. Studying how the roots change as K changes is the subject of the root locus method of the next section.

4-2 The Root Locus Method

Consider the feedback control system of **Fig. 4-1(b)**. The transfer function $G(s)H(s)$, which we write as $GH(s)$, is called the *open-loop transfer function*. Let $GH(s) = K L(s)$ where $K \geq 0$ is a constant and

$$L(s) = \frac{\prod_{i=1}^{m}(s-z_i)}{\prod_{i=1}^{n}(s-p_i)} = \frac{(s-z_1)(s-z_2) \times \cdots \times (s-z_m)}{(s-p_1)(s-p_2) \times \cdots \times (s-p_n)}, \qquad m \leq n. \qquad (4.3)$$

$L(s)$ has m zeros at $z_1, z_2, \ldots, z_m$ and n poles at $p_1, p_2, \ldots, p_n$. They are called the open-loop zeros and poles, respectively, because they are the zeros and poles of the open-loop transfer function $GH(s)$. Note that $L(s)$ is a ratio of two monic polynomials; that is, polynomials where the coefficient of the highest power of s is one. There is no loss of generality in this representation because if the numerator or denominator polynomial of $GH(s)$ is not monic, the coefficient of the highest power of s can be factored out and included in the definition of the constant K, as it was done in Example 4-1.

The roots of the characteristic equation

$$1 + K L(s) = 0 \qquad (4.4)$$

vary as K varies. The root locus method provides simple rules that allow us to sketch the locus in the complex plane of each root of the characteristic equation as K varies from zero to infinity. The rules are derived from two criteria, known as the magnitude criterion and the phase (or angle) criterion. Let the complex number s be a root of the characteristic equation represented by Eq. (4.4). Then, for $K > 0$,

$$K L(s) = -1. \qquad (4.5)$$

For the complex number $K L(s)$ to be equal to -1, its magnitude must be one and its angle an odd multiple of $180°$; that is,

$$K|L(s)| = 1, \qquad (4.6)$$
$$\angle L(s) = k \times 180°, \qquad k = \pm 1, \pm 3, \pm 5, \ldots \qquad (4.7)$$

Equation (4.6) is the magnitude criterion and Eq. (4.7) is the phase (angle) criterion. In writing the foregoing two equations, we used the fact that K is a real positive constant so that $|K L(s)| = K|L(s)|$ and $\angle K L(s) = \angle L(s)$. Using the form of $L(s)$ from Eq. (4.3), we see that the magnitude and phase criteria can be written as

$$K \frac{\prod_{i=1}^{m} |s - z_i|}{\prod_{i=1}^{n} |s - p_i|} = 1 \qquad (4.8)$$

4-2. THE ROOT LOCUS METHOD

and

$$\sum_{i=1}^{m} \angle(s - z_i) - \sum_{i=1}^{n} \angle(s - p_i) = k \times 180°, \qquad k = \pm 1, \pm 3, \pm 5, \ldots \tag{4.9}$$

These equations can be represented graphically in the complex plane. Recall that if A and B are complex numbers represented by points A and B in the complex plane, then the complex number $A - B$ is represented by a vector originating at B and terminating at A. The length of this vector is $|A - B|$ and its angle with respect to the real axis is $\angle(A - B)$. With this representation, if s is a root of the characteristic equation, we draw the lines connecting s to the open-loop poles and zeros. Figure 4-2 shows a case with three poles and one zero. From the graph, the length and angle of each line can be determined. Then, Eq. (4.8) shows that

$$K = \frac{\text{Product of distances from poles}}{\text{Product of distances from zeros}}, \tag{4.10}$$

and Equation (4.9) shows that

sum of angles from zeros − sum of angles from poles = $k \times 180°$, $\quad k = \pm 1, \pm 3, \pm 5, \ldots$
$$\tag{4.11}$$

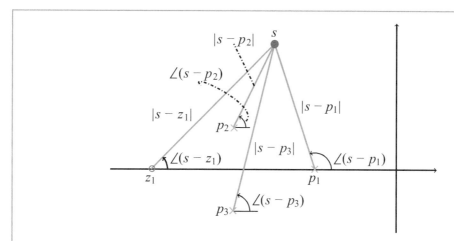

Figure 4-2: Graphical representation in the complex plane of a root s for a case with three poles and one zero.

4-2.1 Rules for Constructing the Root Locus

Multiplying the characteristic equation $1 + K L(s) = 0$ by the denominator of $L(s)$ yields

$$\prod_{i=1}^{n}(s - p_i) + K \prod_{i=1}^{m}(s - z_i) = 0. \tag{4.12}$$

This is a polynomial of degree n; hence, there are n roots. As K varies, each root varies along a certain curve; hence the root locus has n branches. At $K = 0$ the roots are p_1 to p_n.

Rule #1: There are n branches of the root locus and they start at the poles of $L(s)$ at $K = 0$.

The polynomial equation given by Eq. (4.12) has real coefficients; hence, its complex roots must be in conjugate pairs.

Rule #2: The root locus is symmetric with respect to the real axis.

Consider two points s_1 and s_2 on the real axis, as shown in Fig. 4-3. The figure shows also the locations of the zeros and poles of $L(s)$ for a case of two zeros and six poles. The point s_1 is to the left of an odd number of poles and zeros on the real axis, counted together (three in this case), while the point s_2 is to the left of an even number (two in this case). For s_1,

$$\angle(s_1 - p_1) = \angle(s_1 - p_2) = \angle(s_1 - z_1) = 180°,$$
$$\angle(s_1 - z_2) = \angle(s_1 - p_3) = \angle(s_1 - p_4) = 0,$$
$$\angle(s_1 - p_5) = \phi, \quad \angle(s_1 - p_6) = 360° - \phi.$$

Therefore,

$$\sum_{i=1}^{2} \angle(s_1 - z_i) - \sum_{i=1}^{6} \angle(s_1 - p_i) = 180 - (180 + 180 + \phi + 360 - \phi) = -3 \times 180.$$

From the angle criterion, as stated in Eq. (4.11), we see that s_1 lies on the root locus. For s_2,

$$\angle(s_2 - p_1) = \angle(s_2 - p_2) = 180°,$$
$$\angle(s_2 - z_1) = \angle(s_2 - z_2) = \angle(s_2 - p_3) = \angle(s_2 - p_4) = 0,$$
$$\angle(s_2 - p_5) = \theta, \quad \angle(s_2 - p_6) = 360° - \theta.$$

Therefore,

$$\sum_{i=1}^{2} \angle(s_2 - z_i) - \sum_{i=1}^{6} \angle(s_2 - p_i) = 0 - (180 + 180 + \theta + 360 - \theta) = -4 \times 180.$$

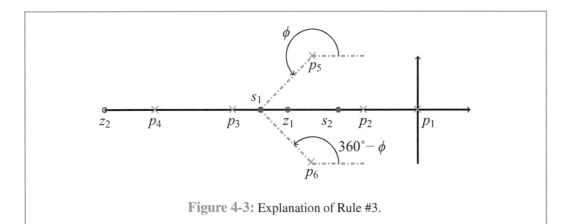

Figure 4-3: Explanation of Rule #3.

4-2. THE ROOT LOCUS METHOD

The point s_2 does not satisfy the angle criterion; it is not on the root locus. Repeating these calculations for other points on the real axis, we arrive at the same conclusions.

> **Rule #3:** Any segment of the real axis to the left of an odd number of poles and zeros, *counted together*, lies on the root locus.

Next, we study the asymptotic behavior of the roots of the characteristic equation as $K \to \infty$. We start with a second-order example.

> **Example 4-2: Root Locus of a Second-Order System**

Compute the roots of the characteristic equation

$$1 + K \frac{s+1}{s(s+2)} = 0 .$$

Solution:

$$1 + K \frac{s+1}{s(s+2)} = 0 \quad \rightarrow \quad s^2 + (2+K)s + K = 0 .$$

There are two roots at

$$s_{1,2} = \frac{-(2+K) \pm \sqrt{(2+K)^2 - 4K}}{2} = -\frac{2+K}{2}\left(1 \mp \sqrt{1 - \frac{4K}{(2+K)^2}}\right) .$$

For small $|x|$, $\sqrt{1+x} \approx 1 + x/2$. Therefore, as $K \to \infty$,

$$s_{1,2} \approx -\frac{2+K}{2}\left[1 \mp \left(1 - \frac{2K}{(2+K)^2}\right)\right] .$$

Thus

$$\lim_{K \to \infty} s_1 = \lim_{K \to \infty} -\frac{K}{2+K} = -1 ,$$

$$\lim_{K \to \infty} s_2 = \lim_{K \to \infty} -(2+K) + \frac{K}{2+K} = -\infty .$$

As $K \to \infty$, one root approaches the open-loop zero at -1 and the other root tends to ∞ along the negative real axis.

Now consider the general case. Dividing Eq. (4.12) by K results in

$$\frac{1}{K}\prod_{i=1}^{n}(s-p_i) + \prod_{i=1}^{m}(s-z_i) = 0.$$

As $K \to \infty$, m roots approach the open-loop zeros z_1 to z_m. If $n = m$, this covers all the roots. If $n > m$, the remaining $n - m$ roots tend to infinity. Let us now reveal the pattern by which these $n-m$ roots approach infinity. As $|s| \to \infty$, the terms with higher power of s dominate. We shall approximate $L(s)$ by keeping the two most dominant terms.

$$L(s) = \frac{\prod_{i=1}^{m}(s-z_i)}{\prod_{i=1}^{n}(s-p_i)} = \frac{s^m - s^{m-1}\sum_{i=1}^{m}z_i + \cdots}{s^n - s^{n-1}\sum_{i=1}^{n}p_i + \cdots} \approx \frac{s^m - s^{m-1}\sum_{i=1}^{m}z_i}{s^n - s^{n-1}\sum_{i=1}^{n}p_i}.$$

Multiplying the numerator and denominator by s^{-m} yields

$$L(s) \approx \frac{1 - s^{-1}\sum_{i=1}^{m}z_i}{s^{n-m} - s^{n-m-1}\sum_{i=1}^{n}p_i}.$$

For small $|x|$, $1 - x \approx 1/(1+x)$. So we can approximate $L(s)$ further as

$$L(s) \approx \frac{1}{\left(s^{n-m} - s^{n-m-1}\sum_{i=1}^{n}p_i\right)\left(1 + s^{-1}\sum_{i=1}^{m}z_i\right)}$$
$$= \frac{1}{s^{n-m} - s^{n-m-1}\left(\sum_{i=1}^{n}p_i - \sum_{i=1}^{m}z_i\right) - s^{n-m-2}\sum_{i=1}^{m}z_i\sum_{i=1}^{n}p_i}.$$

Once again we keep the two most dominant terms in the denominator, to obtain

$$L(s) \approx \frac{1}{s^{n-m} - s^{n-m-1}\left(\sum_{i=1}^{n}p_i - \sum_{i=1}^{m}z_i\right)}.$$

Define

$$\sigma = \frac{\sum_{i=1}^{n}p_i - \sum_{i=1}^{m}z_i}{n-m}$$

and rewrite $L(s)$ as

$$L(s) \approx \frac{1}{s^{n-m} - s^{n-m-1}(n-m)\sigma}.$$

Alternatively,

$$(s-\sigma)^{n-m} = s^{n-m} - (n-m)\sigma s^{n-m-1} + \cdots \approx s^{n-m} - (n-m)\sigma s^{n-m-1}.$$

Hence, as $|s| \to \infty$,

$$L(s) \approx \frac{1}{(s-\sigma)^{n-m}}.$$

Substitution of this approximation in the equation $KL(s) = -1$ yields

$$(s-\sigma)^{n-m} = -K.$$

4-2. THE ROOT LOCUS METHOD

This equation represents $n-m$ straight lines that intersect the real axis at σ and have angles given by

$$\angle(s-\sigma)^{n-m} = (n-m)\angle(s-\sigma) = k \times 180°, \qquad k = \pm 1, \pm 3, \pm 5, \ldots$$

Rule #4: As $K \to \infty$, m branches approach the zeros of $L(s)$. If $n > m$, the remaining $n-m$ branches tend to infinity along asymptotes. The asymptotes intersect the real axis at

$$\sigma = \frac{\text{sum of poles of } L(s) - \text{sum of zeros of } L(s)}{n-m}, \qquad (4.13)$$

and their angles with respect to the real axis are given by

$$\frac{k \times 180°}{n-m}, \qquad k = \pm 1, \pm 3, \pm 5, \ldots$$

For $n-m = 1, 2, 3, 4$, the angles are shown in Table 4-1 and the asymptotes are shown in Fig. 4-4.

Table 4-1: **Angles of the asymptotes.**

$n-m$	1	2	3	4
Angles	180°	±90°	±60°, 180°	±45°, ±135°

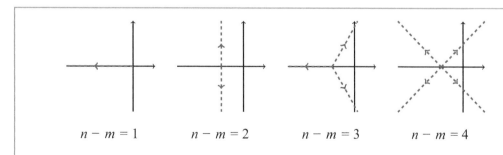

Figure 4-4: Asymptotes.

With the four rules we have developed so far we can sketch the root locus for a wide class of systems. We proceed in steps:

> **Step 1:** Locate the poles and zeros of $L(s)$ in the complex plane.
>
> **Step 2:** Determine segments of the real axis that belong to the root locus. To do this you might start at the extreme right of the real axis beyond all poles and zeros. Move to the left and as you cross poles and zeros mark the segments that are to the left of an odd number of poles and zeros counted together. Continue this motion until you cross all poles and zeros.
>
> **Step 3:** Determine the number of asymptotes, their angles, and their intersection with the real axis. Draw the asymptotes.
>
> **Step 4:** Sketch the root locus with each branch starting at a pole and approaching a zero or tending to infinity along one of the asymptotes. Note that when two branches starting on the real axis meet at a point, they break into complex branches and the meeting point is called a *breakaway point*. Similarly, when two conjugate complex branches reach the real axis at a point, they move on the real axis in opposite directions and the point is called a *break-in point*.

4-2. THE ROOT LOCUS METHOD

Example 4-3: Root Locus of a Second-Order Transfer Function

Sketch the root locus of

$$GH(s) = \frac{K}{s(s+2)}.$$

Solution:

- Open-loop poles: $0, -2$
- Open-loop zeros: None
- Real axis: $[-2, 0]$
- Two asymptotes $\pm 90°$
- $\sigma = \frac{0-2}{2} = -1$

The sketch of the root locus is shown in Fig. 4-5. Two branches start at the open-loop poles 0 and -2. They move on the real axis until they meet at the breakaway point -1. Then they move as conjugate complex roots along a vertical line $s = -1$ towards $\pm\infty$. The rules of the root locus do not require in general that the complex parts of the root locus move exactly along the asymptotes all the way, but for this specific example it can be seen by calculating the roots of the characteristic equation $s^2 + 2s + K = 0$ that the complex parts are exactly on the asymptotes.

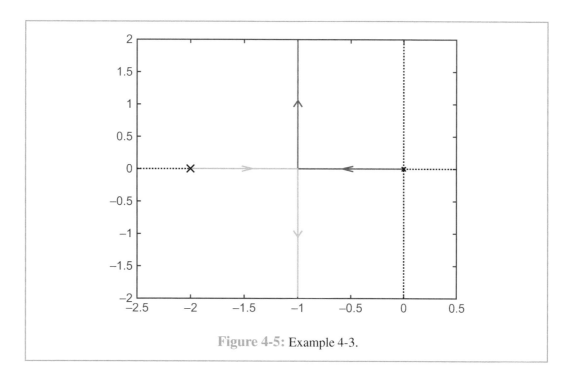

Figure 4-5: Example 4-3.

> Example 4-4: Root Locus of a Third-Order Transfer Function

Sketch the root locus of
$$GH(s) = \frac{K}{(s+1)(s+2)(s+3)}.$$

Solution:

- Open-loop poles: $-1, -2, -3$
- Open-loop zeros: None
- Real axis: $[-2,-1]$ and $(-\infty,-3]$
- Three asymptotes $\pm 60°$, $180°$
- $\sigma = \frac{-1-2-3}{3} = -2$

The sketch of the root locus is shown in **Fig. 4-6**. The three asymptotes intersect the real axis at -2. Two branches start at the open-loop poles -1 and -2. They move on the real axis until they meet at the breakaway point. Then they move as conjugate complex roots approaching infinity along the $\pm 60°$ asymptotes. The third branch starts at the open-loop pole -3 and moves along the real axis towards $-\infty$.

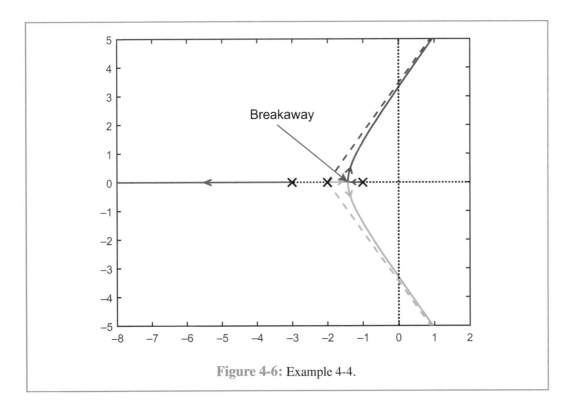

Figure 4-6: Example 4-4.

Example 4-5: Root Locus of a Second-Order Transfer Function with a Zero

Sketch the root locus of

$$GH(s) = \frac{K(s+2)}{s(s+1)}.$$

Solution:

- Open-loop poles: $0, -1$
- Open-loop zeros: -2
- Real axis: $[-1, 0]$ and $(-\infty, -2]$
- One asymptote $180°$

The sketch of the root locus is shown in Fig. 4-7. Two branches start at the open-loop poles 0 and -1. They move on the real axis until they meet at the breakaway point. Then they move as conjugate complex roots. The complex branches reach the real axis at the break-in point, beyond which the two branches move on the real axis, one towards the open-loop zero -2 and the other towards $-\infty$. It is interesting to note that the complex part of the root locus is a circle centered at the open-loop zero -2. This observation is not obvious through the rules of the root locus.

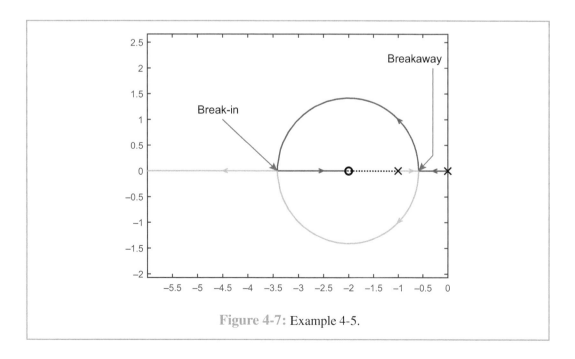

Figure 4-7: Example 4-5.

Example 4-6: Root Locus of a Third-Order Transfer Function with a Zero

Sketch the root locus of

$$GH(s) = \frac{K(s+1)}{s(s+2)(s+3)}.$$

Solution:

- Open-loop poles: $0, -2, -3$
- Open-loop zeros: -1
- Real axis: $[-1, 0]$ and $[-3, -2]$
- Two asymptotes $\pm 90°$
- $\sigma = \frac{(0-2-3)-(-1)}{2} = -2$

The sketch of the root locus is shown in Fig. 4-8. The two asymptotes intersect the real axis at -2. One branch moves on the real axis from the open-loop pole 0 to the open-loop zero -1. The other two branches start at the open-loop poles -2 and -3. They move on the real axis until they reach the breakaway point. Then they move as conjugate complex roots approaching infinity along the asymptotes.

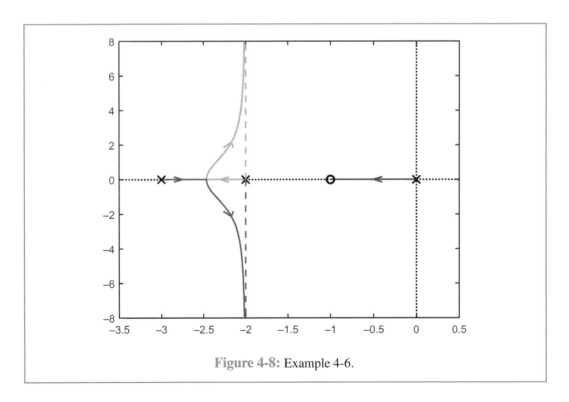

Figure 4-8: Example 4-6.

4-2. THE ROOT LOCUS METHOD

Example 4-7: Root Locus of a Third-Order Transfer Function with a Pair of Complex Zeros

Sketch the root locus of

$$GH(s) = \frac{K(s^2+2s+2)}{s(s+2)(s+3)}.$$

Solution:

- Open-loop poles: $0, -2, -3$
- Open-loop zeros: $-1 \pm j$
- Real axis: $[-2, 0]$ and $(-\infty, -3]$
- One asymptote $180°$

The sketch of the root locus is shown in **Fig. 4-9**. Two branches start at the open-loop poles 0 and -2. They move on the real axis until they meet at the breakaway point. Then they move as conjugate complex roots approaching the open-loop zeros $-1 \pm j$. The third branch starts at the open-loop pole -3 and moves on the real axis towards $-\infty$.

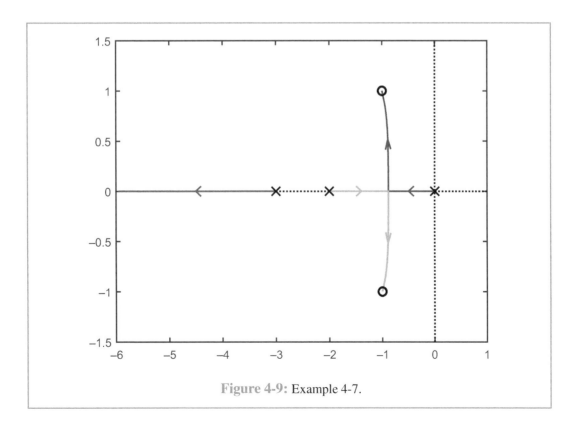

Figure 4-9: Example 4-7.

4-2.2 Refinement of the Root Locus

Next we present two rules to refine the root locus by calculating crossing points of the imaginary axis and breakaway and break-in points.[1]

When the root locus crosses the imaginary axis, as in Example 4-4, the crossing points are the boundary between stability and instability. At these points the closed-loop system is marginally stable with a constant or oscillatory natural response. Since stability can be determined by the Routh-Hurwitz criterion, we can use it to calculate the value of K at which crossing takes place, from which we can calculate the crossing points.

> **Rule #5:** Use the Routh-Hurwitz criterion to determine the crossing points of the imaginary axis.

> **Example 4-8: Crossing Points of the Imaginary Axis**

Find the crossing points of the imaginary axis in Example 4-4.

Solution: The characteristic equation $1 + K L(s) = 0$ is given by

$$(s+1)(s+2)(s+3) + K = 0 \quad \rightarrow \quad s^3 + 6s^2 + 11s + (6+K) = 0.$$

The Routh table is

s^3	1	11
s^2	6	$(6+K)$
s	$(60-K)/6$	
1	$(6+K)$	

The system is stable for $-6 < K < 60$. It is marginally stable at $K = 60$. The closed-loop poles on the imaginary axis at $K = 60$ are the roots of the equation

$$6s^2 + 66 = 0 \quad \rightarrow \quad s = \pm\sqrt{11}j.$$

Hence, the root locus crosses the imaginary axis at $\pm\sqrt{11}j$. At $K = 60$, the natural response of the closed-loop system oscillates at the natural frequency $\omega_n = \sqrt{11}$.

For the real portion of the root locus, $K = -1/L(s)$. Near a breakaway point, s moves towards that point as K increases and when it reaches the breakaway point K is maximum. On the other hand, at a break-in point, s moves away from the point as K increases; hence K is

---

[1]Control textbooks, for example [28] or [11], usually include a third rule to calculate angles of departures and arrival from complex poles and zeros. This rule is needed if the manually generated root locus plot is used in design. However, as we shall see shortly, design calculations are typically done using MATLAB-generated root locus plots. Therefore this third refinement rule is not included here.

4-2. THE ROOT LOCUS METHOD

minimum at the break-in point. Therefore, break-in and breakaway points can be determined by finding the minimum and maximum points of $-1/L(s)$, which are the roots of

$$\frac{d}{ds}\left(-\frac{1}{L(s)}\right) = 0 \quad \longleftrightarrow \quad \frac{d}{ds}\left(\frac{1}{L(s)}\right) = 0.$$

Rule #6: The break-in and breakaway points are located at the roots of

$$\frac{d}{ds}\left(\frac{1}{L(s)}\right) = 0. \tag{4.14}$$

Example 4-9: Breakaway Point

Find the breakaway point of Example 4-4.

Solution:

$$\frac{d}{ds}\left(\frac{1}{L(s)}\right) = \frac{d}{ds}(s^3 + 6s^2 + 11s + 6) = 0,$$

$$3s^2 + 12s + 11 = 0 \quad \longrightarrow \quad s = -2 \pm \frac{1}{\sqrt{3}} = -1.432, -2.577.$$

From the root locus sketch we know that the breakaway point is between -1 and -2. Hence, the root -2.577 is rejected and the breakaway point is at -1.432.

Example 4-10: Break-In and Breakaway Points

Find the break-in and breakaway points of Example 4-5.

Solution:

$$\frac{d}{ds}\left(\frac{1}{L(s)}\right) = \frac{d}{ds}\left(\frac{s(s+1)}{s+2}\right) = 0,$$

$$\frac{(s+2)(2s+1) - s(s+1)}{(s+2)^2} = 0,$$

$$s^2 + 4s + 2 = 0 \quad \longrightarrow \quad s = -2 \pm \sqrt{2} = -0.586, -3.414.$$

From the root locus sketch we see that the breakaway point is at -0.586 and the break-in point is at -3.414.

Our discussion of the root locus so far has been for the closed-loop system of Fig. 4-1(b), where $GH(s) = K L(s)$; that is, the constant K is the gain of the open-loop transfer function. We can also use the root locus to study how the closed-loop poles vary with some other constant in the transfer function $GH(s)$. This requires us to rewrite the characteristic equation in the form $1 + \alpha L(s) = 0$, where α includes the constant to be investigated, while $L(s)$ does not depend on that constant. We illustrate this process by reconsidering the speed control problem of Example 4-1.

Example 4-11: Rewriting the Characteristic Equation in the Standard Form

We saw in Example 4-1 that the characteristic equation of a dc motor controlled by a PI controller is given by

$$s^2 + \left(\frac{1+K_P K_m}{\tau_m}\right) s + \frac{K_I K_m}{\tau_m} = 0 .$$

Suppose we fix the controller gains K_P and K_I so we may study how the closed-loop poles will vary as τ_m varies. Rewrite the characteristic equation as

$$1 + \left(\frac{1+K_P K_m}{\tau_m}\right) \frac{s + K_I K_m/(1+K_P K_m)}{s^2} = 0 .$$

Setting $\alpha = (1+K_P K_m)/\tau_m$, the equation takes the standard form

$$1 + \alpha L(s) = 0, \quad \text{where } L(s) = \frac{s + K_I K_m/(1+K_P K_m)}{s^2} .$$

The root locus is sketched in **Fig. 4-10**. Suppose that for the nominal value of τ_m, the controller gains are designed to locate the closed-loop poles at the red squares. If τ_m changes over the range $[a,b]$, α will change over the range $[(1+K_P K_m)/b,\ (1+K_P K_m)/a]$. For a limited variation of τ_m, the closed-loop poles will change over an arc like the one marked by the red marks. If, however, τ_m becomes very small, α will be very large, which might push the closed-loop poles into the real part of the root locus. This will cause a significant change in the performance of the system because a system that is designed to be underdamped will become overdamped.

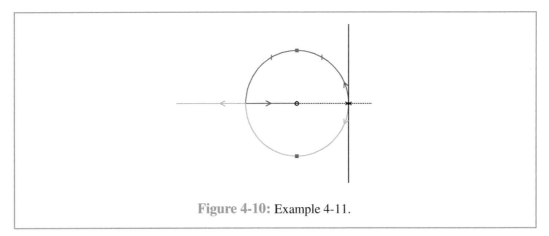

Figure 4-10: Example 4-11.

4-2. THE ROOT LOCUS METHOD

In Appendix D it is shown how to plot the root locus using MATLAB. A question that comes to the mind of many students: If we can plot the root locus exactly using MATLAB, why do we have to learn all these rules to draw a sketch of it? The answer consists of a few points, all of them building on the fact that drawing a sketch of the root locus not only can be done easily in a few minutes, but also it gives us confidence in the plot we draw using MATLAB. Pressing a wrong key while entering the data will usually generate a plot that is quite different from the sketch. Second, the sketch, as approximate as it might be, still offers a good idea about the behavior of the system as the gain K changes. Third, the root locus sketch can help the designer make quick decisions about some parameter choices without detailed calculations. We illustrate this point by once again recalling Example 4-1.

Example 4-12: Making Design Decisions by Sketching the Root Locus

The PI controller of Example 4-1 has two parameters, K_P and K_I. A typical procedure to design these parameters is to fix the ratio K_I/K_P and then investigate how the closed-loop poles vary with K_P. The characteristic equation $1 + KL(s) = 0$ has $K = K_P K_m/\tau_m$, and

$$L(s) = \frac{s + K_I/K_P}{s(s + 1/\tau_m)}$$

has two poles at 0 and $-1/\tau_m$ and a zero at $-K_I/K_P$. The choice of the ratio K_I/K_P determines the location of the zero. Choosing $K_I/K_P < 1/\tau_m$ locates the zero between the two poles, while $K_I/K_P > 1/\tau_m$ locates the zero to the left of the poles. Sketches of the root locus for these two cases are shown in Fig. 4-11. For $K_I/K_P < 1/\tau_m$, the closed-loop system will be overdamped, while for $K_I/K_P > 1/\tau_m$, the closed-loop system can be designed to be underdamped by locating the poles on the complex part of the root locus.

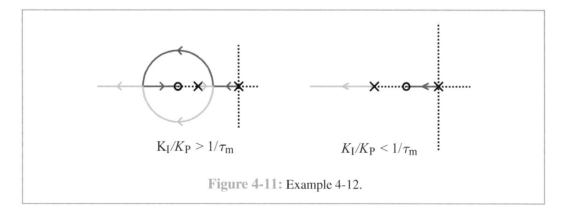

Figure 4-11: Example 4-12.

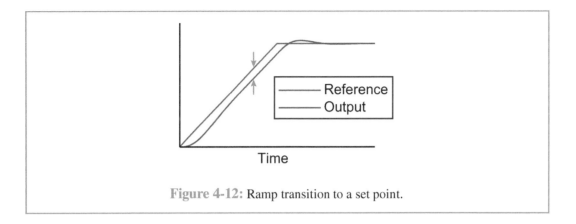

Figure 4-12: Ramp transition to a set point.

4-3 Steady-State Error

The accuracy of the feedback control system is determined by its ability to track the desired reference signal. This accuracy is measured after the decay of the transient response. In other words, we examine the steady-state of the error $e = r - y$, where r is the reference input and y is the output.

In many control systems, the goal is to regulate the output to a constant value, usually called the *set point*. In such cases we study the steady-state error to a step input $r(t) = A\, 1(t)$, where $1(t)$ is the unit step function and A is the amplitude of the step. In some control problems the output is required to track a target moving with constant velocity. An example is an antenna tracking a satellite orbiting at constant velocity. Another example is the speed control of a motor where the measured output is the angular position θ. To regulate the angular velocity $\omega = \dot{\theta}$ to a constant ω_r, θ tracks $\omega_r t$. In such cases the steady-state accuracy of the system is tested by studying the steady-state error to a ramp input $r(t) = At\, 1(t)$, where A is the slope of the ramp. Testing the steady-state error of the response to a ramp input might be useful even when the ultimate goal is to regulate the output to a constant position. Figure 4-12 shows an example of a reference signal that moves the output from zero to the desired position over a time period by first applying a ramp followed by a constant. The figure shows the response of a system whose transfer functions is $100/(s^2 + 10s + 100)$. The error during the transition period is determined by examining the steady-state error to a ramp input. Such a ramp transition of the output from zero to the desired set point is needed if a step input would cause the control signal to become very large, exceeding its maximum permissible value.

In some control systems, the output is required to track a target moving with constant acceleration, as in the example of an antenna tracking an accelerating missile. In such systems we study the steady-state error of the response to a parabola $r(t) = \frac{1}{2} At^2\, 1(t)$.[2] In this section we characterize the steady-state error for these three test signals: the step, the ramp, and the parabola.

---

[2] The factor $\frac{1}{2}$ is included so that $R(s) = A/s^3$.

4-3.1 Unity Feedback System

We start by considering the feedback control system of Fig. 4-1(b) in the special case when $H = 1$. We assume that the closed-loop system is stable; that is, all poles of the closed-loop transfer function

$$T(s) = \frac{G(s)}{1 + G(s)}$$

have negative real parts. The steady-state tracking error is given by

$$e(\infty) = \lim_{t \to \infty} e(t) = \lim_{t \to \infty} [r(t) - y(t)],$$

provided the limit exists. We calculate $e(\infty)$ using the final value theorem of the Laplace transform:

$$e(\infty) = \lim_{s \to 0} sE(s),$$

provided all poles of $sE(s)$ have negative real parts. Since

$$E(s) = R(s) - Y(s) = R(s) - \frac{G(s)}{1 + G(s)} R(s) = \frac{1}{1 + G(s)} R(s),$$

it follows that

$$e(\infty) = \lim_{s \to 0} \frac{s}{1 + G(s)} R(s). \tag{4.15}$$

(a) Step input:

For

$$r(t) = A\, 1(t) \quad \longrightarrow \quad R(s) = \frac{A}{s} \quad \longrightarrow \quad E(s) = \frac{A}{s[1 + G(s)]},$$

$$e(\infty) = \lim_{s \to 0} sE(s) = \lim_{s \to 0} \frac{A}{1 + G(s)} = \frac{A}{1 + \lim_{s \to 0} G(s)}.$$

Define the position error constant K_p as

$$K_p = \lim_{s \to 0} G(s).$$

Then the steady-state error is given by

$$e(\infty) = \frac{A}{1 + K_p}. \tag{4.16}$$

If $G(s)$ has no pole at $s = 0$, $K_p = G(0)$ is finite and $e(\infty) \neq 0$. If $G(s)$ has one or more poles at $s = 0$, $K_p = \infty$ and $e(\infty) = 0$.

(b) Ramp input:

For

$$r(t) = At\, 1(t) \quad \longrightarrow \quad R(s) = \frac{A}{s^2} \quad \longrightarrow \quad E(s) = \frac{A}{s^2[1 + G(s)]},$$

$$e(\infty) = \lim_{s \to 0} sE(s) = \lim_{s \to 0} \frac{A}{s[1+G(s)]} = \frac{A}{\lim_{s \to 0} s\, G(s)}\ .$$

Define the velocity error constant K_v by

$$K_v = \lim_{s \to 0} s\, G(s)\ .$$

Then the steady-state error is given by

$$e(\infty) = \frac{A}{K_v}\ .$$

If $G(s)$ has no pole at $s = 0$, $K_v = 0$ and $e(\infty) = \infty$. If $G(s)$ has one pole at $s = 0$, K_v is finite and different from zero. In this case $e(\infty)$ is finite and different from zero. If $G(s)$ has two or more poles at $s = 0$, $K_v = \infty$ and $e(\infty) = 0$.

(c) Parabola input:

For

$$r(t) = \tfrac{1}{2} A\, t^2\, 1(t) \quad \to \quad R(s) = \frac{A}{s^3} \quad \to \quad E(s) = \frac{A}{s^3[1+G(s)]}\ ,$$

$$e(\infty) = \lim_{s \to 0} sE(s) = \lim_{s \to 0} \frac{A}{s^2[1+G(s)]} = \frac{A}{\lim_{s \to 0} s^2\, G(s)}\ .$$

Define the acceleration error constant K_a by

$$K_a = \lim_{s \to 0} s^2\, G(s)\ .$$

Then the steady-state error is given by

$$e(\infty) = \frac{A}{K_a}\ . \tag{4.17}$$

If $G(s)$ has less than two poles at $s = 0$, $K_a = 0$ and $e(\infty) = \infty$. If $G(s)$ has two poles at $s = 0$, K_a is finite and different from zero. In this case $e(\infty)$ is finite and different from zero. If $G(s)$ has three or more poles at $s = 0$, $K_a = \infty$ and $e(\infty) = 0$.

Because the closed-loop system is stable, in all three cases where $e(\infty)$ is finite, $sE(s)$ has poles with negative real parts. Therefore, the application of the final value theorem to calculate $e(\infty)$ is valid.

Warning: We cannot use the foregoing expressions to calculate the steady-state error unless the closed-loop system is stable.

Definition: The unity feedback control system is said to be of Type N if $G(s)$ has N poles at $s = 0$.

4-3. STEADY-STATE ERROR

Table 4-2: **Summary of the steady-state errors.**

Input	Type 0 $e(\infty)$	Type 1 $e(\infty)$	Type 2 $e(\infty)$
Step, $r(t) = A\,1(t)$	$\dfrac{A}{1+K_p}$	0	0
Ramp, $r(t) = At\,1(t)$	∞	$\dfrac{A}{K_v}$	0
Parabola, $r(t) = \tfrac{1}{2}A\,t^2\,1(t)$	∞	∞	$\dfrac{A}{K_a}$

With this definition, the foregoing results are summarized in Table 4-2.

Example 4-13: Step Input to Type 0 Feedback System

Consider the feedback control system of Fig. 4-1(b) with a plant transfer function

$$G_p(s) = \frac{1}{(s+1)(s+2)(s+3)}$$

and a proportional controller $G_c(s) = K > 0$. Find the steady-state error to a unit step input. Can you design K such that $e(\infty) < 0.01$?

Solution:

$$G(s) = \frac{K}{(s+1)(s+2)(s+3)}\ .$$

The system is Type 0.

$$K_p = G(0) = \frac{K}{6} \quad \longrightarrow \quad e(\infty) = \frac{1}{1+K_p} = \frac{6}{6+K}\ .$$

To answer the posed question take

$$\frac{6}{6+K} < 0.01 \quad \longrightarrow \quad K > 594\ .$$

At this point it is tempting to answer the question with YES. But this answer is premature because we did not check whether the closed-loop system is stable for $K > 594$. Recall that the expression for $e(\infty)$ is valid only if the system is stable. Now let us use the Routh-Hurwitz criterion to check stability. The characteristic equation of the closed-loop system is given by

$$(s+1)(s+2)(s+3) + K = s^3 + 6s^2 + 11s + 6 + K = 0.$$

s^3	1	11
s^2	6	$(6+K)$
s	$(60-K)/6$	
1	$(6+K)$	

The system is stable for $K < 60$. Hence we cannot achieve $e(\infty) < 0.01$ by taking $K > 594$.

Example 4-14: Step Input to Type 1 Feedback System

Reconsider the previous example and change the controller to a PI controller

$$G_c(s) = K_P + K_I/s.$$

Can you design K_P and K_I so that $e(\infty) < 0.01$?

Solution:

$$G(s) = G_p(s)\, G_c(s) = \frac{K_P(s + K_I/K_P)}{s(s+1)(s+2)(s+3)}.$$

The system is Type 1. Therefore, if K_P and K_I are designed such that the closed-loop system is stable, $e(\infty) = 0$. It is not hard to see that we can design such K_P and K_I. For example, taking $K_I/K_P = 1$ so that the zero of $G(s)$ cancels the pole at -1, the characteristic equation is

$$s(s+2)(s+3) + K_P = s^3 + 5s^2 + 6s + K_P = 0.$$

s^3	1	6
s^2	5	K_P
s	$(30-K_P)/5$	
1	K_P	

The system is stable for $0 < K_P < 30$.

4-3.2 Non-Unity Feedback System

Now consider the feedback control system of Fig. 4-1(b) when $H(s) \neq 1$. We can represent the system by an equivalent unity feedback system. The process is illustrated in Fig. 4-13. Figure 4-13(a) is the original non-unity feedback system. In Fig. 4-13(b) we add and subtract Y at the summation node. In Fig. 4-13(c) we replace the parallel connection of $H(s)$ and the branch from Y that enters the summation node with a positive sign by the equivalent transfer function $H(s) - 1$. In Fig. 4-13(d) we replace the inner feedback loop by its equivalent transfer function

$$G_e(s) = \frac{G(s)}{1 + G(s)[H(s) - 1]} \ .$$

This process does not change the stability of the system, which can be checked from the original configuration (a) or the equivalent configuration (d).

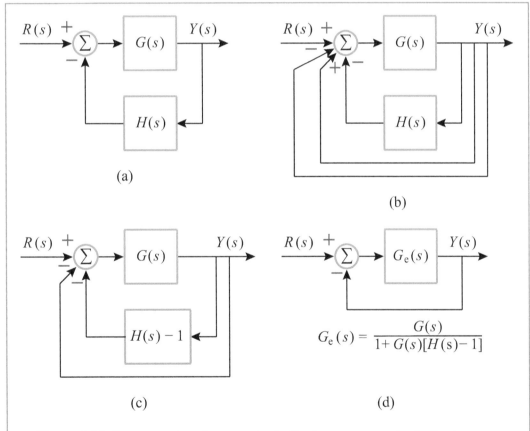

Figure 4-13: Representation of a non-unity feedback system as a unity feedback system.

Example 4-15: Steady-State Error of Non-Unity Feedback System

Consider the feedback control system of Fig. 4-1(b) with

$$G(s) = \frac{K}{s(s+1)}, \qquad H(s) = H \neq 1,$$

where K and H are positive constants. Find the steady-state error to a unit step input.

Solution: The characteristic equation of the closed-loop system is

$$1 + G(s)H(s) = 0 \quad \longrightarrow \quad s^2 + s + KH = 0.$$

The system is stable for all positive K and H. This can be seen by calculating the roots of the equation, sketching the root locus, or applying the Routh-Hurwitz criterion. The equivalent transfer function $G_e(s)$ is given by

$$G_e(s) = \frac{G(s)}{1 + G(s)[H(s) - 1]}$$

$$= \frac{\frac{K}{s(s+1)}}{1 + \frac{K}{s(s+1)}(H-1)}$$

$$= \frac{K}{s^2 + s + K(H-1)}.$$

$G_e(s)$ has no poles at $s = 0$; hence, the system is Type 0.

$$K_p = G_e(0) = \frac{1}{H-1} \quad \longrightarrow \quad e(\infty) = \frac{1}{1 + K_p} = \frac{H-1}{H}.$$

Notice that $G_e(s)$ has no poles at $s = 0$ even though $G(s)$ has a pole at $s = 0$.

Example 4-16: A Non-Unity Feedback System with $H(0) = 1$

Reconsider the previous example with

$$H(s) = \frac{10}{s+10}.$$

Solution: The characteristic equation of the closed-loop system is

$$1 + G(s)H(s) = 0 \quad \longrightarrow \quad s^3 + 11s^2 + 10s + 10K = 0.$$

By application of the Routh-Hurwitz criterion it can be seen that the system is stable for $0 < K < 11$. We calculate the steady-state error for K in this range. The equivalent transfer function $G_e(s)$ is given by

$$G_e(s) = \frac{G(s)}{1 + G(s)[H(s) - 1]}$$

$$= \frac{\frac{K}{s(s+1)}}{1 + \frac{K}{s(s+1)}\left[\frac{10}{s+10} - 1\right]}$$

$$= \frac{\frac{K}{s(s+1)}}{1 + \frac{K}{s(s+1)}\left[\frac{-s}{s+10}\right]}$$

$$= \frac{K(s+10)}{s(s^2 + 11s + 10 - K)}.$$

The system is Type 1; hence, the steady-state error to a step input is zero.

4-4 Effect of Disturbance

Control systems are usually subject to disturbances, which are input signals that cannot be manipulated. To motivate our discussion, let us consider a position servo control problem where the angular position of a motor is to be regulated to a set point. The motor shaft can be subject to torque disturbance $T_d(t)$. In this case, the torque balance equation that results from applying Newton's second law is given by

$$T_m(s) = J_m s^2\, \theta_m(s) + b_m s\, \theta_m(s) + T_d(s), \qquad (4.18)$$

where $T_d(s)$ is the Laplace transform of $T_d(t)$ and the other variables are defined in Chapter 2. Assuming an armature-controlled dc motor with negligible armature inductance L_a and friction coefficient b_m, the system is modeled by

$$\frac{K_t[V_a(s) - K_b s\, \theta_m(s)]}{R_a} = J_m s^2\, \theta_m(s) + T_d(s).$$

Hence

$$\theta_m(s) = \frac{1}{s(J_m s + K_t K_b/R_a)}\left(\frac{K_t}{R_a} V_a(s) - T_d(s)\right)$$

$$= \frac{K_m}{s(\tau_m s + 1)}\left(V_a(s) - \frac{R_a}{K_t} T_d(s)\right), \qquad (4.19)$$

where $K_m = 1/K_b$ and $\tau_m = J_m R_a/(K_t K_b)$. Define the normalized disturbance

$$d(t) = \frac{-R_a\, T_d(t)}{K_t};$$

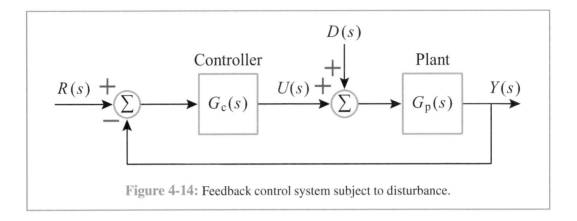

Figure 4-14: Feedback control system subject to disturbance.

then the foregoing equation is rewritten as

$$\theta_m(s) = \frac{K_m}{s(\tau_m s + 1)} \left[V_a(s) + D(s) \right],$$

where $D(s)$ is the Laplace transform of $d(t)$. In this representation, the disturbance d enters the system at the same point as the control input v_a. Note that d has the same units as v_a.

Figure 4-14 provides a general representation of a control system where the disturbance D enters the system at the same point as the control signal U, which is generated by the controller. The output is given by

$$Y(s) = \frac{G_p(s) G_c(s)}{1 + G_p(s) G_c(s)} R(s) + \frac{G_p(s)}{1 + G_p(s) G_c(s)} D(s)$$

$$= \frac{G(s)}{1 + G(s)} R(s) + \frac{G_p(s)}{1 + G(s)} D(s),$$

where $G(s) = G_p(s) G_c(s)$. The expression for Y is obtained by superposition. First set $D = 0$ and find the output due to R; then set $R = 0$ and find the output due to D. If the closed-loop system is stable; that is, all the roots of the characteristic equation $1 + G(s) = 0$ have negative real parts, then both transfer functions $G(s)/[1 + G(s)]$ and $G_p(s)/[1 + G(s)]$ are stable. The tracking error E is given by

$$E(s) = R(s) - Y(s)$$

$$= \frac{1}{1 + G(s)} R(s) - \frac{G_p(s)}{1 + G(s)} D(s). \tag{4.20}$$

Attenuating the effect of the disturbance is achieved by designing $G_c(s)$ so that $G_p(s)/[1 + G(s)]$ is small. This is best done in the frequency domain where $G_c(s)$ is designed such that $|G(j\omega)| \gg 1$ over the frequency band of the disturbance. If this condition is also done over the frequency band of the reference, we will be attenuating the error due to the reference. We shall consider frequency domain design in Chapter 7.

4-4. EFFECT OF DISTURBANCE

Steady-state error due to constant disturbance

An important special case arises when the disturbance is constant. In this case the error due to the disturbance converges to a constant value, which we can determine by applying a step input at the disturbance and calculating the steady-state error using the final-value theorem of the Laplace transform. Let $d(t) = B\,1(t)$, where B is the amplitude of the constant disturbance. The Laplace transform of the error due to the disturbance is given by

$$E_d(s) = -\frac{G_p(s)}{1+G(s)} D(s) = -\frac{G_p(s)}{1+G(s)} \frac{B}{s}$$

and

$$e_d(\infty) = \lim_{s \to 0} s\, E_d(s) = -\lim_{s \to 0} \frac{G_p(s)\, B}{1+G(s)}. \tag{4.21}$$

Example 4-17: Steady-State Error Due to Constant Disturbance

Consider Fig. 4-14 where

$$G_p(s) = \frac{K_m}{s(\tau_m s + 1)}$$

and the disturbance is a unit step. Find the steady-state error due to the disturbance when the controller is

(a) Proportional: $G_c(s) = K$;

(b) Proportional-Integral (PI): $G_c(s) = K_P + K_I/s$.

Solution: It can be seen that in both cases the controller parameters can be designed so that the closed-loop system is stable.

(a)

$$e_d(\infty) = -\lim_{s \to 0} \frac{\dfrac{K_m}{s(\tau_m s + 1)}}{1 + \dfrac{K K_m}{s(\tau_m s + 1)}} = -\lim_{s \to 0} \frac{K_m}{s(\tau_m s + 1) + K K_m} = -\frac{1}{K}.$$

(b)

$$e_d(\infty) = -\lim_{s \to 0} \frac{\dfrac{K_m}{s(\tau_m s + 1)}}{1 + \dfrac{(K_P + K_I/s) K_m}{s(\tau_m s + 1)}} = -\lim_{s \to 0} \frac{K_m s}{s^2(\tau_m s + 1) + K_m(s K_P + K_I)} = 0.$$

It is important to note that with the proportional controller the steady-state error due to the disturbance is not zero, even though when we study the steady-state error due to a step reference signal the error would be zero because the system is Type 1. Including the integrator ensures zero steady-state error due to the step disturbance. Because constant disturbances are common in control systems, integral action via PI or PID (Proportional-Integral-Derivative) controllers is used frequently.

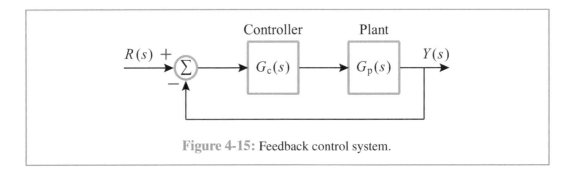

Figure 4-15: Feedback control system.

4-5 Plant Parameter Uncertainty

It is not uncommon for a control engineer to have to design a controller for a plant whose parameters are not exactly known. Even if the parameters were exactly known at the time of the design, they do change during the operation of the system and it is not practical to stop operation to redesign the controller; think of a control system in an automobile that is driven for tens of thousands of miles. Therefore the control design should accommodate plant parameter uncertainty.

There are two approaches to deal with parameter uncertainty. For small parameter changes, we study their effect on the performance of the closed-loop system by *sensitivity analysis*. For large parameter changes, special design methods are developed in an area known as *robust control*, which is not covered in this book.[3]

Consider the feedback control system of Fig. 4-15, where $G_p(s)$ is the nominal transfer function of the plant; that is, the transfer function that is available at the time of the design. Let $G_c(s)$ be the transfer function of the controller that is designed using the nominal $G_p(s)$, resulting in the nominal closed-loop transfer function

$$T(s) = \frac{G_p(s)\,G_c(s)}{1 + G_p(s)\,G_c(s)}. \tag{4.22}$$

Let $G'_p(s)$ be the actual transfer function of the plant, after parameter changes, with

$$T'(s) = \frac{G'_p(s)\,G_c(s)}{1 + G'_p(s)\,G_c(s)}$$

as the actual closed-loop transfer function. In the forthcoming analysis, we relate the relative error in $T(s)$ to the relative error in $G_p(s)$. For convenience, we omit writing the argument s of the transfer functions. The deviation is

---

[3] See, for example, [15, 16, 35, 36].

4-5. PLANT PARAMETER UNCERTAINTY

$$\Delta T = T' - T$$
$$= \frac{G'_p G_c}{1 + G'_p G_c} - \frac{G_p G_c}{1 + G_p G_c}$$
$$= \frac{G'_p G_c (1 + G_p G_c) - G_p G_c (1 + G'_p G_c)}{(1 + G'_p G_c)(1 + G_p G_c)}$$
$$= \frac{(G'_p - G_p) G_c}{(1 + G'_p G_c)(1 + G_p G_c)} = \frac{\Delta G_p \, G_c}{(1 + G'_p G_c)(1 + G_p G_c)},$$

where $\Delta G_p = G'_p - G_p$. The relative error $\Delta T / T$ is given by

$$\frac{\Delta T}{T} = \frac{\Delta G_p \, G_c}{(1 + G'_p G_c)(1 + G_p G_c)} \times \frac{1 + G_p G_c}{G_p G_c} = \frac{1}{1 + G'_p G_c} \frac{\Delta G_p}{G_p}.$$

Define the *sensitivity transfer function* for the actual system as

$$S'(s) = \frac{1}{1 + G'_p(s) G_c(s)}$$

and the sensitivity transfer function for the nominal system with $G(s) = G_p(s) \, G_c(s)$ as

$$S(s) = \frac{1}{1 + G_p(s) \, G_c(s)} = \frac{1}{1 + G(s)}.$$

Hence,

$$\frac{\Delta T(s)}{T(s)} = S'(s) \frac{\Delta G_p(s)}{G_p(s)}.$$

The sensitivity function $S'(s)$ depends on $G'_p(s)$, which is not known. It is better to relate $\Delta T / T$ to $\Delta G_p / G_p$ by a known quantity. For small ΔG_p, we can use the approximation $S'(s) \approx S(s)$, to arrive at

$$\frac{\Delta T(s)}{T(s)} = S(s) \frac{\Delta G_p(s)}{G_p(s)}, \qquad \text{where } S(s) = \frac{1}{1 + G(s)}. \tag{4.23}$$

Designing $G_c(s)$ such that $G(s)$ is large results in a small sensitivity function so that the relative error in the closed-loop transfer function is much smaller than the relative error in the plant transfer function. This is best achieved in the frequency domain where $G_c(s)$ is designed such that $|G(j\omega)| \gg 1$ over the frequency domain of interest, which is the frequency band of the reference signal. This is consistent with the design requirement we saw in the previous section to attenuate the effect of disturbance.

Example 4-18: Sensitivity to Parameter Uncertainty

Consider a feedback control system where the plant is a dc motor represented by

$$G_p(s) = \frac{K_m}{s(\tau_m s + 1)}$$

and the controller is $G_c(s) = K > 0$. Recall that $K_m = 1/K_b$ and $\tau_m = J_m R_a/(K_t K_b)$. Find the relative error in the closed-loop transfer function $\Delta T/T$ in terms of the relative error in the moment of inertia $\Delta J_m/J_m$.

Solution: To start with, let us note that the closed-loop system is stable for all $K > 0$. The change in the plant transfer function is

$$\Delta G_p = \frac{K_m}{s[1 + s(J_m + \Delta J_m)R_a/(K_t K_b)]} - \frac{K_m}{s[1 + sJ_m R_a/(K_t K_b)]}$$
$$= -\frac{s(K_m R_a/K_t K_b)\Delta J_m}{s[1 + s(J_m + \Delta J_m)R_a/(K_t K_b)][1 + sJ_m R_a/(K_t K_b)]},$$

and

$$\frac{\Delta G_p}{G_p} = -\frac{s(J_m R_a/K_t K_b)}{[1 + s(J_m + \Delta J_m)R_a/(K_t K_b)]} \frac{\Delta J_m}{J_m}.$$

Using the approximation $J_m + \Delta J_m \approx J_m$, the preceding expression simplifies to

$$\frac{\Delta G_p}{G_p} = -\frac{s(J_m R_a/K_t K_b)}{[1 + sJ_m R_a/(K_t K_b)]} \frac{\Delta J_m}{J_m} = -\frac{\tau_m s}{\tau_m s + 1} \frac{\Delta J_m}{J_m}.$$

The sensitivity transfer function is given by

$$S(s) = \frac{1}{1 + KK_m/[s(\tau_m s + 1)]} = \frac{s(\tau_m s + 1)}{\tau_m s^2 + s + KK_m}.$$

Therefore,

$$\frac{\Delta T}{T} = S \frac{\Delta G_p}{G_p} = -\frac{\tau_m s^2}{\tau_m s^2 + s + KK_m} \frac{\Delta J_m}{J_m}.$$

An increase in the controller gain K reduces the sensitivity of the closed-loop transfer function to errors in the moment of inertia J_m.

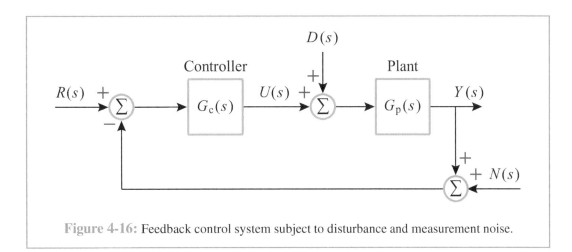

Figure 4-16: Feedback control system subject to disturbance and measurement noise.

4-6 Effect of Measurement Noise

Figure 4-16 shows a feedback control system with three inputs: the reference R, the disturbance D, and the measurement noise N, which corrupts the measurement of the output Y. The output is given by

$$Y(s) = \frac{G_p(s)\,G_c(s)}{1+G_p(s)\,G_c(s)}\,R(s) + \frac{G_p(s)}{1+G_p(s)\,G_c(s)}\,D(s) - \frac{G_p(s)\,G_c(s)}{1+G_p(s)\,G_c(s)}\,N(s)\,.$$

This expression can be written by superposition, taking one input at a time, or can be written directly from the block diagram. Note that the denominator of all three transfer functions is the same, namely 1 plus the loop gain. The numerator of each transfer function is the forward path from the input to the output. The tracking error is given by

$$\begin{aligned}E(s) = R(s) - Y(s) &= \frac{1}{1+G(s)}\,R(s) - \frac{G_p(s)}{1+G(s)}\,D(s) + \frac{G(s)}{1+G(s)}\,N(s) \\ &= S(s)\,R(s) - S(s)\,G_p(s)\,D(s) + [1-S(s)]\,N(s)\,,\end{aligned} \qquad (4.24)$$

where $G(s) = G_p(s)\,G_c(s)$ is the loop gain and $S(s) = 1/[1+G_p(s)\,G_c(s)]$ is the sensitivity transfer function. To reduce the error due to R and D, we want $|G| \gg 1$ so that $|S| \ll 1$. This design, however, makes $|1-S| \approx 1$, which allows the noise N to affect the error without attenuation. Attenuating the effect of the measurement noise requires $|G| \ll 1$ so that $|S| \approx 1$ and $|1-S| \approx 0$. It is clear that we cannot achieve both requirements at the same time. This trade-off is usually handled in the frequency domain by designing the system to have $|G(j\omega)| \gg 1$ over the frequency bands of R and D and then rolling off $|G(j\omega)|$ to a small value for higher frequencies; thereby attenuating the effect of measurement noise at those frequencies.

4-7 Control Constraints

While the objectives in control design are to reduce the steady-state tracking error and to shape the transient response of the system, the design has to cope with the fact that the control input of the plant cannot exceed its maximal permissible value determined by the capacity of the plant. For example, the armature voltage of an armature-controlled dc motor cannot exceed a certain value beyond which the volage can damage the coils of the motor.

There are two approaches to coping with control constraints. The first approach considers the magnitude of the control in design, either explicitly or implicitly. In the *explicit approach*, the control is designed to meet the constraint $|u(t)| \leq U_{\max}$, where $U_{\max}$ is the maximum permissible value of the control input.[4] This approach leads to a complicated design procedure, which usually uses nonlinear analysis. In contrast, the *implicit approach* takes the control magnitude into consideration by formulating an optimal control problem where the control task is to minimize a performance index of the form $\int_0^\infty [e^2(t) + \rho\, u^2(t)]\, dt$, where e is the tracking error, u is the control input, and ρ is a positive factor that weighs minimizing the error versus minimizing the control.[5] Increasing ρ puts a heavier penalty on the control and results in a smaller control magnitude. By iterating the design for different values of ρ, we arrive at a design that meets the control constraint. This approach is pursued using state-space models like the ones we will see later in Chapter 8.

The second approach does not take the control constraint into consideration during the design, but uses simulation and/or analysis to determine whether the design meets the control constraint. This is the approach we follow in this book, although we rarely go the extra step of checking the control constraint. Meeting the control constraint may require adjusting the reference signal.[6] In Example 4-19, a step input is smoothed out through a prefilter used to limit the magnitude of the control input during the transient period.

> Example 4-19: Smoothing a Step Reference by a Prefilter

Consider the double-integrator plant $G_p(s) = 1/s^2$ and suppose the control is constrained to $|u(t)| \leq 5$. A controller $G_c(s) = 20(s+1)/(s+10)$ is designed that assigns the closed-loop poles at $-1.1234 \pm 1.1478j$ and -7.7531. The dc gain of the closed-loop transfer function $T(s) = 20(s+1)/(s^3 + 10s^2 + 20s + 20)$ is 1; hence the steady-state error to step inputs is zero. The reference is a step input that sets the output at a desired set point. For small reference signals, the control constraint is met, but for large reference signals it could be violated.

---

[4] See, for example, [18].
[5] See, for example, [23, 24].
[6] Read about the reference governor approach in [14].

Parts (a) and (b) of Fig. 4-17 show the output and control for a unit step reference. The control exceeds 15 during the transient period. The level of the control can be reduced by smoothing the reference signal by passing the step input through a prefilter and using the output of the filter as the reference. The prefilter $F(s) = 5/(s+5)$ is used. It has a dc gain of 1 so that $T(s)F(s)$ still has a dc gain of 1, leading to a zero steady-state error for step inputs. Parts (c) and (d) of Fig. 4-17 compare the output and control signals with and without the prefilter. The use of prefilter reduces the control during the transient period at the expense of increasing the rise time of the output response.

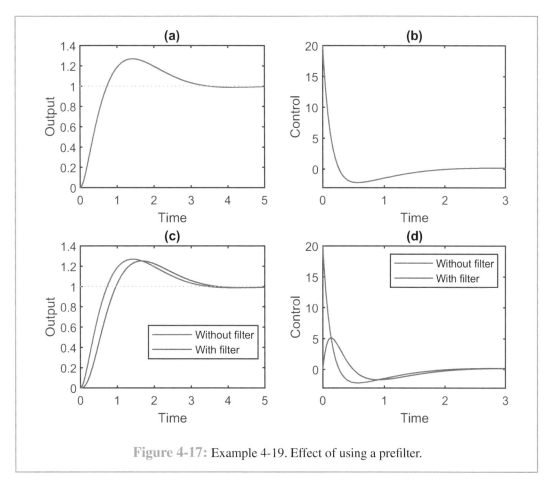

Figure 4-17: Example 4-19. Effect of using a prefilter.

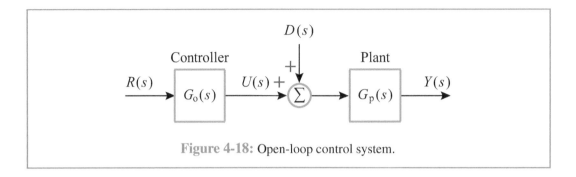

Figure 4-18: Open-loop control system.

4-8 Why Use Closed-Loop Instead of Open-Loop?

The two fundamental structures for controlling the behavior of a system are the open-loop control of **Fig. 4-18** and the closed-loop control of **Fig. 4-16**. In both cases the driving input is the reference signal R and a disturbance signal D is present at the input of the plant.

For the open-loop system, the reference drives the controller $G_o(s)$, which generates the control signal U. The transfer function from R to Y is

$$T_o(s) = G_p(s)\, G_o(s)\,,$$

and the tracking error is given by

$$\begin{aligned} E_o(s) &= R(s) - Y(s) \\ &= [1 - G_p(s)\, G_o(s)]R(s) - G_p(s)\, D(s)\,. \end{aligned} \qquad (4.25)$$

For the closed-loop system, the output Y is measured by a sensor, whose measurement is corrupted by noise N. The measured output is subtracted from the reference signal, resulting in the error, which drives the controller that generates the control signal U. The transfer function from R to Y is

$$T(s) = \frac{G(s)}{1 + G(s)}\,,$$

and the tracking error is given by

$$E(s) = \frac{1}{1 + G(s)}\, R(s) - \frac{G_p(s)}{1 + G(s)}\, D(s) + \frac{G(s)}{1 + G(s)}\, N(s)\,, \qquad (4.26)$$

where $G(s) = G_p(s)\, G_c(s)$. In this section we look at the pros and cons of the closed-loop control system versus the open-loop control system.

The first advantage of feedback is that it can stabilize an unstable plant because the control design can make the closed-loop transfer $T(s)$ stable even when the plant transfer function $G_p(s)$ is unstable. The ability to assign the closed-loop poles also means that the transient response of the closed-loop system can be greatly improved over the transient response of the open-loop system. It might appear that unstable poles of $G_p(s)$ can be cancelled by zeros of the controller transfer function $G_o(s)$, thus resulting in a stable transfer function $T_o(s)$. Such unstable pole-zero cancellation is not allowed because it does not work. The effect of

4-8. WHY USE CLOSED-LOOP INSTEAD OF OPEN-LOOP?

the unstable poles of $G_p(s)$ appears in the response to non-zero initial conditions and the response to the disturbance D. Similarly, stable but undesired poles should not be canceled.

The transfer function from the disturbance D to the tracking error E is $-G_p(s)$ for the open-loop system and $-G_p(s)/[1+G(s)]$ for the closed-loop system. By designing the controller $G_c(s)$ such that $G(s)$ is large, the feedback controller can greatly attenuate the effect of the disturbance, compared with the open-loop system.

We saw before that in a feedback control system the relative error in the closed-loop transfer is related to the relative error in the plant transfer function by the expression

$$\frac{\Delta T(s)}{T(s)} = \frac{1}{1+G(s)} \frac{\Delta G_p(s)}{G_p(s)}.$$

Once again, by designing $G(s)$ to be large, $\Delta T/T$ can be made much smaller than $\Delta G_p/G_p$. On the other hand, for the open-loop system

$$\frac{\Delta T_o(s)}{T_o(s)} = \frac{\Delta G_p(s)}{G_p(s)}.$$

Thus, plant parameter changes have full effect on the transfer function from R to Y.

Are there any situations where the open-loop system might be similar to or better than the closed-loop system? If the plant transfer function is stable and the open-loop controller $G_o(s)$ is designed such that $T_o(s) = T(s)$, the steady-state tracking error due to the reference R will be the same as in the closed-loop system. To design such a $G_o(s)$, we note that for the closed-loop system, the transfer function from R to U is given by

$$\frac{G_c(s)}{1+G(s)}.$$

Setting

$$G_o(s) = \frac{G_c(s)}{1+G(s)}$$

results in $T_o(s) = T(s)$.

The main drawback of the closed-loop system is that a sensor is needed to measure the output and measurement noise affects the performance of the system. The output is not measured in the open-loop system. Table 4-3 offers a comparison between closed-loop and open-loop control systems.[7]

---

[7] Table 4-3 is taken from a similar table in [23].

Table 4-3: **Comparison of closed-loop and open-loop control systems.**

Feature	Closed-loop control	Open-loop control
Stability	Unstable plant can be stabilized	Unstable plant cannot be stabilized
Transient response	Great improvement in response to initial conditions is possible	No improvement in response to initial conditions is possible
Steady-state error due to the reference		Can be made the same as closed-loop if the plant is stable
Effect of disturbance	Effect can be greatly reduced	Effect cannot be reduced
Effect of plant parameter variations	Effect can be greatly reduced	Effect cannot be reduced
Effect of measurement noise	It increases the tracking error	No effect

Summary

Concepts
- Closed-loop poles are the roots of the characteristic equation
- Using knowledge of the open-loop poles and zeros, the root locus method allows us to draw loci of the closed-loop poles as the loop gain changes.
- The rules of the root locus method allow us to sketch the root locus quickly.
- Exact drawing of the root locus can be done using MATLAB.
- For unity feedback systems, the steady-state errors in the responses to step, ramp, and parabola reference inputs are given in terms of the position, velocity, and acceleration error constants, respectively.
- The tracking error due to the reference and disturbance signals can be reduced by increasing the loop gain.
- The sensitivity transfer function relates the relative error in the closed-loop transfer function to the relative error in the open-loop transfer function.
- The sensitivity transfer function can be reduced by increasing the loop gain.
- The effect of measurement noise can be reduced by decreasing the loop gain.
- There is a design trade-off in choosing the loop gain to meet the foregoing conflicting requirements. The trade-off is better handled in the frequency domain.
- Meeting control constraints may require smoothing the reference signal.
- Closed-loop control has several advantages over open-loop control.

Mathematical Models

Characteristic equation: $\quad 1 + GH(s) = 0$

Steady-state error:
- Step input $\quad e(\infty) = A/(1+K_p)$
- Ramp input $\quad e(\infty) = A/K_v$
- Parabola input $\quad e(\infty) = A/K_a$

Sensitivity transfer function: $\quad S = 1/(1+G_p G_c)$

Tracking error due to reference R: $\quad SR$

Tracking error due to disturbance D: $\quad -SG_p D$

Tracking error due to measurement noise N: $\quad (1-S)N$

Important Terms Provide definitions or explain the meaning of the following terms:

acceleration error constant	control constraint	position error constant
asymptotes	disturbance	reference
breakaway point	measurement noise	root locus
break-in point	non-unity feedback system	root locus branch
characteristic equation	open-loop control	type of feedback system
closed-loop control	open-loop poles	unity feedback system
closed-loop poles	plant parameter uncertainty	velocity error constant

PROBLEMS

4.1 Consider the feedback control system of Fig. 4-1(b). Sketch the root locus for each of the following transfer functions as K varies from zero to infinity. Drawing the sketch should start by finding:

(i) the segments of the real axis that belong to the root locus;

(ii) the number of asymptotes, their angles and their intersection with real axis.

(a) $GH(s) = \dfrac{K}{s(s+1)(s+3)}$

(b) $GH(s) = \dfrac{K}{s(s+2)(s+4)(s+6)}$

(c) $GH(s) = \dfrac{K(s+7)}{(s+1)(s+2)}$

(d) $GH(s) = \dfrac{K(s+2)}{s^2+1}$

4.2 Repeat Problem 4.1 for the following transfer functions.

(a) $GH(s) = \dfrac{K}{(s+1)(s+2)(s+6)}$

(b) $GH(s) = \dfrac{K}{s(s+2)^2(s+4)}$

(c) $GH(s) = \dfrac{K(s+2)}{(s+1)(s+4)}$

(d) $GH(s) = \dfrac{K(s+2)(s+4)}{(s^2+1)}$

4.3 Repeat Problem 4.1 for the following transfer functions.

(a) $GH(s) = \dfrac{K(s-1)}{(s^2+2s+2)}$

(b) $GH(s) = \dfrac{K(s^2+6s+18)}{(s+1)(s^2+2s+2)}$

(c) $GH(s) = \dfrac{K(s+2)}{s^2(s+4)}$

(d) $GH(s) = \dfrac{K(s+1)(s+2)(s+4)}{(s+3)(s^2-2s+2)}$

4.4 Repeat Problem 4.1 for the following transfer functions.

(a) $GH(s) = \dfrac{K(s^2+2s+2)}{(s-1)(s+4)}$

(b) $GH(s) = \dfrac{K(s+4)(s^2+4s+13)}{(s^2+2s+2)(s^2+2s+10)}$

(c) $GH(s) = \dfrac{K(s+3)}{s^2(s+1)}$

(d) $GH(s) = \dfrac{K(s+1)(s+4)}{(s+3)(s^2-2s+2)}$

4.5 For each of the cases of Problem 4.1, refine the sketch of the root locus by finding:

(i) crossing points of the imaginary axis (if any);

(ii) break-in and breakaway points (if any).

4.6 For each of the cases of Problem 4.2, refine the sketch of the root locus by finding:

(i) crossing points of the imaginary axis (if any);

(ii) break-in and breakaway points (if any).

4.7 For each of the cases of Problem 4.3, refine the sketch of the root locus by finding:

(i) crossing points of the imaginary axis (if any);

(ii) break-in and breakaway points (if any).

4.8 For each of the cases of Problem 4.4, refine the sketch of the root locus by finding:

(i) crossing points of the imaginary axis (if any);

(ii) break-in and breakaway points (if any).

4.9 Consider the feedback control system of Fig. 4-1(b). Sketch the root locus for each of the following transfer functions as K varies from zero to infinity.

(a) $GH(s) = \dfrac{s}{s^3 + s^2 + K}$

(b) $GH(s) = \dfrac{5(s+K)}{s(s+1)}$

(c) $GH(s) = \dfrac{\frac{1}{4}(s+1)}{s^2(s+K)}$

(d) $GH(s) = \dfrac{s+K}{s^2(s+1)}$

4.10 For each of the transfer functions of Problem 4.1, plot the root locus using MATLAB and compare with the sketch developed in that problem and refined in Problem 4.5.

4.11 For each of the transfer functions of Problem 4.2, plot the root locus using MATLAB and compare with the sketch developed in that problem and refined in Problem 4.6.

4.12 For each of the transfer functions of Problem 4.3, plot the root locus using MATLAB and compare with the sketch developed in that problem and refined in Problem 4.7.

4.13 For each of the transfer functions of Problem 4.4, plot the root locus using MATLAB and compare with the sketch developed in that problem and refined in Problem 4.8.

4.14 For each of the MATLAB plots of Problem 4.10,

(a) find a pair of dominant closed-loop poles with $\zeta = 0.7$;

(b) using MATLAB, find the value of K at this point and determine all the closed-loop poles;

(c) for comparison, find the value of K using the magnitude criterion Eq. (4.8).

4.15 Consider a unity feedback control system. For each of the following transfer functions, determine the type of the feedback system and calculate the steady-state error to the step input $r(t) = 10\,1(t)$, the ramp input $r(t) = 10t\,1(t)$, and the parabola input $r(t) = 10t^2\,1(t)$.

(a) $G(s) = \dfrac{10}{s(s+1)(s+3)}$

(b) $G(s) = \dfrac{20}{s(s+1)(s+3)}$

(c) $G(s) = \dfrac{100}{s(s+2)(s+4)(s+6)}$

(d) $G(s) = \dfrac{10(s+7)}{(s+1)(s+2)}$

(e) $G(s) = \dfrac{5(s+1)}{s^2(s+3)}$

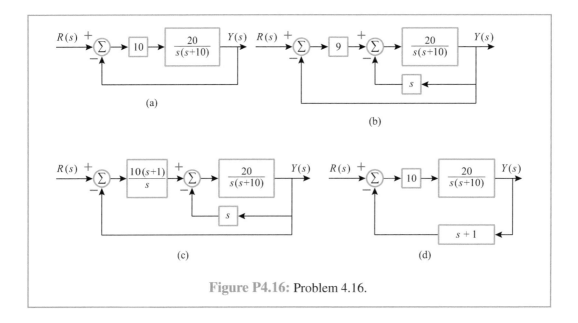

Figure P4.16: Problem 4.16.

4.16 For each of the systems in Fig. P4.16, find the type of the feedback system and calculate the corresponding error constant (K_p for Type 0, K_v for Type 1, and K_a for Type 2).

4.17 Consider a unity feedback control system with

$$G(s) = \frac{K}{s(s+3)}, \qquad K > 0.$$

(a) Choose K such that $K_v > 10$ and the natural frequency of the closed-loop transfer function $\omega_n < 10$.

(b) What is the steady-state error to the step input $r(t) = 1(t)$?

(c) What is the steady-state error to the ramp input $r(t) = t\,1(t)$?

4.18 Consider a unity feedback control system with

$$G(s) = \frac{K}{(s+1)(s+2)}, \qquad K > 0.$$

(a) Can you choose K such that the steady-state error to a unit step input is less than 0.01?

(b) Can you choose K such that the steady-state error to a unit step input is less than 0.01 while keeping the natural frequency of the closed-loop transfer function $\omega_n < 10$?

4.19 Consider a unity feedback control system with

$$G_p(s) = \frac{10}{(s+1)(s+10)}, \qquad G_c(s) = K_P + \frac{K_I}{s}, \qquad K_P > 0, \quad K_I > 0.$$

Can you choose the PI controller gains K_P and K_I such that

(a) the steady-state error to a unit step input is less than 0.01?
(b) the steady-state error to the ramp input $r(t) = t\, 1(t)$ is less than 0.01?
(c) the steady-state error to the parabola input $r(t) = t^2\, 1(t)$ is less than 0.01?

4.20 Consider a unity feedback control system with

$$G(s) = \frac{K}{s(s+1)(s+2)}, \qquad K > 0.$$

Can you choose K such that

(a) the steady-state error to a unit step input is less than 0.1?
(b) the steady-state error to the ramp input $r(t) = t\, 1(t)$ is less than 0.1?

4.21 Consider a non-unity feedback control system where $G(s)$ has one or more poles at $s = 0$ and $H(s) \neq 1$ but $H(0) = 1$. Show that

$$G_e(s) = \frac{G(s)}{1 + G(s)[H(s) - 1]}$$

has a pole at $s = 0$.

4.22 A cruise-control system is modeled by

$$m\dot{v} = au - bv - mg\sin\theta,$$

where v is the velocity of the vehicle, m is its mass, u is the throttle angle (control input), a is the engine gain, bv is the force due to road friction, and θ is the inclination of the road. The controller is driven by the tracking error $e = v_{\text{des}} - v$, where v_{des} is the velocity set point. Find the steady-state tracking error for $\theta = 30°$ when the controller is

(a) a proportional controller: $G_c(s) = K$;
(b) a PI controller: $G_c(s) = K_P + K_I/s$.

4.23 Consider a unity feedback control system with

$$G(s) = \frac{K}{s^3 + 13s^2 + 32s + 20}.$$

(a) Find the range of K for stability.
(b) Find the frequency of oscillation when the system is marginally stable.
(c) Find the position error constant K_p and the steady-state error to a unit step input. Can you make the error less than 0.1?

4.24 Consider a unity feedback control system with

$$G(s) = \frac{K}{s^3 + 10s^2 + 25s + 10}.$$

(a) Find the range of K for stability.

(b) Find the frequency of oscillation when the system is marginally stable.

(c) Find the position error constant K_p and the steady-state error to a unit step input. Can you make the error less than 0.01?

4.25 Consider a unity feedback control system with

$$G(s) = \frac{K}{s^4 + 7s^3 + 11s^2 + 7s}.$$

(a) Find the range of K for stability.

(b) Find the value of K at which the system oscillates and the frequency of oscillation.

(c) Calculate the steady-state error $e(\infty)$ for a unit step input. Can you choose K to make $e(\infty) < 0.1$?

(d) Calculate the steady-state error $e(\infty)$ for a unit ramp input. Can you choose K to make $e(\infty) < 0.1$?

4.26 Consider the idle speed control of Problem 3.16 together with the linearized model of Fig. P3.16. The time delay is modeled by the transfer function

$$H(s) = \frac{1 - (0.04/2)s + (0.04)^2 s^2/12}{1 + (0.04/2)s + (0.04)^2 s^2/12}.$$

Suppose the feedback control is taken as $\Delta u = K(\Delta N_r - \Delta N)$, where N_r is the desired engine speed and ΔN_r is the change from the nominal value.

(a) Draw the root locus as K changes from 0 to ∞ and find the maximum value of K that brings the system to verge of instability.

(b) Find K at which the real part of dominant poles is about -2.

(c) What are the closed-loop poles for the value of K determined in (b).

(d) What are the overshoot and settling time for the value of K of determined in (b).

4.27 A riderless bicycle can be stabilized by equipping it with a steering mechanism and feedback control from the measured lean angle. In [1], the dynamics of the bicycle, in constant forward velocity, from the steering angle to the lean angle is modeled by

$$G_1(s) = 3.373 \left(\frac{s+8}{s^2 - 18.9} \right)$$

and the steering mechanism is modeled by

$$G_2(s) = H(s) \frac{(33.9)^2}{s^2 + 40.68s + (33.9)^2},$$

where $H(s)$ is the transfer function of 0.015 s time delay. Using Padé approximation, the time delay is modeled by

$$H(s) = \frac{1 - (0.015/2)s + (0.015)^2 s^2/12}{1 + (0.015/2)s + (0.015)^2 s^2/12}.$$

The transfer function from the input of the steering mechanism to the lean angle is $G_1(s)\,G_2(s)$. A PID controller is designed in [1] as

$$G_c(s) = K_P\left(1 + \frac{1.5431}{s} + 0.075s\right),$$

with $K_P = 2.5177$.

(a) Draw the root locus of the system as K_P changes from 0 to ∞.
(b) Show that the system is stable for $K_1 < K_P < K_2$ and find K_1 and K_2.

For the rest of this problem, $K_P = 2.5177$.

(c) Find the closed-loop poles.
(d) What is the steady-state error to a unit step reference?
(e) Plot the step response of the closed-loop system. Determine the overshoot and the settling time.

4.28 A process in the pulp and paper industry is modeled by the transfer function [5]

$$G_p(s) = \frac{K_1 e^{-\tau s}}{\tau_1 s + 1} \approx \left(\frac{K_1}{\tau_1 s + 1}\right)\left(\frac{1 - (\tau s)/2 + (\tau s)^2/12}{1 + (\tau s)/2 + (\tau s)^2/12}\right),$$

where the nominal values of the parameters are $K_1 = 1$, $\tau_1 = 3$ s, and $\tau = 0.2$ s. It is common in this industry to design the controller using the Internal Model Principle, where the controller cancels the process dynamics, process poles with controller zeros and process zeros with controller poles. For the given process, a PI controller is taken as

$$G_c(s) = \frac{K(\tau_1 s + 1)}{s}.$$

(a) Draw the root locus as K varies from 0 to ∞.
(b) Choose K such that the settling time of the step response is about 2 sec.
(c) Compare the open-loop and closed-loop step responses.
(d) Using simulation, study the effect of $\pm 20\%$ change in each of K_1, τ_1, and τ on the closed-loop step response. Take the changes in the parameters one at a time. The controller parameters are kept at the nominal values.
(e) Comment on the results.

4.29 A process in the pulp and paper industry is modeled by the transfer function [5]

$$G_p(s) = \frac{K_1(\beta s + 1)}{s(\tau_1 s + 1)},$$

where the nominal values of the parameters are $K_1 = 0.005$, $\tau_1 = 20$ s, and $\beta = 300$ s. It is common in this industry to design the controller using the Internal Model Principle, where the controller cancels the process dynamics, process poles with controller zeros and process zeros with controller poles. For the given process, a PID controller plus filter is taken as

$$G_c(s) = \left(\frac{K(\tau_1 s + 1)(s + z)}{s}\right)\left(\frac{1}{\beta s + 1}\right).$$

(a) Sketch the root locus as K varies from 0 to ∞.

(b) Choose z and K such that the settling time of the step response is about 2 s.

(c) Plot the step response of the closed-loop system.

(d) Using simulation, study the effect of $\pm 20\%$ change in each of K_1, τ_1 and β on the closed-loop step response. Take the changes in the parameters one at a time. The controller parameters are kept at the nominal values.

(e) Comment on the results.

4.30 For people with type 1 diabetes, the pancreatic β-cells do not produce endogenous insulin, which may lead to high glucose levels. Treatment with exogenous insulin is needed to avoid extended periods of high glucose. An automated system for insulin delivery (closed-loop insulin pump) has been the subject of research for many years and continues to be an active area of research [6, 19, 34].

Figure P4.30 shows the block diagram of a typical closed-loop insulin pump system [34]. Glucose levels are measured by subcutaneous continuous glucose monitor (CGM) and insulin is delivered by a continuous subcutaneous insulin infusion (CCII) pump. The insulin dose is calculated by the controller, which determines the input voltage to the pump. A mathematical model from the insulin delivery rate to the measured glucose level is given by the transfer function [19, 34]

$$G_p(s) = \frac{K_o}{(\tau_1 s + 1)(\tau_2 s + 1)^2}.$$

The time constants $\tau_1 = 247$ (min) and $\tau_2 = 17$ (min) are uniform across patients while the gain K_o is individualized to the patient. In particular, $K_o = -12000/\text{TDI}$, where TDI is the total daily insulin dose of the patient. A PID controller is taken as

$$G_c(s) = K \frac{(\tau_1 s + 1)(\tau_2 s + 1)}{s},$$

where the zeros of $G_c(s)$ are chosen to cancel two poles of $G_p(s)$.

(a) Choose K in terms of K_o such that the closed-loop step response has overshoot less than 5%. Determine the rise and settling times.

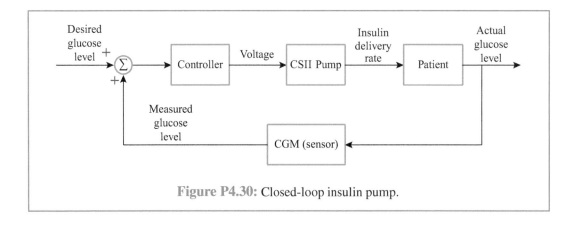

Figure P4.30: Closed-loop insulin pump.

(b) Using simulation, study the effect of $\pm 20\%$ change in K_o and $\pm 10\%$ change in each of τ_1 and τ_2 on the closed-loop step response. Take the changes in the parameters one at a time. The controller parameters are kept at the nominal values.

(c) Comment on the results.

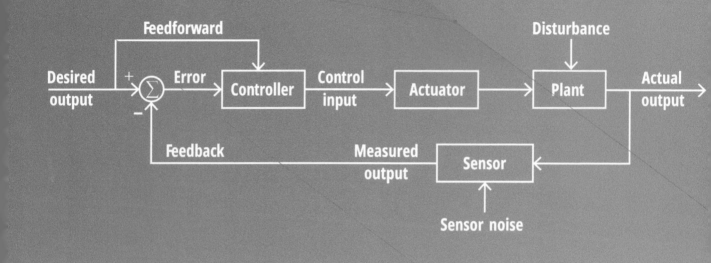

Chapter 5
Time-Domain Design

Chapter Contents

     Overview, 155
5-1 Design Specifications, 155
5-2 Design by Gain Adjustment, 156
5-3 Design by Cascade Compensation, 163
5-4 Lead and PD Controller, 165
5-5 Lag and PI Controller, 179
5-6 Lead-Lag and PID Controller, 186
5-7 Tuning PID Controllers and Anti-Windup Schemes, 194
     Chapter Summary, 201
     Problems, 202

Objectives

Upon learning the material presented in this chapter, you should be able to:

1. Know the time-domain design specifications.
2. Design by adjustment of a constant gain.
3. Design by cascade compensation.
4. Design lead, lag, and lead-lag compensators.
5. Design PI (Proportional-Integral), PD (Proportional-Derivative), and PID (Proportional-Integral-Derivative) controllers.
6. Compensate for the effect of a controller's zero by prefilter or set-point weighting.
7. Tune PID controllers.
8. Understand integrator windup and the use of anti-windup schemes.

Overview

The basic feedback control system is shown in Fig. 5-1. Given the plant transfer function $G_p(s)$, the objective of the control design problem is to design the controller transfer function $G_c(s)$ so that the closed-loop system meets certain design specifications. Time-domain design specifications are stated in Section 5-1. Sections 5-2 to 5-6 describe different procedures, increasing in complexity, for designing $G_c(s)$.

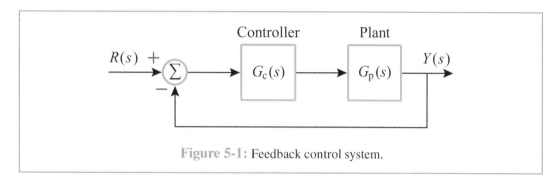

Figure 5-1: Feedback control system.

One of the most used controllers in industry is the proportional-integral-derivative (PID) controller. Section 5-7 describes two practical issues in implementing PID controllers, tuning their gains and preventing integrator drift during actuator saturation.

5-1 Design Specifications

Time-domain design specifications are classified into two groups:

- Steady-state error specifications, which take the form of requirements on the steady-state error $e(\infty)$ to step, ramp, or parabola inputs. Equivalently, they may take the form of requirements on the error constants, K_p (for Type 0 systems), K_v (for Type 1 systems), or K_a (for Type 2 systems).

- Transient-response specifications, which are requirements imposed on the step response of the closed-loop system. They constitute:

  - Requirement on the percent overshoot (*PO*).
  - Requirement on the rise time (T_r), the peak time (T_p), or the settling time (T_s).

When the closed-loop system has a pair of dominant complex poles, which are the roots of $s^2 + 2\zeta\omega_n s + \omega_n^2$, the transient response specifications are translated into requirements on the damping ratio ζ and the natural frequency ω_n, using relations that were developed in Chapter 3. Those relations are exact only when the closed-loop transfer function $T(s)$ is given by the transfer function defined by Eq. (3.9):

$$T(s) = \frac{K\omega_n^2}{s^2 + 2\zeta\omega_n s + \omega_n^2} . \tag{5.1}$$

For other cases we should ensure that the closed-loop transfer function can be approximated by the above second-order transfer function. We have seen in Chapter 3 that such an approximation is valid when other poles are far to the left in the complex plane relative to the dominant complex poles, or when poles and zeros appear in pairs where the pole and zero are very close to each other. In any case, the transient response should be checked by simulation to make sure that the specifications are met.

5-2 Design by Gain Adjustment

Our first attempt to meet the design specifications is to use a *proportional controller* $G_c(s) = K$, where K is a positive constant. The gain K is adjusted to meet the requirements. Since we choose only one parameter, we can adjust it to meet only one of the specifications. We can then check whether or not the other specifications are met by the same gain. There are two reasons for starting the design process by adjusting the gain of a proportional controller. If the chosen gain happens to meet all specifications, we end up with the simplest controller. If the adjusted gain does not meet all specifications, our calculations usually guide us to the next type of controller to try.

Choosing K to meet a requirement on the steady-state error is realized by calculating the appropriate error constant, K_p, K_v, or K_a, in terms of K, using formulas from Section 4-3. Choosing K to meet a requirement on the transient response is accomplished by first translating the transient response requirement to a requirement on a pair of complex dominant poles. This step can take one of four forms: a requirement on ζ (for overshoot), a requirement on ω_n, a requirement on the real part of the poles $-\zeta\omega_n$ (for settling time), or a requirement on the imaginary parts of the poles $\omega_d = \omega_n\sqrt{1-\zeta^2}$.

The choice of K is realized in three steps:

1. Draw the root locus of $G_p(s)$.

2. Draw a line or a curve in the complex plane that has the desired property of the closed-loop poles and find its intersection with the root locus. The intersection point is the desired closed-loop pole.

3. Find K at the desired closed-loop pole by applying the magnitude criterion of the root locus method.

The second of the above three steps is illustrated in **Fig. 5-2**:

(a) For a desired ζ, find the intersection of the root locus with a line from the origin that makes angle ϕ with the negative real axis, where $\cos\phi = \zeta$.

(b) For a desired ω_n, find the intersection of the root locus with the semicircle centered at the origin with radius ω_n.

5-2. DESIGN BY GAIN ADJUSTMENT

(c) For a desired $-\zeta\omega_n$, find the intersection of the root locus with the vertical line drawn at $-\zeta\omega_n$.

(d) For a desired ω_d, find the intersection of the root locus with the horizontal line drawn at ω_d.

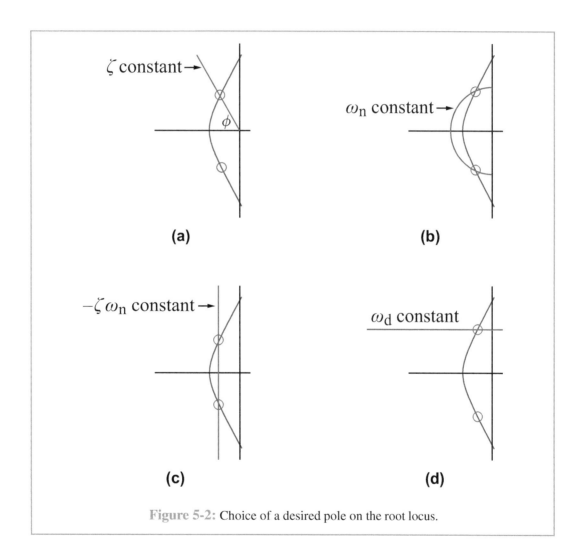

Figure 5-2: Choice of a desired pole on the root locus.

Example 5-1: Gain Design for Desired Overshoot

Given the plant transfer function

$$G_p(s) = \frac{1}{(s+1)(s+3)},$$

design K such that PO (percent overshoot) $\leq 5\%$.

Solution: For $PO \leq 5\%$, set $\zeta = 1/\sqrt{2}$, for which $PO = 4.3\%$, which is based on Eq. (3.22) in Section 3-4.2, where it was shown that $\zeta = 1/\sqrt{2} \approx 0.7$ leads to $PO = 4.3\%$. Now proceed as follows:

Step 1: Draw the root locus of $G_p(s)$ (**Fig. 5-3**).

Step 2: Draw a line from the origin at an angle $\phi = \cos^{-1}(1/\sqrt{2}) = 45°$ with the negative real axis. The line intersects the root locus at $s^\star = -2 + 2j$.

Step 3: Apply the magnitude criterion at $s^\star$ to find K.

$$|KG_p(s^\star)| = 1 \quad \rightarrow \quad \frac{K}{|(-2+2j+1)|\,|(-2+2j+3)|} = 1 \quad \rightarrow \quad K = 5.$$

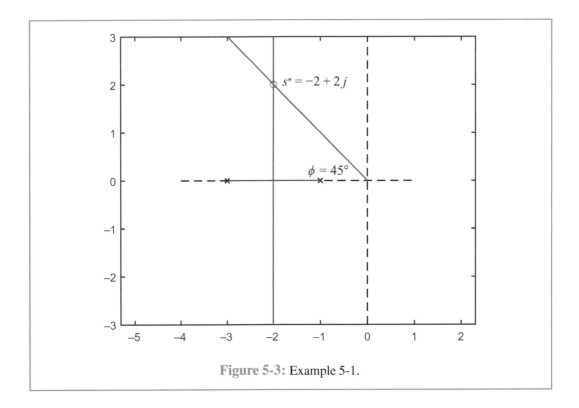

Figure 5-3: Example 5-1.

5-2. DESIGN BY GAIN ADJUSTMENT

Example 5-2: Gain Design for Desired Steady-State Error

Given the plant transfer function

$$G_p(s) = \frac{1}{(s+1)(s+3)},$$

design K such that the steady-state error to a unit step input is less than 5%.

Solution: From the root locus in the previous example, the system is stable for all $K \geq 0$. Hence,

$$K_p = KG_p(0) = \frac{K}{3}$$

and

$$e(\infty) = \frac{1}{1+K_p} < 0.05 \quad \rightarrow \quad \frac{3}{3+K} < 0.05 \quad \rightarrow \quad K > 57.$$

Example 5-3: Gain Design for Desired Overshoot and Steady-State Error

Given the plant transfer function

$$G_p(s) = \frac{1}{(s+1)(s+3)},$$

design K such that

(a) $PO \leq 5\%$ and

(b) the steady-state error to a unit step input is less than 5%.

Solution: From Example 5-1, the first requirement is satisfied for $K \leq 5$ and from Example 5-2, the second requirement is satisfied for $K > 57$. Hence, there is no value of K that meets both requirements.

Example 5-4: Gain Design for Desired Overshoot

Given the plant transfer function

$$G_p(s) = \frac{10}{s(s+1)(s+10)},$$

design K such that the overshoot of the step response is less than 20%. Find the settling time.

Solution: For $PO \leq 20\%$, set $\zeta = 0.5$, for which $PO = 16.3\%$ (according to Section 3-4.2). Now proceed as follows:

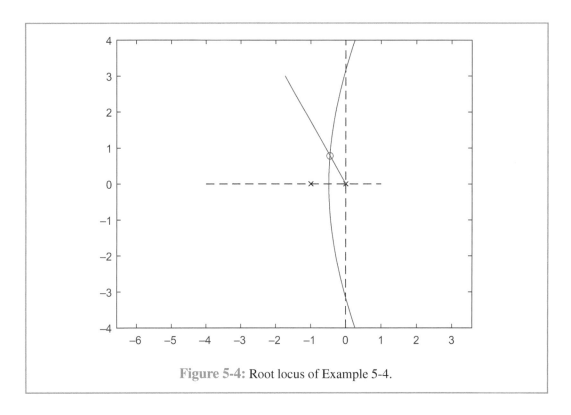

Figure 5-4: Root locus of Example 5-4.

Step 1: Draw the root locus of $G_p(s)$ (Fig. 5-4 shows only the complex part of the root locus).

Step 2: Draw a line from the origin at an angle $\phi = \cos^{-1}(0.5) = 60°$ with the negative real axis. The line intersects the root locus at the desired closed-loop pole $s^\star$. If the root locus is drawn to scale, we can use a ruler to determine the real part of $s^\star$; we can then calculate the imaginary part from the slope of the line. Alternatively, we may approximate the real part of $s^\star$.

The point of intersection lies slightly to the right of the breakaway point. Find the breakaway point from

$$\frac{d}{ds}(s^3 + 11s^2 + 10s) = 3s^2 + 22s + 10 = 0 \,.$$

The roots of this equation are $s = -0.49$ and $s = -6.85$. From the root locus, the breakaway point is between $s = 0$ and $s = -1$. Hence, it is $s = -0.49$. Approximate the real part of $s^\star$ by -0.4. Then, $\text{Im}[s^\star] = 0.4\tan(60°) = 0.4\sqrt{3}$. Hence, $s^\star \approx -0.4 + 0.4\sqrt{3}j$.

Step 3: Apply the magnitude criterion at $s^\star$ to find K:

$$\left| \frac{10K}{s(s+1)(s+10)} \right|_{s=s^\star} = 1 \,,$$

$$\rightarrow \quad K = \frac{1}{10}|(-0.4+0.4\sqrt{3}j)||(-0.4+0.4\sqrt{3}j+1)||(-0.4+0.4\sqrt{3}j+10)| \approx 0.71$$

5-2. DESIGN BY GAIN ADJUSTMENT

and

$$T_s \approx \frac{4}{\zeta\omega_n} = \frac{4}{0.5|-0.4+0.4\sqrt{3}j|} = \frac{4}{0.5 \times 0.8} = 10 \text{ s}.$$

Step 4: Simulate the step response of the closed-loop system with $K = 0.71$; the MATLAB code is displayed in Fig. 5-5. The step response parameters are summarized in Table 5-1 for three values of K, including $K = 0.71$, for which $PO = 12.6\%$ and $T_s = 7$ s. The requirement on the overshoot is met. The actual $T_s = 7$ is lower than the estimated value $T_s = 10$.

```
>> G1 = zpk([],[0 -1 -10],10)
G1 =
        10
    --------------
    s (s+1) (s+10)
>> T1 = feedback(0.71*G1,1);
>> stepinfo(T1)
ans =
  struct with fields:
         RiseTime: 2.0826
     SettlingTime: 7.0474
      SettlingMin: 0.9099
      SettlingMax: 1.1262
        Overshoot: 12.6225
       Undershoot: 0
             Peak: 1.1262
         PeakTime: 4.5932
```

Figure 5-5: MATLAB code for Example 5-4.

Table 5-1: Example 5-4: comparison of the step response parameters for three gains.

K	PO	T_r (s)	T_p (s)	T_s (s)
0.71	12.6%	2.08	4.59	7.05
0.822	15.9%	1.84	4.15	9
1	20.1%	1.56	3.62	8.49

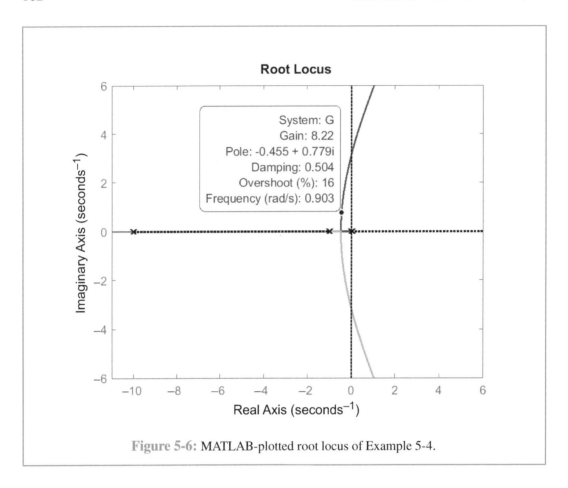

Figure 5-6: MATLAB-plotted root locus of Example 5-4.

The root locus can be plotted using MATLAB, which is shown in Fig. 5-6. Moving the cursor over the root locus we can find a point where $\zeta = 0.5$. We reached a point where $\zeta = 0.504$ at which $K = 0.822$. The step response of the closed-loop system with $K = 0.822$ was simulated in MATLAB and the results are summarized in Table 5-1; $PO = 15.9\%$ and $T_s = 9$ s. The requirement on the overshoot is met. Notice that with the exact calculations of MATLAB, the PO is closer to the expected $PO = 16.3\%$, which corresponds to $\zeta = 0.5$.

Now, let us take another crack at the design of K. The poles of $G_p(s)$ are 0, -1, and -10. The pole at -10 is far to the left relative to the other two poles. Therefore we may approximate $G_p(s)$ as

$$G_p(s) = \frac{1}{s(s+1)(0.1s+1)} \approx \frac{1}{s(s+1)} .$$

The complex part of the root locus of the approximate transfer function is a vertical line at $s = -0.5$. The intersection of this line with the $\zeta = 0.5$ line is at $s^\star = -0.5 + 0.5\sqrt{3}j$. Application of the magnitude criterion yields

$$\frac{K}{|(-0.5+0.5\sqrt{3}j)||(-0.5+0.5\sqrt{3}j+1)|} = 1 \quad \rightarrow \quad K = 1 .$$

5-3. DESIGN BY CASCADE COMPENSATION

The step response of the closed-loop system with the actual $G_p(s)$ was simulated using MATLAB and the results appear in Table 5-1. The requirement on the overshoot is almost satisfied even with the approximation since $PO = 20.1\%$. This is helped by the fact that choosing $\zeta = 0.5$ was an over-design since $\zeta = 0.5$ yields $PO = 16.3\%$ while the requirement is $PO = 20\%$. From the table we see also that as the gain K increases, the overshoot increases but the rise and peak times decrease.

5-3 Design by Cascade Compensation

To improve the performance of the control system beyond the proportional controller, the controller transfer function $G_c(s)$ is chosen from a family of transfer functions that include *Lead compensator*, *Lag compensator*, *Lead-Lag compensator*, *PI (Proportional-Integral) controller*, *PD (Proportional-Derivative) controller*, and *PID (Proportional-Integral-Derivative) controller*. All these compensators, as well as the simple P (Proportional) controller, can be realized by the circuit of Fig. 5-7.

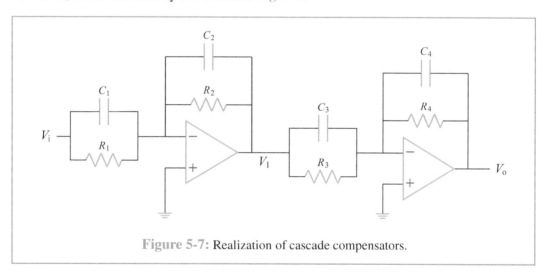

Figure 5-7: Realization of cascade compensators.

It is shown in Example 2-9 that the transfer function of the circuit is given by Eq. (2.20) as

$$\frac{V_o}{V_i} = \frac{C_1 C_3}{C_2 C_4} \left(\frac{s + \dfrac{1}{C_1 R_1}}{s + \dfrac{1}{C_2 R_2}} \right) \left(\frac{s + \dfrac{1}{C_3 R_3}}{s + \dfrac{1}{C_4 R_4}} \right). \qquad (5.2)$$

Specific choices of the R and C components result in the aforementioned transfer functions, as shown in Table 5-2. Note that removing one of the capacitors is equivalent to setting $C = 0$, while removing one of the resistors is equivalent to setting $R = \infty$.

The derivative component of the PD and PID controllers is obtained by approximation. In particular, starting with the lead compensator with $C_3 = C_4 = 0$, Eq. (5.2) simplifies to

$$\frac{R_2 R_4}{R_1 R_3} \left(\frac{C_1 R_1 s + 1}{C_2 R_2 s + 1} \right), \qquad (5.3)$$

Table 5-2: **Realization of different compensators using the circuit of** Fig. 5-7.

Compensator	Condition	Transfer Function
P	$C_1 = C_2 = C_3$ $= C_4 = 0$	$G_c(s) = K$ $K = \dfrac{R_2 R_4}{R_1 R_3}$
Lead	$C_3 = C_4 = 0$ $C_1 R_1 > C_2 R_2$	$G_c(s) = K\left(\dfrac{s+z}{s+p}\right)$ $K = \dfrac{C_1 R_4}{C_2 R_3}, \quad z = \dfrac{1}{C_1 R_1}, \quad p = \dfrac{1}{C_2 R_2} > z$
Lag	$C_3 = C_4 = 0$ $C_1 R_1 < C_2 R_2$	$G_c(s) = K\left(\dfrac{s+z}{s+p}\right)$ $K = \dfrac{C_1 R_4}{C_2 R_3}, \quad z = \dfrac{1}{C_1 R_1}, \quad p = \dfrac{1}{C_2 R_2} < z$
Lead-Lag	$C_1 R_1 > C_2 R_2$ $C_3 R_3 < C_4 R_4$	$G_c(s) = K\left(\dfrac{s+z_1}{s+p_1}\right)\left(\dfrac{s+z_2}{s+p_2}\right)$ $K = \dfrac{C_1 C_3}{C_2 C_4}, \; z_1 = \dfrac{1}{C_1 R_1}, \; p_1 = \dfrac{1}{C_2 R_2} > z_1$ $z_2 = \dfrac{1}{C_3 R_3}, \quad p_2 = \dfrac{1}{C_4 R_4} < z_2$
PI	$R_2 = \infty, \; C_3 = C_4 = 0$	$G_c(s) = K_1 + \dfrac{K_2}{s}$ $K_1 = \dfrac{C_1 R_4}{C_2 R_3}, \quad K_2 = \dfrac{R_4}{C_2 R_1 R_3}$
PD	$C_2 = C_3 = C_4 = 0$	$G_c(s) = K_1 + K_2 s$ $K_1 = \dfrac{R_2 R_4}{R_1 R_3}, \quad K_2 = \dfrac{C_1 R_2 R_4}{R_3}$
PID	$R_2 = \infty, \; C_4 = 0$	$G_c(s) = K_1 + \dfrac{K_2}{s} + K_3 s$ $K_1 = \dfrac{R_4(C_1 R_1 + C_3 R_3)}{C_2 R_1 R_3}$ $K_2 = \dfrac{R_4}{C_2 R_1 R_3}, \quad K_3 = \dfrac{C_1 C_3 R_4}{C_2}$

which has a pole at $-1/(C_2 R_2)$ and a zero at $-1/(C_1 R_1)$. Also, choosing $C_2 R_2 \ll C_1 R_1$ locates the pole far to the left in the complex plane with respect to the zero. Hence, the transfer function can be approximated by neglecting the pole; that is, by setting $C_2 = 0$. The resulting transfer function is

$$\frac{R_2 R_4}{R_1 R_3}(C_1 R_1 s + 1) \stackrel{\text{def}}{=} K_1 + K_2 s ,$$

with
$$K_1 = \frac{R_2 R_4}{R_1 R_3}, \quad K_2 = \frac{C_1 R_2 R_4}{R_3}.$$

The condition $C_2 = 0$ can be achieved by removing the capacitor C_2, but in reality the parasitic capacitance across R_2 will induce C_2 with a small value.

When cascade compensation is used, the system with proportional controller is referred to as the *uncompensated system*. The six types of cascade compensators are grouped into three groups depending on their functionality:

(a) To stabilize the system and improve the transient response using:
- Lead compensator or
- PD controller (special case of the lead compensator).

(b) To improve the steady-state error using:
- Lag compensator or
- PI controller (special case of the lag compensator).

(c) To improve the transient response and the steady-state error using:
- Lead-Lag compensator or
- PID controller (special case of the lead-lag compensator).

The suggested approach to design cascade compensation is to start with the uncompensated system where $G_c(s) = K$ and then choose the gain K to meet one of the design specifications. If the design does not meet all design specifications, the next step will depend on what needs to be improved. If we need to improve the transient response, we bring in a lead or PD controller. If we need to improve the steady-state error, we bring in a lag or PI controller. If we need to do both, we bring in a lead-lag or PID controller.

The design of the last case is usually realized in two steps. For example, in a lead-lag compensator, we may start first by designing the lead component to improve the transient response, and then design the lag component to improve the steady-state error.

5-4 Lead and PD Controller

The lead compensator and PD controller are used to stabilize the system and improve the transient response. We start with the lead compensator.

5-4.1 Lead Compensator

The transfer function of the lead compensator is given by

$$G_c(s) = K \left(\frac{s+z}{s+p} \right), \quad K > 0, \quad p > z > 0. \tag{5.4}$$

$G_c(s)$ has a pole at $-p$ and a zero at $-z$. They lie on the negative real axis of the s-plane with the zero to the right of the pole.

The addition of a lead compensator shifts the root locus of the uncompensated system to the left, which allows for stabilization and improvement of the transient response. This property is illustrated by the following example.

Example 5-5: The Lead Compensator Shifts the Root Locus to the Left

Sketch the root loci for the uncompensated system

$$KG_p(s) = \frac{K}{s^2}$$

and the compensated system

$$G_p(s)\,G_c(s) = K\left(\frac{s+z}{s^2(s+p)}\right), \quad p > z > 0 .$$

Solution: The root loci are shown in **Fig. 5-8**. The system cannot be stabilized by a proportional controller but can be stabilized by a lead compensator.

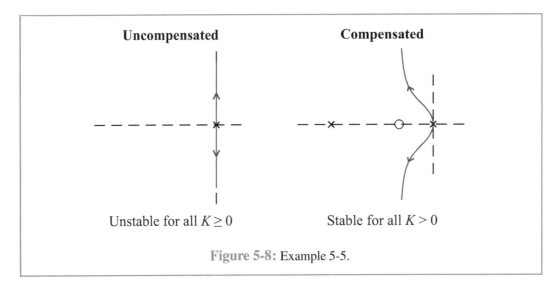

Figure 5-8: Example 5-5.

Design Procedure:

Step 1: From the transient response specifications, choose the desired dominant complex closed-loop poles or the area of the s-plane where the poles should be located. A requirement

5-4. LEAD AND PD CONTROLLER

on PO takes the form $PO \leq \alpha\%$ for some number $\alpha < 100$. Requirements on T_s, T_p or T_r take the form $T_s \leq \beta$, $T_p \leq \gamma$, and $T_r \leq \delta$ for some specified numbers β, γ, and δ. From

$$PO \leq \alpha\% \quad \rightarrow \quad e^{-\zeta\pi/\sqrt{1-\zeta^2}} \leq 0.01\alpha \quad \rightarrow \quad \zeta \geq \frac{[\ln(100/\alpha)/\pi]}{\sqrt{1+[\ln(100/\alpha)/\pi]^2}} \stackrel{\text{def}}{=} \zeta_o.$$

Recall that for $PO \leq 5\%$ set $\zeta \geq 1/\sqrt{2}$ and for $PO \leq 20\%$ set $\zeta \geq 0.5$.

Note that the desired closed-loop pole $s^\star$ lies to the left of a line from the origin that makes an angle ϕ with the negative real axis, where $\cos\phi = \zeta_o$. Also,

$$T_s \leq \beta \quad \rightarrow \quad \frac{4}{\zeta\omega_n} \leq \beta \quad \rightarrow \quad \zeta\omega_n \geq \frac{4}{\beta} \quad \rightarrow \quad \text{Re}[s^\star] \leq -\frac{4}{\beta},$$

$$T_p \leq \gamma \quad \rightarrow \quad \frac{\pi}{\omega_d} \leq \gamma \quad \rightarrow \quad \omega_d \geq \frac{\pi}{\gamma} \quad \rightarrow \quad \text{Im}[s^\star] \geq \frac{\pi}{\gamma},$$

$$T_r \leq \delta \quad \rightarrow \quad \frac{2.16\zeta + 0.6}{\omega_n} \leq \delta \quad \rightarrow \quad \omega_n \geq \frac{2.16\zeta + 0.6}{\delta}.$$

After choosing ζ (within the range $0.3 \leq \zeta \leq 0.75$), ω_n should lie outside a semicircle centered at the origin of radius $(2.16\zeta + 0.6)/\delta$.

Step 2: Sketch the root locus of the uncompensated system; that is, the system with $G_c(s) = K$. If the desired closed-loop poles can be chosen on the complex part of the root locus, stop. You do not need a lead compensator. Otherwise, proceed to the next step.

Step 3: The design of the lead compensator involves choosing three parameters: z, p, and K. Start by choosing z, which is equivalent to locating the compensator's zero on the negative real axis. Two suggestions for locating the zero are:

- Choose the zero to cancel one of the poles of $G_p(s)$. Because the pole of the compensator is to the left of its zero, the net effect amounts to replacing one pole of $G_p(s)$ by another pole to the left, which shifts the root locus to the left. Warning: Do not cancel poles with non-negative real parts ($\text{Re}[p_i] \geq 0$). We will say more about this point later.

- Locate the zero under the desired closed-loop pole (move it to the right or left as needed).

Step 4: Apply the angle criterion at the desired closed-loop pole $s^\star$ to find the location of the compensator's pole $-p$.

Step 5: Apply the magnitude criterion at $s^\star$ to find K.

Step 6: Simulate the step response of the closed-loop system to make sure the transient response specifications are satisfied. You will need to redo the design if the specifications are not satisfied.

Example 5-6: Lead Compensator Design for Desired Settling Time

Given the plant transfer function

$$G_p(s) = \frac{1}{s(s+1)},$$

design a compensator $G_c(s)$ so that $T_s \leq 1$ s.

Solution: For

$$T_s \leq 1 \quad \rightarrow \quad \frac{4}{\zeta \omega_n} \leq 1 \quad \rightarrow \quad \zeta \omega_n \geq 4 \quad \rightarrow \quad \text{Re}[s^\star] \leq -4.$$

The uncompensated root locus is shown in Fig. 5-9. The complex part of the root locus is a vertical line at $\text{Re}[s] = -\frac{1}{2}$. Use a lead compensator to shift the root locus to the left. Choose the zero of the compensator to cancel the plant's pole at -1; hence $z = 1$. The compensated root locus is shown in Fig. 5-9. The design specification will be met with $p \geq 8$ so that $p/2 \geq 4$. Choose $p = 8$. Any closed-loop pole on the vertical line at $\text{Re}[s] = -4$ will meet the requirement $T_s \leq 1$.

We have freedom in choosing the imaginary part of $s^\star$. We choose it such that $s^\star$ has $\zeta = 1/\sqrt{2}$. That places $s^\star$ on a line from the origin that makes 45° angle with the negative real axis. Therefore, $s^\star = -4 + 4j$. Apply the magnitude criterion at $s^\star$ to find K:

$$\left|\frac{K}{s(s+8)}\right|_{s=-4+4j} = 1 \quad \rightarrow \quad \frac{K}{|(-4+4j)||(-4+4j+8)|} = 1 \quad \rightarrow \quad K = 32.$$

Thus,

$$G_c(s) = \frac{32(s+1)}{s+8} \quad \rightarrow \quad G_p(s)G_c(s) = \frac{32}{s(s+8)} \quad \rightarrow \quad T(s) = \frac{32}{s^2+8s+32}.$$

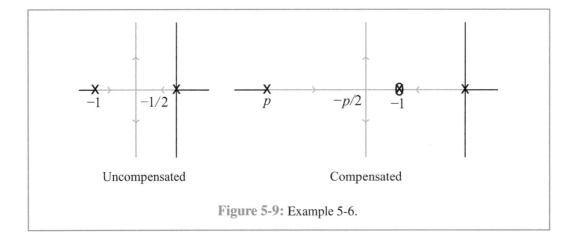

Figure 5-9: Example 5-6.

5-4. LEAD AND PD CONTROLLER

Because $T(s)$ is a second-order transfer function, $T_s = 4/(\zeta\omega_n) = 4/4 = 1$ s.

Example 5-7: Lead Compensator Design for Desired Overshoot and Peak Time

Given the plant transfer function

$$G_p(s) = \frac{0.5}{s(s+2.5)},$$

design a compensator $G_c(s)$ such that $PO \leq 20\%$ and $T_p \leq 0.5$ s.

Solution: For $PO \leq 20\%$, take $\zeta = 0.5$.

$$T_p \leq 0.5 \quad \rightarrow \quad \frac{\pi}{\omega_d} \leq 0.5 \quad \rightarrow \quad \omega_d \geq 2\pi.$$

The uncompensated root locus is shown in Fig. 5-10. The complex part of the root locus is a vertical line at $\text{Re}[s] = -1.25$. The point on that line that has $\zeta = 0.5$ is at the intersection with a line originating at the origin that makes an angle of $60°$ with the negative real axis. At this point $\omega_d = 1.25\tan(60°) = 1.25\sqrt{3} \approx 2.165 < 2\pi$. The design specifications cannot be met by the uncompensated system. Hence, use a lead compensator to shift the root locus to the left. On the shifted root locus, if $\text{Re}[s^\star] = -\alpha$, $\omega_d = \alpha\sqrt{3}$. For $\omega_d \geq 2\pi$, we need $\alpha \geq 2\pi/\sqrt{3} = 3.63$. Set $\alpha = 4$ so that the desired closed-loop pole is $s^\star = -4 + 4\sqrt{3}j$. The design of the lead compensator starts by locating its zero.

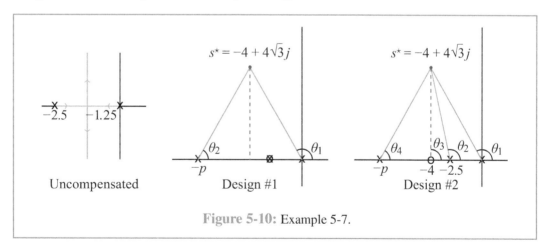

Figure 5-10: Example 5-7.

We consider two designs, where in the first design the zero is chosen to cancel a pole of $G_p(s)$ while in the second design it is located under the desired closed-loop pole.

Design #1: Set $z = 2.5$ to cancel the plant's pole at -2.5. Recall that this is the only pole we can cancel because we cannot cancel the pole at the origin. To find the location of the compensator's pole, apply the angle criterion. From Fig. 5-10,

$$-[\theta_1 + \theta_2] = -180°, \quad \theta_1 = 180 - 60 = 120° \quad \rightarrow \quad \theta_2 = 60°,$$

and
$$\tan\theta_2 = \frac{4\sqrt{3}}{p-4} = \tan 60° = \sqrt{3} \quad \rightarrow \quad p = 8.$$

Apply the magnitude criterion to find K:
$$\left|\frac{0.5K}{s(s+8)}\right|_{s=-4+4\sqrt{3}j} = 1 \quad \rightarrow \quad K = 128.$$

Hence,
$$G_c(s) = \frac{128(s+2.5)}{s+8} \quad \rightarrow \quad T(s) = \frac{64}{s^2+8s+64}.$$

Design #2: Locate the zero under the desired closed-loop pole $s^\star = -4 + 4\sqrt{3}j$ by setting $z = 4$. Apply the angle criterion to find the location of the pole $-p$. From **Fig. 5-10**,
$$\theta_3 - (\theta_1 + \theta_2 + \theta_4) = -180°,$$

$$\theta_1 = 120°, \quad \theta_2 = 180° - \tan^{-1}\left(\frac{4\sqrt{3}}{1.5}\right) = 102.2°, \quad \theta_3 = 90°,$$

$$90 - (120 + 102.2 + \theta_4) = -180 \quad \rightarrow \quad \theta_4 = 47.8°,$$

and
$$\tan\theta_4 = \frac{4\sqrt{3}}{p-4} = \tan 47.8 = 1.1 \quad \rightarrow \quad p = 10.28.$$

Apply the magnitude criterion to find K:
$$\left|\frac{0.5K(s+4)}{s(s+2.5)(s+10.28)}\right|_{s=-4+4\sqrt{3}j} = 1 \quad \rightarrow \quad K = 153.$$

Hence,
$$G_c(s) = \frac{153(s+4)}{s+10.28} \quad \rightarrow \quad T(s) = \frac{76.5(s+4)}{(s+4.78)(s^2+8s+64)}.$$

Figure 5-11 compares the step responses of the closed-loop transfer functions for the two designs. Design #2 has a larger overshoot and a smaller rise time due to the zero of the closed-loop transfer function. In particular, for Design #1, $PO = 16.3\%$, $T_r = 0.2$ s, and $T_p = 0.45$ s, while for Design #2, $PO = 23.9\%$, $T_r = 0.18$ s, and $T_p = 0.43$ s. The difference is not significant because the effect of the zero at -4 in the second design is reduced by the closed-loop pole at -4.78, which is fairly close to the zero.

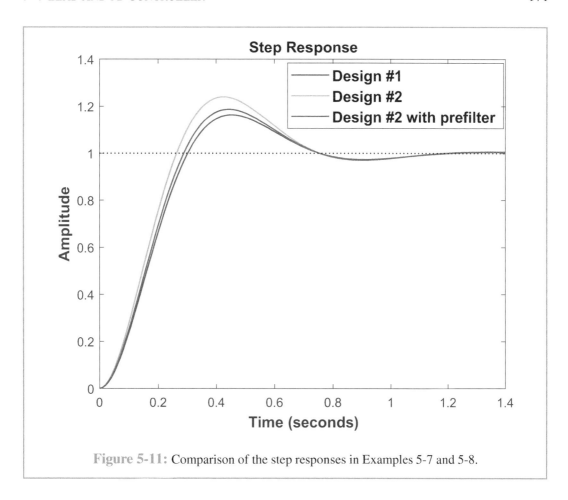

Figure 5-11: Comparison of the step responses in Examples 5-7 and 5-8.

Prefilter

In the previous example, we can reduce the overshoot in Design #2 by using a prefilter to move the zero of the closed-loop transfer function, as shown in Fig. 5-12. The filter transfer function is

$$F(s) = \frac{bs+z}{s+z}, \quad b \geq 0, \tag{5.5}$$

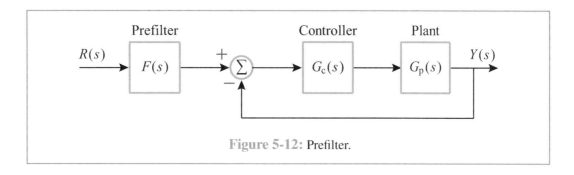

Figure 5-12: Prefilter.

and $-z$ is a zero of the closed-loop transfer function that we want to move to $-z/b$. The filter has a dc gain of 1; that is, $F(0) = 1$, so that it does not change the dc gain of the closed loop transfer function. The filtered closed-loop transfer function is $T(s) F(s)$. If $b = 0$, the filter cancels the zero of the plant. We have seen in Section 4-7 that a prefilter $z/(s+z)$ can be used to smooth the reference input and reduce the amplitude of the control signal.

Example 5-8: Lead Compensator with Prefilter

Reconsider Design #2 of the previous example and use a prefilter to reduce the overshoot.

Solution: We set $b = 8/9$ to replace the zero at -4 by a zero at -4.5, which is closer to the pole at -4.78. The step response of the filtered closed-loop transfer function is shown in Fig. 5-11, and the step response parameters are summarized in Table 5-3.

Table 5-3: **Examples 5-7 and 5-8: comparison of the step responses.**

	PO	T_r (s)	T_p (s)	T_s (s)
Design #1	16.3%	0.2	0.45	1
Design #2	23.38%	0.18	0.43	1.02
Design #2 with prefilter	18.6%	0.196	0.45	1.01

Warning about Pole-Zero Cancelation: Don't cancel a pole with $\text{Re}[s] \geq 0$ because such a pole will be a hidden unstable pole. It will appear in the natural response of the closed-loop system due to non-zero initial conditions and the response due to disturbances. The latter case is illustrated in the next example.

Example 5-9: Unstable Hidden Pole

Reconsider Example 5-7 and suppose that we design the lead compensator by canceling the pole of $G_p(s)$ at $s = 0$.

Solution: For

$$G_c(s) = \frac{Ks}{s+p} \quad \longrightarrow \quad G_p(s)G_c(s) = \frac{0.5K}{(s+2.5)(s+p)}.$$

It can be verified that closed-loop poles at $-4 \pm 4\sqrt{3}j$ are achieved with $p = 5.5$ and $K = 100.5$. The closed-loop transfer function from R to Y is given by

$$\frac{Y}{R} = \frac{50.25}{s^2 + 8s + 64}.$$

5-4. LEAD AND PD CONTROLLER

Note that the dc gain is not 1 because the system now is Type 0. However, this is not the issue. As far as the transient response specifications are concerned, they are met. Suppose that a disturbance D enters the system, as shown in Fig. 5-13. The transfer function from D to Y is given by

$$\frac{Y}{D} = \frac{0.5(s+5.5)}{s(s^2+8s+64)}.$$

The hidden pole at $s = 0$ appears in this transfer function; hence, it is unstable. A constant disturbance would cause a ramp signal at the output.

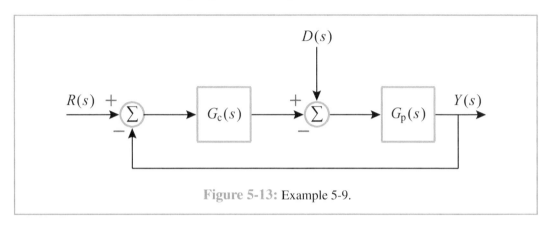

Figure 5-13: Example 5-9.

5-4.2 Proportional Derivative (PD) Controller

The transfer function of the PD controller is

$$G_c(s) = K_P + K_D s \stackrel{\text{def}}{=} K(s+z), \quad \text{where } K = K_D, \quad z = \frac{K_P}{K_D}.$$

The PD controller is a special case of the lead compensator as the compensator pole tends to $-\infty$. Starting with the lead compensator transfer function

$$G_c(s) = K_1 \left(\frac{s+z}{s+p} \right) = \frac{K_1}{p} \left(\frac{s+z}{\frac{1}{p}s+1} \right), \tag{5.6}$$

we set $K_1 = Kp$, which leads to

$$G_c(s) = K \left(\frac{s+z}{\frac{1}{p}s+1} \right) \to K(s+z) \text{ as } p \to \infty.$$

Since differentiation of signals amplifies noise, the derivative term s is typically implemented as $s/(\tau s + 1)$ with very small τ.

The addition of a PD controller shifts the root locus of the uncompensated system to the left, which allows for stabilization and improvement of the transient response. This property is illustrated by the following example.

Example 5-10: The PD Controller Shifts the Root Locus to the Left

Sketch the root locus for the uncompensated system

$$KG_p(s) = \frac{K}{s(s+a)}, \quad a > 0,$$

and the root locus for the compensated system

$$G_p(s)\,G_c(s) = K\,\frac{s+z}{s(s+a)}, \quad z > 0.$$

Solution: The root loci are shown in **Fig. 5-14**.

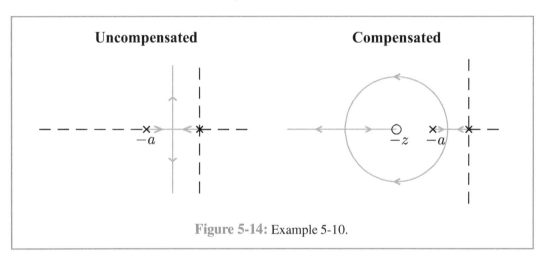

Figure 5-14: Example 5-10.

Example 5-11: PD Controller Design to Reduce the Settling Time

Given the plant transfer function

$$G_p(s) = \frac{1}{s(s+1)},$$

(a) Design $G_c = K$ to have $\zeta = 0.5$. What is T_s?
(b) Design $G_c = K(s+z)$ to have $\zeta = 0.5$ and reduce T_s to $1/4$ of the uncompensated value.
(c) Compare the step responses.

5-4. LEAD AND PD CONTROLLER

Solution:
(a) For $G_c = K$,

$$T(s) = \frac{\frac{K}{s(s+1)}}{1 + \frac{K}{s(s+1)}} = \frac{K}{s^2 + s + K} \quad \rightarrow \quad 2\zeta\omega_n = 1, \quad \omega_n^2 = K,$$

$$\zeta = 0.5 \quad \rightarrow \quad \omega_n = 1 \quad \rightarrow \quad K = 1,$$

$$T_s = \frac{4}{\zeta\omega_n} = 8 \text{ s}.$$

(b) For $G_c = K(s+z)$

$$T(s) = \frac{\frac{K(s+z)}{s(s+1)}}{1 + \frac{K(s+z)}{s(s+1)}} = \frac{K(s+z)}{s^2 + (K+1)s + Kz} \quad \rightarrow \quad 2\zeta\omega_n = K+1, \quad \omega_n^2 = Kz.$$

For $\zeta = 0.5$, $\omega_n = K+1$. For $T_s = 2$ s,

$$T_s = \frac{4}{\zeta\omega_n} = \frac{8}{\omega_n} = 2 \quad \rightarrow \quad \omega_n = 4 \quad \rightarrow \quad K = 3 \text{ and } z = \frac{16}{3},$$

$$T(s) = \frac{3s + 16}{s^2 + 4s + 16}.$$

(c) The results of the MATLAB simulation are summarized in Table 5-4.

Table 5-4: **Example 5-11: comparison of the step responses.**

G_c	PO	T_r (s)	T_s (s)
1	16.3%	1.64	8.1
$3s + 16$	23.38%	0.29	1.89

The PD controller reduces the settling time as desired but the presence of a zero in the closed-loop transfer function causes an increase in the overshoot and reduction in the rise time.

The procedure for designing the PD controller using the root locus is similar to the procedure for the lead compensator except that steps 3 and 4 for choosing the zero and pole of the compensator are replaced with a single step to choose the zero of the PD controller:

▶ Apply the angle criterion at the desired closed-loop pole $s^\star$ to find the location of the PD controller's zero at $-z$. ◀

Example 5-12: PD Controller Design for Desired Overshoot and Peak Time

Reconsider Example 5-7. Given the plant transfer function

$$G_p(s) = \frac{0.5}{s(s+2.5)},$$

design a compensator $G_c(s)$ so that $PO \leq 20\%$ and $T_p \leq 0.5$ s.

Solution: Proceeding as in Example 5-7, the desired closed-loop pole is chosen as

$$s^\star = -4 + 4\sqrt{3}j.$$

We need to shift the uncompensated root locus to the left to have this desired pole. Use a PD controller $G_c(s) = K(s+z)$. To find z, apply the angle criterion at $s^\star$. From **Fig. 5-15**,

$$\theta_3 - (\theta_1 + \theta_2) = -180°,$$

$$\theta_1 = 120°, \quad \theta_2 = 180° - \tan^{-1}\left(\frac{4\sqrt{3}}{1.5}\right) = 102.2° \quad \rightarrow \quad \theta_3 = 42.2°,$$

$$\tan\theta_3 = \frac{4\sqrt{3}}{z-4} = \tan 42.2 = 0.907 \quad \rightarrow \quad z = 11.64.$$

Apply the magnitude criterion at $s^\star$ to find K.

$$\left|\frac{0.5K(s+11.64)}{s(s+2.5)}\right|_{s=-4+4\sqrt{3}j} = 1 \quad \rightarrow \quad K \approx 11.$$

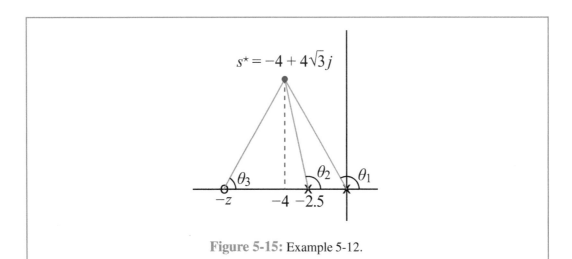

Figure 5-15: Example 5-12.

5-4. LEAD AND PD CONTROLLER

Hence,

$$G_c = 11(s+11.64) \quad \rightarrow \quad T(s) = \frac{5.5(s+11.64)}{s^2+8s+64} .$$

The step response of the PD controller is compared with the two lead compensator designs of Example 5-7. The results are summarized in Table 5-5. Similar to Design #2 of the lead compensator, the presence of a zero in the closed-loop transfer function under the PD controller causes an increase in the overshoot and reduction in the rise time. In the case of the lead compensator, the effect of the zero at -4 is moderated by the pole at -4.78. In the case of the PD controller, there is no pole that moderates the effect of the zero, but the zero is farther to the left, at -11.64 compared with -4 for the lead compensator. That is why the overshoot of the PD controller is comparable to the overshoot of Design #2 of the lead compensator.

Table 5-5: **Example 5-12: comparison of the step responses.**

G_c	PO	T_r (s)	T_p (s)	T_s (s)
Lead: $G_c = 128(s+2.5)/(s+8)$	16.3%	0.2	0.45	1
Lead: $G_c = 153(s+4)/(s+10.28)$	23.98%	0.18	0.43	1.02
PD: $G_c = 11(s+11.64)$	22.1%	0.15	0.35	0.95

Set-Point Weighting

As it was done with the lead compensator, the overshoot of the closed-loop transfer function of the PD controller can be reduced by a prefilter that changes the location of the transfer function's zero. In this case, the filter can be realized by weighting the reference input to the controller, as shown in Fig. 5-16. The closed-loop transfer function is

$$Y = G_p(s)\left[K_P(R-Y) + K_D s(bR-Y)\right] ,$$

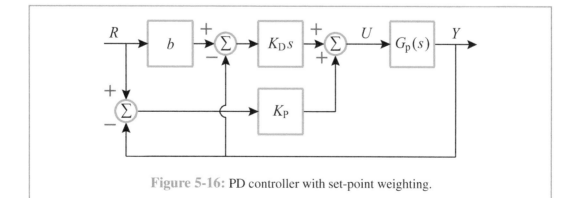

Figure 5-16: PD controller with set-point weighting.

$$\frac{Y}{R} = \frac{(K_D bs + K_P)\, G_p(s)}{1 + (K_D s + K_P)\, G_p(s)}$$

$$= \frac{K(bs + z)\, G_p(s)}{1 + K(s + z)\, G_p(s)}, \quad \text{where } K = K_D, \; z = \frac{K_P}{K_D}.$$

We can rewrite this transfer function as

$$\frac{Y}{R} = \underbrace{\left(\frac{bs + z}{s + z}\right)}_{\text{Prefilter}} \underbrace{\left(\frac{K(s + z)\, G_p(s)}{1 + K(s + z)\, G_p(s)}\right)}_{\text{Closed-loop TF under PD control}}, \tag{5.7}$$

which shows that set-point weighting is equivalent to a prefilter that moves the zero from $-z$ to $-z/b$. When $b = 0$, set-point weighting cancels the zero. The case $b = 0$ is interesting in another aspect that can be seen in the time domain. When $b \neq 0$ the control signal is given by

$$u(t) = K_P[r(t) - y(t)] + K_D \frac{d}{dt}[br(t) - y(t)],$$

while when $b = 0$, the control signal is given by

$$u(t) = K_P[r(t) - y(t)] + K_D \frac{d}{dt}[-y(t)].$$

A step jump in the reference signal causes a spike in the control signal when $b \neq 0$ but not with $b = 0$ because $r(t)$ is not differentiated.

Example 5-13: PD Controller with Set-Point Weighting

Reconsider the previous example and implement the PD controller with set-point weighting $b = 0$.

Solution: The closed-loop transfer function under PD control is

$$T(s) = \frac{5.5(s + 11.64)}{s^2 + 8s + 64},$$

and under the PD control with $b = 0$ set-point weighting is

$$T(s) = \frac{64}{s^2 + 8s + 64}.$$

The results of the step-response simulation are shown in Table 5-6.

Table 5-6: **Example 5-13: comparison of the step responses.**

PD	PO	T_r (s)	T_p (s)	T_s (s)
PD: $G_c = 11(s+11.64)$	22.1%	0.15	0.35	0.95
PD with $b=0$ set-point weighting	16.3%	0.2	0.45	1

5-5 Lag and PI Controller

The lag compensator and PI controller are used to reduce the steady-state error. We start with the lag compensator.

5-5.1 Lag Compensator

The transfer function of the lag compensator is

$$G_c(s) = K\left(\frac{s+z}{s+p}\right), \qquad K > 0, \quad z > p > 0. \tag{5.8}$$

$G_c(s)$ has a pole at $-p$ and a zero at $-z$. They lie on the negative real axis of the s-plane with the zero to the left of the pole.

The lag compensator increases the error constant (K_p, K_v, or K_a) because the dc gain of $(s+z)/(s+p) = z/p > 1$. The increase in the error constant is to be achieved while maintaining the transient response of the uncompensated system. This is done by locating the pole and zero of the compensator very close to the origin, far from the closed-loop poles of the uncompensated system. In particular, if $s^\star$ is the dominant closed-loop pole of the uncompensated system, z and p are chosen small enough so that

$$\angle(s^\star + z) \approx \angle(s^\star + p) \quad \text{and} \quad |s^\star + z| \approx |s^\star + p|. \tag{5.9}$$

Because z and p are very small, the ratio z/p can be chosen high enough to achieve the desired increase in the error constant. The conditions of Eq. (5.9) are illustrated in Fig. 5-17. The magnitude and angle criteria of the root locus are approximately satisfied at $s^\star$ for the compensated system. Therefore the value of K for the compensated system can be approximated by its value for the uncompensated system. The choice of z and p to meet the conditions of Eq. (5.9) is subjective. A rule of thumb is to choose z such that the ratio of $|\text{Re}[s^\star]|$ to z is between 5 and 10.

Design Procedure:

Step 1: Study the uncompensated system and find dominant complex closed-loop poles, $s^\star$, that meet the transient response specifications.

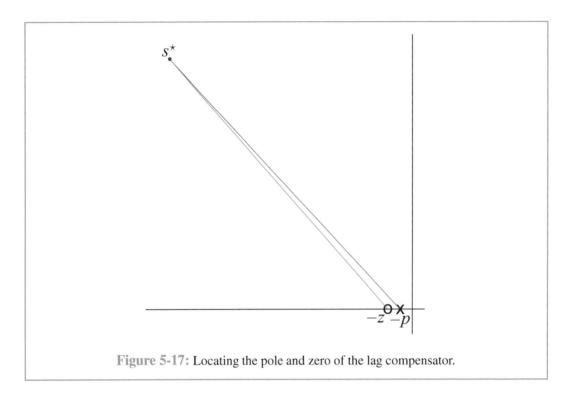

Figure 5-17: Locating the pole and zero of the lag compensator.

Step 2: Find K of the uncompensated system and calculate the error constant (K_p, K_v, or K_a).

Step 3: If the error constant is high enough, stop. Otherwise, use a lag compensator.

Step 4: Choose z such that the ratio $|\text{Re}[s^\star]|/z$ is between 5 and 10.

Step 5: Choose p such that the ratio z/p is greater than the required increase in the error constant.

Step 6: Set K as in the uncompensated system.

Step 7: Simulate the step response of the closed-loop system to make sure the transient response specifications are satisfied.

Example 5-14: Lag Compensator Design for Desired Overshoot and Steady-State Error

---

Given the plant transfer function

$$G_p(s) = \frac{1}{(s+1)(s+3)},$$

design $G_c(s)$ so that: (1) $PO \leq 20\%$ and (2) the steady-state error to step inputs is less than 1%.

Solution: Start with the uncompensated system $G_c = K$ and choose K to meet the overshoot requirement. For $PO \leq 20\%$, set $\zeta = 0.5$. The complex part of the root locus is a vertical line at $\text{Re}[s] = -2$. The desired closed-loop pole is at the intersection of this line with the line from the origin that makes angle $60°$ with the negative real axis. Hence,

$$s^\star = -2 + 2\tan(60°)j = -2 + 2\sqrt{3}j.$$

Application of the magnitude criterion yields

$$K = |(-2+2\sqrt{3}j+1)(-2+2\sqrt{3}j+3)| = 13,$$

$$K_p = KG_p(0) = \frac{K}{3} = \frac{13}{3}.$$

For $e(\infty) < 0.01$,

$$\frac{1}{1+K_p} < 0.01 \quad \rightarrow \quad K_p > 99.$$

Use a lag compensator to increase K_p. With $\text{Re}[s^\star] = -2$, set $z = 0.1 \times 2 = 0.2$. We need

$$\frac{z}{p} > \frac{99}{13/3} = 22.85 \quad \rightarrow \quad p < \frac{0.2}{22.85} = 0.00875.$$

Set $p = 0.008$. Keep $K = 13$. Hence,

$$G_c = 13\left(\frac{s+0.2}{s+0.008}\right),$$

$$K_p = G_p(0)\,G_c(0) = 109.33 \quad \rightarrow \quad e(\infty) = 0.009.$$

The closed-loop poles, computed using MATLAB, are $-1.92 \pm 3.42j$ and -0.171. The complex poles are fairly close to the uncompensated poles $-2 \pm 3.46j$. The closed-loop transfer function has a zero at -0.2, which is close to the pole at -0.171. The step responses of the uncompensated and compensated closed-loop transfer functions are shown in Fig. 5-18.

The compensated system reduces the steady-state error, but it approaches the steady-state in a sluggish way, resulting in a large settling time; $T_s = 12.6$ s compared with $T_s = 2.02$ s for the uncompensated system. The slow convergence is the effect of the pole at -0.171, even though its effect is reduced by the zero at -0.2.

5-5.2 Proportional-Integral (PI) Controller

The transfer function of the PI controller is

$$G_c(s) = K_P + \frac{K_I}{s} \stackrel{\text{def}}{=} K\left(\frac{s+z}{s}\right), \quad (5.10)$$

$$\text{where } K = K_P \text{ and } z = \frac{K_I}{K_P}.$$

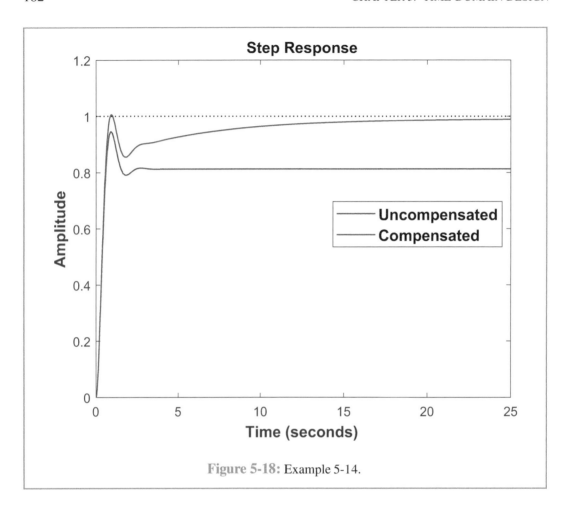

Figure 5-18: Example 5-14.

The PI controller is a special case of the lag compensator as the compensator's pole tends to zero. Because the PI controller has a pole at the origin, it increases the type of the uncompensated, from Type 0 to Type 1, from Type 1 to Type 2, and so on. The increase in the type usually takes care of the requirement on the steady-state error. For example, if the uncompensated system is Type 0 and the steady-state error to a step input is required to be less than a certain value, the use of a PI controller results in zero steady-state error to step inputs. The design of the PI controller involves the choice of z and K.

Design Procedure:

Step 1: Study the uncompensated system and find dominant complex closed-loop poles, $s^\star$, that meet the transient response specifications.

Step 2: Find K of the uncompensated system and calculate the error constant (K_p, K_v, or K_a).

Step 3: If the error constant is high enough, stop. Otherwise, use a PI controller.

Step 4: Choose z to meet the transient response specifications. There are three approaches to choose z:

1. Choose small z so that the zero is close to the origin and far from the uncompensated closed-loop poles. This approach is similar to the design of the lag compensator. The constant K can be approximated by its uncompensated value and the settling time is expected to increase.

2. Choose z to reshape the uncompensated root locus by shifting it to the left. Then find a point on the reshaped root locus that meets the design specifications.

3. Choose a desired closed-loop pole $s^\star$ that meets the design specifications and apply the angle criterion to locate the controller's zero such that $s^\star$ lies on the compensated root locus.

Step 5: Apply the magnitude criterion at $s^\star$ to find K.

Step 6: Simulate the step response of the closed-loop system to make sure the transient response specifications are satisfied.

Example 5-15: PI Controller Design for Desired Overshoot and Steady-State Error

---

Reconsider the plant transfer function

$$G_p(s) = \frac{1}{(s+1)(s+3)}$$

from Example 5-14, where the goal is to design $G_c(s)$ such that: (1) $PO \leq 20\%$ and (2) the steady state error to step inputs is less than 1%.

Solution: We have already seen from Example 5-14 that the design specifications cannot be met with a choice of a gain K in $G_c = K$. So, use a PI controller:

$$G_p(s)\,G_c(s) = K\left(\frac{s+z}{s(s+1)(s+3)}\right).$$

Design #1: Choose the controller's zero to cancel one of the plants's poles. Which pole to cancel, -1 or -3? Cancelling the pole at -3 leaves two poles at 0 and -1, while canceling the pole at -1 leaves two poles at 0 and -3. The root locus in the second case will be further to the left. So we choose $z = 1$ and the compensated transfer function becomes

$$G_p(s)\,G_c(s) = K\left(\frac{1}{s(s+3)}\right).$$

The complex part of the root locus is a vertical line at $\text{Re}[s] = -1.5$. The requirement on the overshoot is met with $\zeta = 0.5$. Hence, the desired closed-loop pole is $s^\star = -1.5 + 1.5\sqrt{3}j$. Application of the magnitude criterion yields

$$K = |(-1.5+1.5\sqrt{3})(-1.5+1.5\sqrt{3}j+3)| = 9\,.$$

Hence,
$$G_c(s) = 9\left(\frac{s+1}{s}\right) \quad \rightarrow \quad T(s) = \frac{9}{s^2+3s+9}.$$

The dc gain of $T(s) = 1$, showing that the steady-state error to step inputs is zero. The second-order transfer function $T(s)$ has $\zeta = 0.5$; hence $PO = 16.3\%$.

Design #2: Choose the controller's zero to locate the desired closed-loop pole at
$$s^\star = -1 + \sqrt{3}j.$$

Apply the angle criterion. From Fig. 5-19,

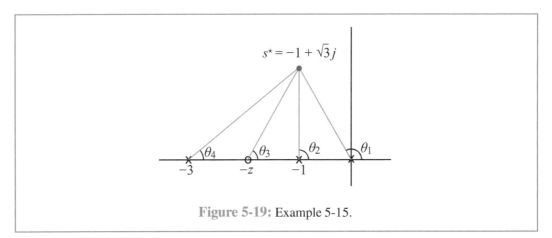

Figure 5-19: Example 5-15.

$$\theta_3 - (\theta_1 + \theta_2 + \theta_4) = -180°,$$

$$\theta_1 = 120°, \quad \theta_2 = 90°, \quad \theta_4 = \tan^{-1}\left(\frac{\sqrt{3}}{2}\right) = 40.9°,$$

$$\theta_3 = 120 + 90 + 40.9 - 180 = 70.9°,$$

$$\tan\theta_3 = \tan 70.9° = 2.888 = \frac{\sqrt{3}}{z-1} \quad \rightarrow \quad z = 1.6.$$

Apply the magnitude criterion to find K:
$$K = \left|\frac{s(s+1)(s+3)}{s+1.6}\right|_{s=-1+\sqrt{3}j} = 5.$$

Hence,
$$G_c(s) = \frac{5(s+1.6)}{s} \quad \rightarrow \quad T(s) = \frac{5(s+1.6)}{(s+2)(s^2+2s+4)}.$$

Table 5-7 compares the closed-loop step responses of the two designs. The increase in the overshoot of Design #2 is due to the zero of the closed-loop transfer function.

5-5. LAG AND PI CONTROLLER

Table 5-7: **Example 5-15. Comparison of the step responses.**

	PO	T_r (s)	T_s (s)
Design #1	16.3%	0.55	2.7
Design #2	21.7%	0.7	4.1

Set-Point Weighting: Figure 5-20 shows a set-point weighting scheme for the PI controller. The closed-loop transfer function is given by

$$Y = G_p(s)[K_P(bR - Y) + \frac{K_I}{s}(R - Y)],$$

$$\frac{Y}{R} = \frac{G_p(s)(K_P b + K_I/s)}{1 + G_p(s)(K_P + K_I/s)}$$

$$= \frac{G_p(s)(K_P bs + K_I)}{s + G_p(s)(sK_P + K_I)}$$

$$= \frac{G_p(s)K(bs + z)}{s + G_p(s)K(s + z)}, \quad \text{where } K = K_P, \; z = \frac{K_I}{K_P}$$

$$= \underbrace{\left(\frac{bs + z}{s + z}\right)}_{\text{Prefilter}} \underbrace{\left(\frac{K(s + z) G_p(s)}{s + K(s + z)G_p(s)}\right)}_{\text{Closed-loop TF under PI control}}. \tag{5.11}$$

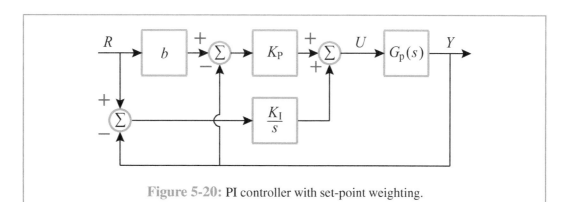

Figure 5-20: PI controller with set-point weighting.

Example 5-16: PI Controller with Set-Point Weighting

Reconsider Design #2 of Example 5-15 and implement the PI controller with set-point weighting. Try $b = 1.6/2$ and $b = 0$.

Solution: With $b = 1.6/2$, the closed-loop transfer function is

$$\frac{4(s+2)}{(s+2)(s^2+2s+4)} = \frac{4}{s^2+2s+4}.$$

With $b = 0$, the closed-loop transfer function is

$$\frac{8}{(s+2)(s^2+2s+4)}.$$

The step responses of the closed-loop transfer functions for $b = 1$, $b = 1.6/2 = 0.8$, and $b = 0$ are shown in Fig. 5-21. The case $b = 1$ is the PI controller with no set-point weighting. As b moves from 1 towards 0, the overshoot decreases and the rise time increases.

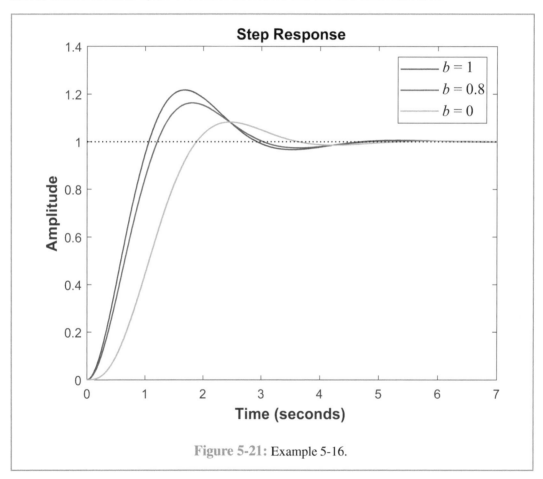

Figure 5-21: Example 5-16.

5-6 Lead-Lag and PID Controller

Improving both the transient response and the steady-state error may need the combination of both the lead and lag compensators, leading to the lead-lag compensator, or the combination the PD and PI controllers, leading to the PID controller.

5-6.1 Lead-Lag Compensator

$$G_c(s) = K\left(\frac{s+z_1}{s+p_1}\right)\left(\frac{s+z_2}{s+p_2}\right), \quad K > 0, \; p_1 > z_1 > 0, \; z_2 > p_2 > 0. \quad (5.12)$$

Design Procedure:

Step 1: Design z_1, p_1 and K of the lead part to meet the transient response specifications following the design procedure of the lead compensator.

Step 2: Choose z_2 and p_2 to increase the error constant while maintaining the transient response, following the design procedure of the lag compensator. Keep K as in the first step.

Step 3: After each step, simulate the step response of the closed-loop system.

Example 5-17: Design of a Lead-Lag Compensator

Given the plant transfer function

$$G_p(s) = \frac{1}{s(s+0.4)}.$$

1. Design a lead compensator to achieve $PO \leq 20\%$ and $T_s \leq 5$ s.

2. Add a lag compensator to increase K_v by a factor of 5.

Solution: To start with, note that the required settling time cannot be met with the uncompensated system because the complex part of the uncompensated root locus is a vertical line at $\text{Re}[s] = -0.2$ so that the settling time for any closed-loop poles on this line is $T_s = 4/0.2 = 20$ s.

To design the lead compensator, choose a desired closed-loop pole that meets the design specifications. To account for the increased overshoot due to the compensator's zero, over-design the system to achieve $PO \leq 5\%$ and $T_s \leq 5$ s. For $PO \leq 5\%$, set $\zeta = 1/\sqrt{2}$. For $T_s = 4/(\zeta\omega_n) \leq 5$, set $\zeta\omega_n \geq 0.8$. These requirements are met by choosing the desired closed-loop pole as $s^* = -1 + j$. Locate the compensator's zero under the desired pole by setting $z_1 = 1$ and apply the angle criterion at s^* to find p_1.

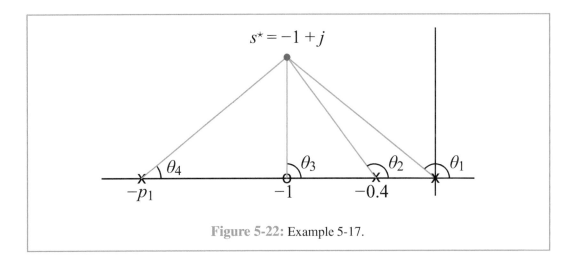

Figure 5-22: Example 5-17.

From Fig. 5-22,
$$\theta_3 - (\theta_1 + \theta_2 + \theta_4) = -180°,$$

$$\theta_1 = 135°, \quad \theta_2 = 180° - \tan^{-1}\left(\frac{1}{0.6}\right) = 121°, \quad \theta_3 = 90° \;\rightarrow\; \theta_4 = 14°,$$

$$\tan\theta_4 = \tan 14° = 0.25 = \frac{1}{p_1 - 1} \;\rightarrow\; p_1 = 5.$$

Apply the magnitude criterion to find K.

$$\left|\frac{K(s+1)}{s(s+0.4)(s+5)}\right|_{s=-1+j} = 1 \;\rightarrow\; K = 6.8.$$

With
$$G_c(s) = \frac{6.8(s+1)}{(s+5)},$$

the closed-loop transfer function is

$$T(s) = \frac{6.8(s+1)}{(s+3.4)(s^2+2s+2)},$$

whose complex poles are $-1 \pm j$. The simulation of $T(s)$ results in $PO = 18.4\%$, $T_r = 0.8$ s, and $T_s = 3.8$ s. Now design the lag part of the compensator to increase K_v by a factor of 5. With $\text{Re}[s^\star] = -1$, set $z_2 = 0.1$. Then, $z_2/p_2 = 5$ yields $p_2 = 0.02$. Keep $K = 6.8$. With

$$G_c(s) = \frac{6.8(s+1)(s+0.1)}{(s+5)(s+0.02)},$$

the closed-loop transfer function is

$$T(s) = \frac{6.8(s+1)(s+0.1)}{(s+3.455)(s+0.102)(s^2+1.863s+1.929)}.$$

Its complex poles are $-0.993 \pm 1.03j$. The simulation of $T(s)$ results in $PO = 23.2\%$, $T_r = 0.77$ s, and $T_s = 4.05$ s. The increase in T_s is small because the pole and zero of the lag part are very close to each other. The overshoot, however, does not meet the design specification. How can we reduce the overshoot?

- Redesign with a larger ζ,
- push z_2 and p_2 closer to the origin, or
- use a prefilter.

We try the first and third approaches. Redesign with $\zeta = 0.75$. Keep $\zeta \omega_n = 1$, as in the previous design. Then $\omega_n = 1/0.75$ and $\omega_d = \omega_n \sqrt{1-\zeta^2} = 0.882$. Thus, $s^\star = -1 + 0.882j$. Set $z_1 = 1$ and apply the angle criterion to find p_1.

$$\theta_3 - (\theta_1 + \theta_2 + \theta_4) = -180°,$$

$$\theta_1 = 180° - \tan^{-1}(0.882) = 138.59°,$$

$$\theta_2 = 180° - \tan^{-1}(0.882/0.6) = 124.23°, \qquad \theta_3 = 90°,$$

$$\rightarrow \theta_4 = 7.18° \quad \rightarrow \quad \tan\theta_4 = 0.126 = \frac{0.882}{p_1 - 1} \quad \rightarrow \quad p_1 = 8.$$

Apply the magnitude criterion to find K.

$$\left|\frac{K(s+1)}{s(s+0.4)(s+8)}\right|_{s=-1+0.882j} = 1 \quad \rightarrow \quad K = 11.38.$$

With

$$G_c(s) = \frac{11.38(s+1)}{(s+8)},$$

the closed-loop transfer function is

$$T(s) = \frac{11.38(s+1)}{(s+6.4)(s^2+2s+1.778)}.$$

Its complex poles are $-1 \pm 0.882j$. The simulation of $T(s)$ results in $PO = 14.5\%$, $T_r = 0.8$ s, and $T_s = 3.9$ s. Adding the lag part from the previous design yields

$$G_c(s) = \frac{11.38(s+1)(s+0.1)}{(s+8)(s+0.02)}$$

and

$$T(s) = \frac{11.38(s+1)(s+0.1)}{(s+6.425)(s+0.102)(s^2+1.893s+1.738)}.$$

The complex poles of $T(s)$ are $-0.947 \pm 0.918j$. The simulation of $T(s)$ results in $PO = 18.66\%$, $T_r = 0.76$ s, and $T_s = 4.25$ s.

Alternatively, we can keep the first design and add the prefilter $F(s) = (0.5s+1)/(s+1)$ to move the zero from -1 to -2. Simulation of the filtered closed-loop transfer function yields $PO = 9.57\%$, $T_r = 1.22$ s, and $T_s = 4.58$ s.

5-6.2 Proportional-Integral-Derivative (PID) Controller

$$G_c(s) = K_P + \frac{K_I}{s} + K_D s = \frac{K_D s^2 + K_P s + K_I}{s} = \frac{K(s+z_1)(s+z_2)}{s},$$

where $K = K_D$, $z_1 z_2 = \frac{K_I}{K_D}$, $z_1 + z_2 = \frac{K_P}{K_D}$,

$$G_c(s) = \underbrace{K(s+z_1)}_{PD} \underbrace{\left(\frac{s+z_2}{s}\right)}_{PI}. \tag{5.13}$$

Design Procedure

Step 1: Design z_1 and K of the PD part to meet the transient response specifications following the design procedure of the PD controller.

Step 2: Choose z_2 small enough to maintain the transient response.

Step 3: After each step, simulate the step response of the closed-loop system.

Example 5-18: Design of a PID Controller

Given the plant transfer function

$$G_p(s) = \frac{(s+7)}{(s+2)(s+5)(s+10)},$$

design a PID controller such that

1. $PO \leq 20\%$;
2. $T_p \leq 0.2$ s;
3. zero steady-state error to step inputs.

5-6. LEAD-LAG AND PID CONTROLLER

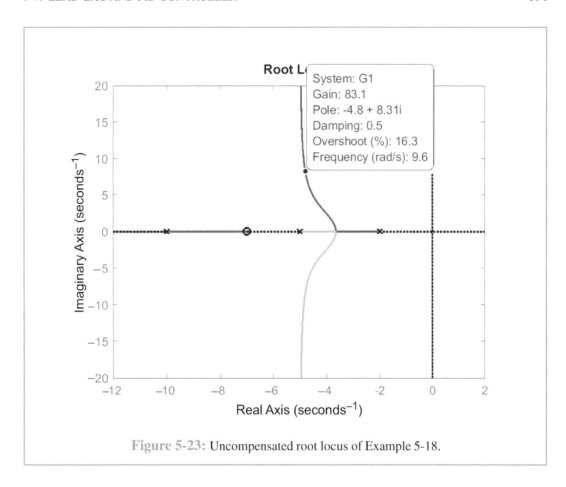

Figure 5-23: Uncompensated root locus of Example 5-18.

Solution: The third requirement is met by the use of the PID controller because it increases the type of the system to Type 1. So we need to work on the first two requirements. Let us start with the uncompensated system $G_c = K$. For $PO \leq 20\%$, set $\zeta = 0.5$. On the root locus of the system, drawn using MATLAB, $\zeta = 0.5$ is achieved at $s \approx -4.8 + 8.3j$ with $K \approx 83$, as shown in Fig. 5-23. The closed-loop transfer function is

$$T(s) = \frac{83(s+7)}{(s+7.41)(s^2 + 9.594s + 91.95)},$$

whose complex poles are $-4.8 \pm 8.3j$. The step-response parameters are $PO = 17.9\%$, $T_r = 0.17$ s, $T_p = 0.37$ s, and $T_s = 0.85$ s. Proceed to design the PD controller to reduce T_p:

$$T_p = \frac{\pi}{\omega_d} \leq 0.2 \quad \rightarrow \quad \omega_d \geq \frac{\pi}{0.2} = 15.7 \ .$$

Set $\omega_d = 16$. With $\zeta = 0.5$,

$$\zeta \omega_n = \frac{\zeta \omega_d}{\sqrt{1-\zeta^2}} = 9.2 \approx 9 \ .$$

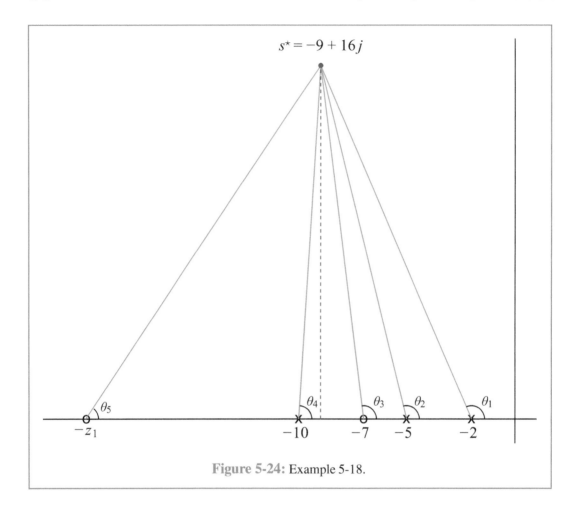

Figure 5-24: Example 5-18.

Set the desired closed-loop pole as $s^\star = -9 + 16j$. Design the PD controller $K(s+z_1)$ to locate the dominant closed-loop pole at $s^\star$. Apply the angle criterion to find z_1. From Fig. 5-24,

$$\theta_3 + \theta_5 - (\theta_1 + \theta_2 + \theta_4) = -180°,$$

$$\theta_1 = 180° - \tan^{-1}\left(\frac{16}{7}\right) = 113.63°, \quad \theta_2 = 180° - \tan^{-1}\left(\frac{16}{4}\right) = 104.04°,$$

$$\theta_3 = 180° - \tan^{-1}\left(\frac{16}{2}\right) = 97.13°, \quad \theta_4 = \tan^{-1}(16) = 86.42°,$$

$$\theta_5 = 113.63 + 104.04 + 86.42 - 97.13 - 180 = 26.96°,$$

$$\tan\theta_5 = \tan 26.96° = 0.509 = \frac{16}{z_1 - 9} \quad \rightarrow \quad z_1 = 40.46 \approx 40 \,.$$

Apply the magnitude criterion to find K.

$$\left| \frac{K(s+7)(s+40)}{(s+2)(s+5)(s+10)} \right|_{s=-9+16j} = 1 \quad \rightarrow \quad K = 8.2\,,$$

5-6. LEAD-LAG AND PID CONTROLLER

$$G_c = 8.2(s+40) \quad \rightarrow \quad T(s) = \frac{8.2(s+7)(s+40)}{(s+7.116)(s^2+18.08s+336.7)} .$$

The complex poles of $T(s)$ are $-9.042 \pm 15.967j$. The step-response parameters are $PO = 20.19\%$, $T_r = 0.075$ s, $T_p = 0.168$ s, and $T_s = 0.421$ s. Finally, add the PI controller $(s+z_2)/s$. The closest closed-loop pole to the imaginary axis is $s = -7.116$. Set $z_2 < 7.116/10 = 0.77116$. Set it $z_2 = 0.5$. Hence,

$$G_c = \frac{8.2(s+40)(s+0.5)}{s} \quad \rightarrow \quad T(s) = \frac{8.2(s+7)(s+40)(s+0.5)}{(s+7.125)(s+0.485)(s^2+17.59s+332.2)} .$$

The complex poles of $T(s)$ are $-8.795 \pm 15.964j$. The step-response parameters are $PO = 17.74\%$, $T_r = 0.078$ s, $T_p = 0.168$ s, and $T_s = 0.866$ s.

Figure 5-25 compares the step responses of the P, PD, and PID controllers. Even though the response of the PID controller reaches within 2% of the steady state at $T_s = 0.886$ s, it moves slowly towards the steady state. This is the effect of setting the zero of the PI controller very close to its pole at the origin.

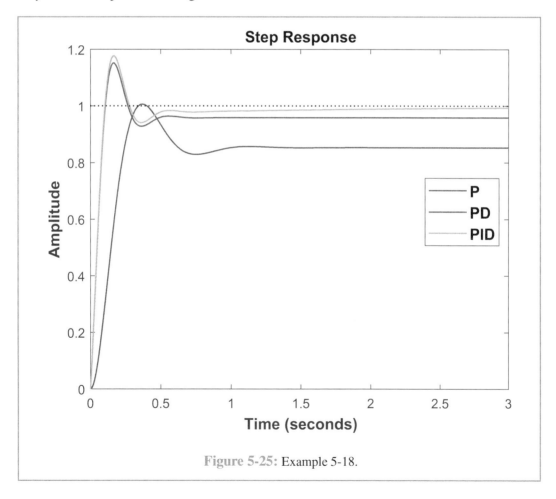

Figure 5-25: Example 5-18.

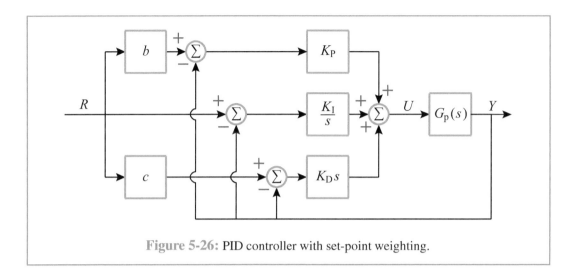

Figure 5-26: PID controller with set-point weighting.

Set-Point Weighting

As in the PD and PI controllers, set-point weighting can be used with the PID controller. The scheme, shown in Fig. 5-26, combines the schemes for the PD and PI controllers. The control signal, in the time-domain, is given by

$$u(t) = K_P[br(t) - y(t)] + K_I \int [r(\tau) - y(\tau)] \, d\tau + K_D \frac{d}{dt}[cr(t) - y(t)] \,.$$

The closed-loop transfer function is given by

$$\frac{Y}{R} = \frac{(cK_D s^2 + bsK_P + K_I) \, G_P(s)}{s + (K_D s^2 + K_P s + K_I) \, G_p(s)} \,. \tag{5.14}$$

The extra closed-loop zeros due to the PID controller are the roots of

$$cK_D s^2 + bsK_P + K_I = 0 \,.$$

The locations of the two zeros can be adjusted by choosing the weights b and c, which is an effective way to reduce the overshoot caused by the presence of the zeros. The choice $c = 0$ is preferred because it avoids the differentiation of the reference signal. This leaves only one zero at $-K_I/(bK_P)$. This zero can be also eliminated by taking $b = 0$.

5-7 Tuning PID Controllers and Anti-Windup Schemes

The PID controller, also known as the three-term controller, is one of the most used controllers in industry because a wide range of systems that are commonly used in industrial processes can be effectively controlled by PID controllers. A main reason for its popularity is that its three gains can be tuned without the knowledge of a mathematical model of the system. Experienced system operators determine the three gains by trial and error using experimental

5-7. TUNING PID CONTROLLERS AND ANTI-WINDUP SCHEMES

Table 5-8: **Effect of changing the three gains of the PDI controller on the step response.**

PID Gain	Overshoot	Settling Time	Steady-State Error
Increase K_P	Increases	Minimal impact	Decreases
Increase K_I	Increases	Increases	Zero error
Increase K_D	Decreases	Decreases	No impact

testing. Such empirical procedures depend on having general understanding of the effect of changing each gain on the step response. Table 5-8 summarizes these effects.

There are several published tuning procedures for PID controllers.[1] There are even industrially available automatic tuning algorithms for PID controllers.[2] One of the manual tuning procedures, known as the Quarter Amplitude Delay Method, is described here:

Step 1: With $K_I = K_D = 0$, adjust K_P until the second peak of the oscillation of the step response is about one-fourth of the first peak.

Step 2: With K_P fixed and $K_I = 0$, slowly increase K_D until you reduce the overshoot to the desired level.

Step 3: With K_P and K_D fixed, slowly increase K_I. The steady-state error will be zero, but initially the settling time will be large. Keep increasing K_I to reduce the settling time to an acceptable level.

Example 5-19: PID Controller Design

In this example we perform a simulation experiment to tune the PID controller for an unknown plant using the aforementioned tuning procedure. We set the goal to achieve 20% overshoot in the step response. The plant transfer function used in the simulation is

$$G_p(s) = \frac{10}{(s+1)(s+3)(s+10)},$$

but this model is not used during the tuning steps.

---
[1] See, for example, Section 4.3 of [11].
[2] See, for example, [17].

Step 1: With $K_I = K_D = 0$, we incremented K_P in steps of 1, starting from $K_P = 1$ while checking the step response for the condition that the second peak of the oscillation is about one fourth of the first peak. At $K_P = 20$ we obtained the response in Fig. 5-27. The first peak $= 1.26 - 0.867 = 0.393$ and the second peak $= 0.96 - 0.867 = 0.093$. The ratio of the first to the second peak is 4.2, which we considered to be close enough to 4. So, we fixed $K_P = 20$.

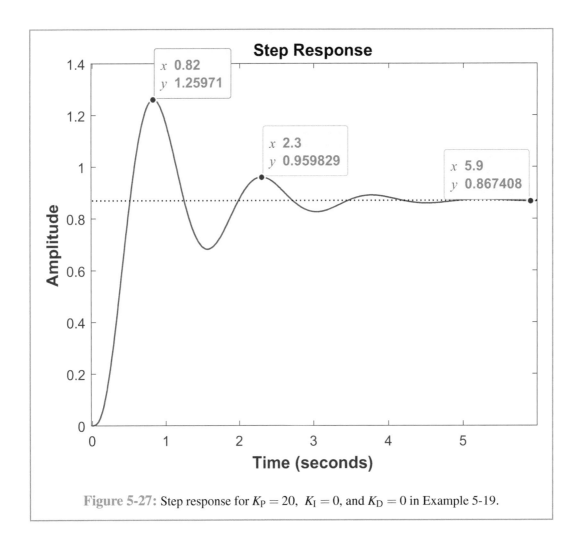

Figure 5-27: Step response for $K_P = 20$, $K_I = 0$, and $K_D = 0$ in Example 5-19.

5-7. TUNING PID CONTROLLERS AND ANTI-WINDUP SCHEMES

Step 2: With $K_P = 20$ and $K_I = 0$, we incremented K_D in steps of 0.5, starting from $K_D = 1$. The step response results are shown in Table 5-9. We set $K_D = 2.5$ because the overshoot is the closest to the goal of 20%.

Table 5-9: **Tuning K_D in Example 5-19.**

K_D	T_r (s)	T_s (s)	PO (%)
1	0.317	2.487	32.45
1.5	0.316	2.136	27.71
2	0.31	1.75	23.65
2.5	0.3	1.66	20.23
3	0.296	1.53	17.34

Step 3: With $K_P = 20$ and $K_D = 2.5$, we incremented K_I in steps of 0.5, starting from $K_I = 0.5$. We went through 30 values. Table 5-10 shows samples of the results. To maintain the overshoot under 20%, we set $K_I = 14.5$.

Table 5-10: **Tuning K_I in Example 5-19.**

K_I	T_r (s)	T_s (s)	PO (%)
0.5	0.37	84.85	4.97
3	0.36	12.9	7.52
5.5	0.352	6.26	10.2
8	0.346	3.65	12.8
10.5	0.341	2.03	15.44
13	0.335	1.92	18.09
14.5	0.333	1.89	19.66
15	0.331	1.88	20.2

Hence the result of the tuning procedure is $K_P = 20$, $K_I = 14.5$, and $K_D = 2.5$. The tuned step response is shown in Fig. 5-28.

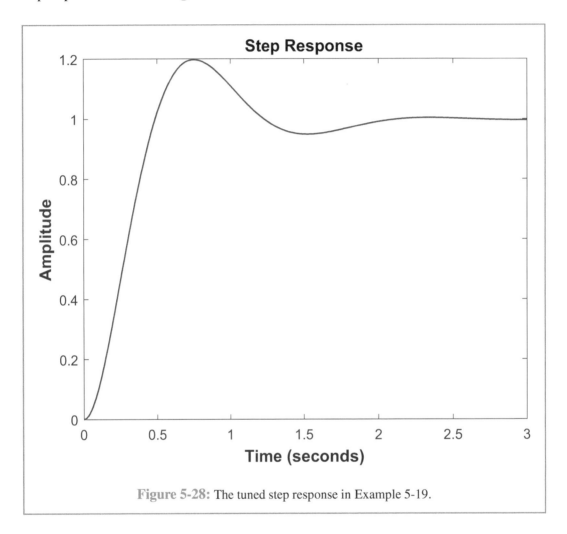

Figure 5-28: The tuned step response in Example 5-19.

Integrator Windup: All actuators have physical limits. Limiters (saturation functions) are used to limit the control signal to its extreme values so as to protect the actuator. This has a serious impact when integral action is used in the control as in PI and PID controllers. The integrator by itself is not a stable system because it has a pole at the origin. This does not cause problems as long as the feedback loop is closed. However when the control saturates, the feedback loop is open because the output of the saturating element does not react to its input. The state of the integrator may then drift to large values. When the controller comes out of saturation, it may take a long time for the system to recover its normal operation. This phenomenon is known as *integrator windup*.

5-7. TUNING PID CONTROLLERS AND ANTI-WINDUP SCHEMES

Anti-windup schemes:

Several anti-windup schemes are used to stop the integrator windup during control saturation.[3] One such scheme is shown in Fig. 5-29. The signal e_s, which is the difference between the input v and the output u of the saturating element, is fed back to the integrator through a positive gain K_w. When the control is not saturating, $u = v$ and $e_s = 0$. In this case the feedback around the integrator has no effect. When the control saturates, $e_s \neq 0$ and the loop around the integrator is closed. The equivalent transfer function from e to v becomes $K_I/(s+K_w)$, which is stable.

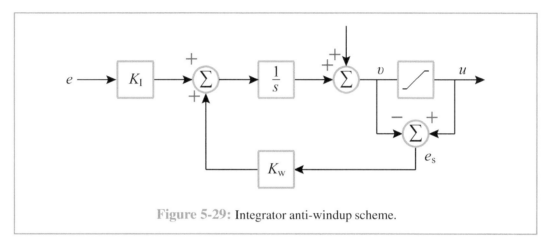

Figure 5-29: Integrator anti-windup scheme.

---

[3] See, for example, [3].

Example 5-20: Anti-Windup Controller

Consider a feedback control system formed of the plant $G_p(s) = 4/(s+1)$ and the PI controller $G_c(s) = 0.25 + 1/s$. The control is saturated at ± 0.3. A unit step input is applied at the reference. Figure 5-30 shows the output, control, and integrator signals without anti-windup scheme and with the anti-windup scheme of Fig. 5-29 with $K_w = 1$. It is clear that the anti-windup scheme reduces the drift of the integrator. The system recovers faster as the gain K_w increases. This is shown in Fig. 5-31 by comparing the responses for $K_w = 1$ and $K_w = 10$.

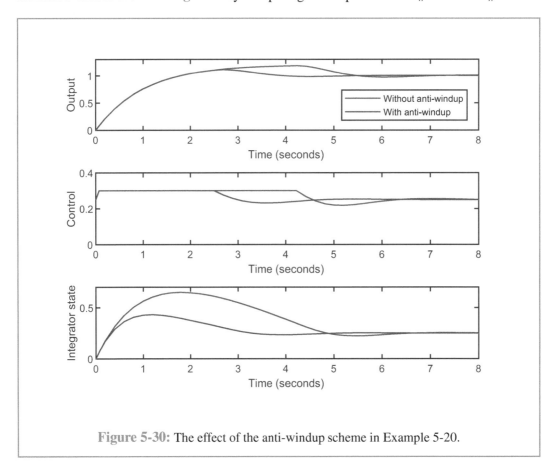

Figure 5-30: The effect of the anti-windup scheme in Example 5-20.

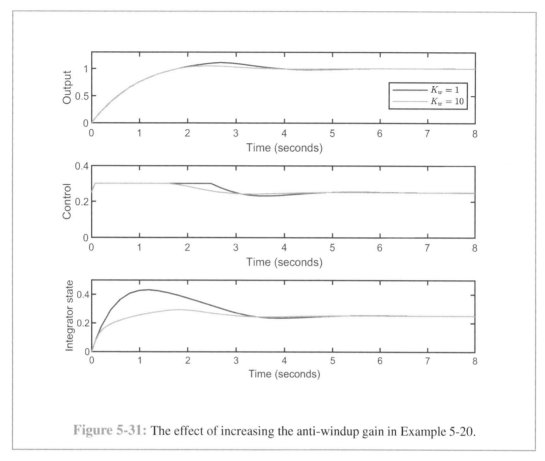

Figure 5-31: The effect of increasing the anti-windup gain in Example 5-20.

Summary

Concepts

- Gain adjustment can meet only one design specification.
- The lead compensator and PD controller can stabilize the system and improve the transient response because they shift the root locus to the left in the complex plane.
- The lag compensator and PI controller can reduce the steady-state error because they increase the error constant K_p, K_v, or K_a.
- The lead-lag compensator combines the functions of the lead and lag compensators.
- The PID controller combines the functions of the PI and PD controllers.
- Control saturation can cause integrator windup, which can be mitigated by anti-windup schemes.
- For a class of systems, the three gains of the PID controller can be tuned without knowing a mathematical model of the system.

Mathematical Models

Lead compensator: $G_c(s) = K\left(\dfrac{s+z}{s+p}\right)$, $K > 0$, $p > z > 0$

PD controller: $G_c(s) = K_P + K_D s$, $K_P > 0$, $K_D > 0$

Lag compensator: $G_c(s) = K\left(\dfrac{s+z}{s+p}\right)$, $K > 0$, $z > p > 0$

PI controller: $G_c(s) = K_P + \dfrac{K_I}{s}$, $K_P > 0$, $K_I > 0$

Lead-lag compensator: $G_c(s) = K\left(\dfrac{s+z_1}{s+p_1}\right)\left(\dfrac{s+z_2}{s+p_2}\right)$, $K > 0$, $p_1 > z_1 > 0$,
$z_2 > p_2 > 0$

PID controller: $G_c(s) = K_P + \dfrac{K_I}{s} + K_D s$, $K_P > 0$, $K_I > 0$, $K_D > 0$

Important Terms Provide definitions or explain the meaning of the following terms:

- anti-windup scheme
- gain adjustment
- integrator windup
- lag compensator
- lead compensator
- lead-lag compensator
- PD controller
- PI controller
- PID controller
- prefilter
- set-point weighting
- tuning PID controller

PROBLEMS

5.1 Consider a unity feedback control system with

$$G(s) = \dfrac{K}{(s+2)(s+4)}$$

Design K such that the system operates at 5% overshoot. Determine the settling time.

5.2 Consider a unity feedback control system with

$$G(s) = \dfrac{K}{s(s+2)}.$$

Design K such that the system operates at 20% overshoot. Determine the settling time.

5.3 Consider a unity feedback control system with

$$G(s) = \dfrac{K(s+3)}{s(s+1)}.$$

(a) Draw the root locus using MATLAB.

(b) Choose K such that $T_s = 2$ s. What is PO?

(c) Simulate the step response of the closed-loop transfer function and determine T_s and PO. If the results are different than part (b), explain why.

5.4 Consider a unity feedback control system with

$$G(s) = \frac{K(s+3)}{s(s+1)}.$$

(a) Draw the root locus using MATLAB.

(b) It is desired to choose K such that $PO \leq 5\%$. Verify that there are two choices of K that meet this requirement.

(c) Simulate the step response of the closed-loop transfer function for each choice of K and comment on the results.

5.5 Consider a unity feedback control system with

$$G(s) = \frac{K}{s(s+1)(s+2)}.$$

(a) Draw the root locus using MATLAB.

(b) Choose K such that $PO = 60\%$. What is T_s?

(c) Simulate the step response of the closed-loop transfer function and determine T_s and PO.

(d) Comments on the results of parts (b) and (c).

5.6 Consider a unity feedback control system with

$$G(s) = \frac{10K}{(s+1)^2(s+10)}.$$

(a) Approximate G by

$$G(s) \approx \frac{K}{(s+1)^2}$$

and use the approximate transfer function to choose K to achieve 5% overshoot.

(b) Simulate the step response of the exact closed-loop transfer function with the choice of K from part (a). Using the results of the simulation, comment on the appropriateness of the approximation.

5.7 Consider a unity feedback control system with

$$G_p(s) = \frac{1}{s(s+1)(s+10)}.$$

(a) Design $G_c = K$ such that the system operates at 5% overshoot. Determine the settling time.

(b) Design a lead compensator to decrease the settling time by a factor of 2 while operating at 5% overshoot.

(c) Using MATLAB simulation, compare the performance of the compensated and uncompensated systems.

5.8 Consider a unity feedback control system with

$$G_p(s) = \frac{1}{(s+1)(s+3)}.$$

(a) Design $G_c = K$ such that the system operates at 10% overshoot. Determine the settling time.

(b) Design a compensator to decrease the settling time by a factor of 3 while operating at 10% overshoot.

(c) Using MATLAB verify that the design specifications of part (b) are satisfied.

5.9 Consider a unity feedback control system with $G_p(s) = 1/s^2$.

(a) Sketch the root locus when $G_c(s) = K \geq 0$.

(b) Design a lead compensator $G_c(s) = K(s+z)/(s+p)$ so that the closed-loop system has complex poles at $-1 \pm j$. Take $z = \frac{1}{2}$.

5.10 Consider a unity feedback control system with

$$G(s) = \frac{K}{(s+1)(s+2)(s+11)}.$$

(a) Design K such that the system operates at 5% overshoot. Determine the settling time.

(b) Design a lead compensator to decrease the settling time by a factor of 2 while operating at 5% overshoot.

(c) Using MATLAB simulation, compare the performance of the compensated and uncompensated systems.

5.11 Consider a unity feedback control system with the plant transfer function $G_p(s) = 1/s^2$. To achieve overshoot under 5%, the desired closed-loop poles are chosen to be $-1 \pm j$.

(a) Design a PD controller to achieve the desired closed-loop poles.

(b) Simulate the step response of the closed-loop transfer function and determine the overshoot.

(c) If the overshoot in (b) is higher than 5%, how would you modify the controller to reduce the overshoot to under 5%? Verify that your modification does indeed bring the overshoot under 5%.

5.12 Consider the plant transfer function

$$G_p(s) = \frac{10}{s(s+1)(s+5)}.$$

(a) Design a compensator $G_c(s)$ such that: (1) $PO \leq 5\%$; (2) $T_s \leq 4$ s.; (3) steady-state error to step inputs $\leq 1\%$.

(b) Simulate the step response of the closed-loop system to verify that all specifications are met.

5.13 The first design of Example 5-7 chooses the zero of the lead compensator to cancel the pole of the plant at -2.5. This of course assumes that the location of the plant pole is known precisely. In this problem we investigate the effect of the uncertainty in the pole's location. Using MATLAB, simulate the step response of the closed-loop transfer function when

(a) $G_p(s) = \dfrac{0.5}{s(s+2.5)}$,   (b) $G_p(s) = \dfrac{0.5}{s(s+3)}$,   (c) $G_p(s) = \dfrac{0.5}{s(s+2)}$.

The compensator is

$$G_c(s) = \dfrac{128(s+2.5)}{s+8}$$

in all cases. Case (a) is the ideal case treated in Example 5-7. Cases (b) and (c) represent $\pm 20\%$ error in the pole's location. Comment on the simulation results.

5.14 Consider a unity feedback control system with the plant transfer function

$$G_p(s) = \dfrac{1}{s(s+2)} .$$

(a) Design $G_c(s) = K$ to operate at 20% overshoot. What are T_s and K_v?

(b) Design $G_c(s)$ to increase K_v by a factor of 2 while maintaining the transient response.

(c) Using simulation, compare the step responses of the uncompensated and compensated systems.

5.15 Consider a unity feedback control system with

$$G_p(s) = \dfrac{1}{(s+3)(s+5)} .$$

(a) Design $G_c = K$ to operate at 5% overshoot. What are T_s and K_p?

(b) Design $G_c(s)$ to increase K_p by a factor of 10 while maintaining the transient response.

(c) Using simulation, compares the step responses of the uncompensated and compensated systems.

5.16 For the plant transfer function

$$G_p(s) = \dfrac{4}{(s+1)(s+4)} ,$$

(a) design a compensator $G_c(s)$ such that: (1): $PO \leq 20\%$; (2) zero steady-state error to step inputs;

(b) verify that the design requirements are met.

5.17 For the plant transfer function

$$G_p(s) = \frac{10}{s(s+10)},$$

(a) design a compensator $G_c(s)$ such that; (1): $PO \leq 20\%$; (2) zero steady-state error to ramp inputs;

(b) simulate the step response of the closed-loop system. If the overshoot is higher than required, modify the controller to bring the overshoot down.

5.18 Consider a unity feedback control system with

$$G_p(s) = \frac{1}{s(s+12)}.$$

(a) Design $G_c = K$ such that the system operates at 20% overshoot.

(b) Redesign the compensator $G_c(s)$ of part (a) to reduce the settling time by a factor of 2 while maintaining the percent overshoot.

(c) Redesign the compensator $G_c(s)$ of part (a) to increase K_v by a factor of 4 while maintaining the dominant closed-loop poles.

5.19 Consider a unity feedback control system with

$$G(s) = \frac{K}{s(s+2)}.$$

(a) Design K such that the system operates at 5% overshoot.

(b) Add a compensator to reduce the settling time of part (a) by a factor of 5.

(c) Add another compensator to increase K_v of part (b) by a factor of 5.

5.20 Consider a unity feedback control system with

$$G_p(s) = \frac{1}{s(s+10)}.$$

(a) Design $G_c = K$ such that the system operates at 20% overshoot.

(b) Redesign the compensator $G_c(s)$ of part (a) to reduce the settling time by a factor of 4 while maintaining the percent overshoot.

(c) Redesign the compensator $G_c(s)$ of part (a) to increase K_v by a factor of 2 while maintaining the dominant closed-loop poles.

5.21 Let

$$G_p(s) = \frac{1}{s(s+0.4)}.$$

Design a lead-lag compensator such that: (1) $PO \leq 8\%$; (2) $T_s \leq 1$ s; (3) $K_v \geq 10$. Use simulation to verify that the design specifications are met.

5.22 Let

$$G_p(s) = \frac{(s+7)}{(s+2)(s+5)(s+10)}.$$

Design a PID controller such that: (1) $PO \leq 8\%$; (2) $T_s \leq 0.5$ s; (3) zero steady-state error to step inputs. Use simulation to verify that the design specifications are met.

5.23 A linearized model of a synchronous generator connected to an infinite bus is shown in Fig. P3.17. Let $b_1 = 34.29$, $b_2 = 0$, $b_3 = 0.149$, $b_4 = 0.3341$, $\sin \delta^\star = 0.78$, and $E_q^\star = 1.0552$.

(a) Find the transfer function $G_p(s) = Y(s)/U(s)$.

(b) Design a PID controller $U = G_c(s)(R-Y)$ such that the overshoot is less than 5%.

(c) Plot the step response.

5.24 Figure P5.24 shows a schematic diagram of an inverted pendulum on a cart. The pivot of the pendulum is mounted on a cart that can move in a horizontal direction. θ is the rotation angle of the pendulum (measured clockwise from the vertical position) and y is the displacement of the cart. Viewing the cart acceleration as the control input to the pendulum, we can model the pendulum by the second-order differential equation [22]

$$(J + mL^2)\ddot{\theta} = mgL\sin\theta - mL\ddot{y}\cos\theta,$$

where J is the moment of inertial of the pendulum with respect to the center of gravity, m the mass of the pendulum, L the distance from the center of gravity to the pivot, and g the acceleration due to gravity. With $\ddot{y} = 0$, the pendulum has equilibrium at the vertical position with $\theta = \dot{\theta} = 0$. Defining the control input as $u = -\ddot{y}/g$, the linearization of the system near the equilibrium point is given by

$$\ddot{\theta} = a(\theta + u), \quad \text{where } a = \frac{mgL}{J + mL^2}.$$

Let $m = 0.1$ kg, $J = 0.008$ kg m², $L = 0.5$ m, and $g = 9.81$ m/s².

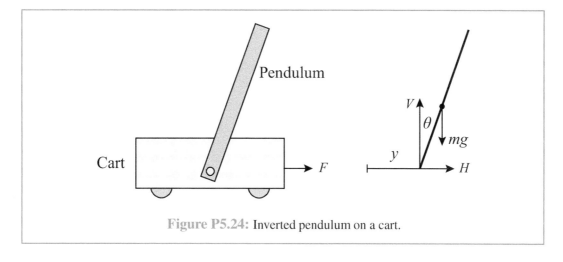

Figure P5.24: Inverted pendulum on a cart.

(a) Find the transfer function θ/U. Show that the system is unstable.

(b) Draw the root locus when $u = -K\theta$, for $K \geq 0$. Can you choose K to stabilize the system?

(c) Design a compensator such that the closed-loop system has a pair of complex poles with $\omega_n = 1$ rad/s and $\zeta = 1/\sqrt{2}$.

5.25 A riderless bicycle can be stabilized by equipping it with a steering mechanism and feedback control from the measured lean angle. In [1], the dynamics of the bicycle, in constant forward velocity, from the steering angle to the lean angle is modeled by

$$G_1(s) = 3.373 \left(\frac{s+8}{s^2 - 18.9} \right),$$

and the steering mechanism is modeled by

$$G_2(s) = H(s) \frac{(33.9)^2}{s^2 + 40.68s + (33.9)^2},$$

where $H(s)$ is the transfer function of a time delay. In this problem we ignore the time delay; that is, $H(s) \approx 1$.

(a) Design a PI controller such that settling time is less than 2 s.

(b) Plot the step response of the closed-loop system. Determine the overshoot and the settling time.

5.26 Consider the idle speed control of Problem 3.16 together with the linearized model of Fig. P3.16. The time delay is modeled by the transfer function

$$H(s) = \frac{1 - (0.04/2)s + (0.04)^2 s^2/12}{1 + (0.04/2)s + (0.04)^2 s^2/12}.$$

Consider the PI controller

$$\Delta U = \left(K_P + \frac{K_I}{s} \right) (\Delta N_r - \Delta N),$$

where N_r is the desired engine speed and ΔN_r is the change from the nominal value. Design the PI controller to achieve settling time less than 2 s. Plot the step response.

5.27 Repeat the PID tuning simulation experiment of Example 5-19. Use

$$G_p(s) = \frac{1}{s(s+1)}$$

in the simulation.

5.28

(a) If you want to increase the error constant (K_p or K_v) of uncompensated system while maintaining the transient response, which compensator would you use and how would you design it?

(b) If you want to reduce the settling time of uncompensated system while maintaining the overshoot, which compensator would you use and how would you design it?

(c) What is set-point weighting? Why and how is it used?

(d) What is a prefilter? Why and how is it used?

(e) In PID control, how would increasing the derivative gain K_D affect the overshoot, settling time, and steady-state error?

(f) In PID control, how would increasing the proportional gain K_P affect the overshoot, settling time, and steady-state error?

(g) In PID control, how would increasing the integration gain K_I affect the overshoot, settling time, and steady-state error?

(h) What is integrator windup and what is an anti-windup scheme?

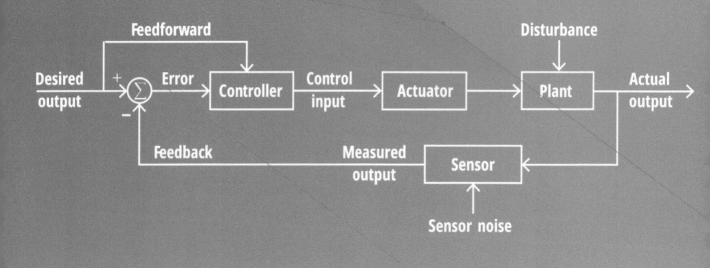

Chapter 6

Frequency-Response Methods

Chapter Contents

      Overview, 211
6-1 Polar Plot, 211
6-2 Bode Plots, 217
6-3 Time-Delay Systems, 233
6-4 Nyquist Criterion, 234
6-5 Relative Stability, 249
6-6 Relations between Frequency-Domain and Time-Domain Responses, 256
      Chapter Summary, 266
      Problems, 267

Objectives

Upon learning the material presented in this chapter, you should be able to:

1. Draw the polar plot of a transfer function.
2. Draw the Bode plots of a transfer function.
3. Apply the Nyquist stability criterion to determine the stability of the feedback control system.
4. Measure relative stability by the gain and phase margins.
5. Find the gain and phase margins and their corresponding crossover frequencies from the Bode plots.
6. Determine the type of a feedback system and the error constant K_p, K_v, or K_a from the Bode plots.
7. Understand the relationship between the open-loop frequency response, the closed-loop frequency response, and the closed-loop step response.

Overview

The *frequency response* of a transfer function $G(s)$ is $G(j\omega)$ where the frequency ω varies from $-\infty$ to ∞ rad/s. In other words, the frequency response is obtained by restricting the complex variable s to the imaginary axis of the complex plane. We have seen in Chapter 3 that the steady-state response of a stable transfer function $G(s)$ to the sinusoidal input $A\sin(\omega t + \theta)$ is a sinusoidal signal of the same frequency with the amplitude and phase modified by the amplitude and phase of the frequency response $G(j\omega)$. The definition of the frequency response, however, is not limited to stable transfer functions.

We can characterize and study the stability of the control system of Fig. 6-1 using the frequency response of $G(s)\,H(s)$, which we write as $GH(s)$. Such characterization uses graphical representations of the frequency response. Two such representations are covered in this chapter: the polar plot of Section 6-1 and the Bode plots of Section 6-2.

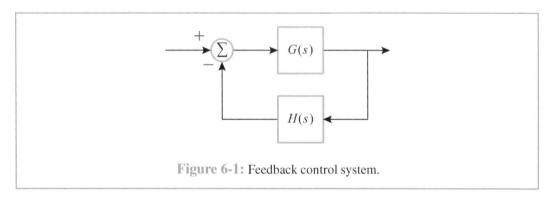

Figure 6-1: Feedback control system.

The polar plot is used in developing the *Nyquist stability criterion* introduced in Section 6-3 and studying *relative stability* in Section 6-4. The notion of relative stability deals with the ability of the closed-loop system to maintain stability in the face of uncertainty in $GH(s)$. Relative stability is quantified by two parameters: the *gain and phase margins*, which measure the strength of the stability of the system.

For transfer functions that have no poles in $\text{Re}[s] > 0$, the gain and phase margins can be easily depicted using Bode plots. Using Bode plots we can also shape the frequency response of the system to meet requirements on the performance of the closed-loop system, such as tracking and the effect of disturbance and measurement noise. These applications of Bode plots are introduced in Chapter 7 in the context of frequency-domain design. Section 6-5 provides certain relations between the frequency-domain and time-domain responses.

6-1 Polar Plot

For each frequency ω, the frequency response $G(j\omega)$ is a complex number that can be written as
$$G(j\omega) = \text{Re}[G(j\omega)] + j\text{Im}[G(j\omega)] = |G(j\omega)|e^{j\angle[G(j\omega)]}\,.$$

The complex number can be represented by a point in a complex plane whose real axis is $\text{Re}[G]$ and imaginary axis is $\text{Im}[G]$. The plane is called the *G-plane*. The polar plot of $G(j\omega)$

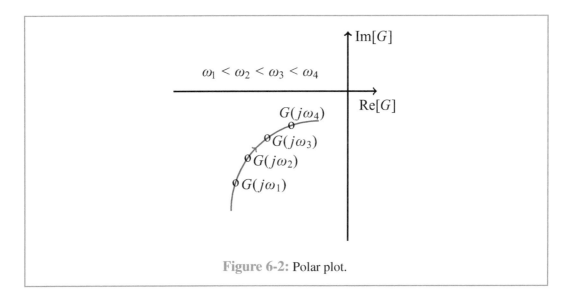

Figure 6-2: Polar plot.

is a plot of $G(j\omega)$ in the G-plane as ω is varied from $-\infty$ to ∞. Figure 6-2 shows a segment of a polar plot with four points corresponding to four frequencies. The arrowhead on the plot shows the direction of increasing frequency.

The polar plot can be drawn using the MATLAB command "*nyquist*", but a sketch that shows the general shape of the plot can be drawn by simple calculations. We shall see later, when we study the Nyquist criterion, that it is often sufficient to have an approximate sketch of the polar plot.

Drawing an approximate sketch of the polar plot can be realized by the following steps:

Step 1: Derive analytic expressions for $\text{Re}[G]$, $\text{Im}[G]$, and $\angle[G]$ in terms of ω. Calculating $\text{Re}[G]$ and $\text{Im}[G]$ is sufficient to determine G, but calculating $\angle[G]$ simplifies drawing the sketch.

Step 2: For $\omega > 0$, calculate the limits of $\text{Re}[G]$, $\text{Im}[G]$, and $\angle[G]$ as:

(a) $\omega \to 0$, and

(b) $\omega \to \infty$.

Step 3: Find points of intersection with

(a) the real axis: $\text{Im}[G] = 0$, and

(b) the imaginary axis: $\text{Re}[G] = 0$.

A rational transfer function $G(s)$ with real coefficients has the property that

$$G(j\omega) = X(\omega) + jY(\omega) \quad \rightarrow \quad G(-j\omega) = X(\omega) - jY(\omega) \ .$$

Therefore, the polar plot is symmetric with respect to the real axis. The plot for $\omega < 0$ is a mirror image of the plot for $\omega > 0$, where the image is reflected off the real axis.

Example 6-1: Polar Plot of a First-Order Transfer Function

Sketch the polar plot of
$$G(s) = \frac{K}{\tau s + 1}, \quad K > 0, \quad \tau > 0.$$

Solution:
$$G(j\omega) = \frac{K}{1 + j\omega\tau} = \frac{K(1 - j\omega\tau)}{1 + \omega^2\tau^2},$$

$$\text{Re}[G] = \frac{K}{1 + \omega^2\tau^2}, \quad \text{Im}[G] = -\frac{K\omega\tau}{1 + \omega^2\tau^2}, \quad \angle[G] = -\tan^{-1}(\omega\tau).$$

Compute the limits:

- As $\omega \to 0$, $\text{Re}[G] \to K$, $\text{Im}[G] \to 0$, $\angle[G] \to 0$,
- As $\omega \to \infty$, $\text{Re}[G] \to 0$, $\text{Im}[G] \to 0$, $\angle[G] \to -\frac{\pi}{2}$.

Intersection with the axes:

- Real axis: $\text{Im}[G] = 0 \;\rightarrow\; \omega = 0 \;\rightarrow\; \text{Re}[G] = K$,
- Imaginary axis: no intersections except at the origin.

Using this information we draw the sketch shown in Fig. 6-3.

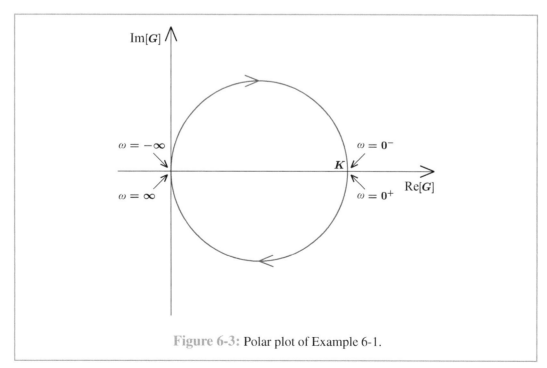

Figure 6-3: Polar plot of Example 6-1.

At $\omega = 0$, the plot starts at the point $(K, 0)$ on the real axis. As ω increases, $\text{Im}[G]$ becomes negative. Hence, the plot moves into the lower half of the plane. There are no intersections with the axes until the plot reaches the origin as ω tends to infinity. Because the limit of $\angle[G]$ is $-\pi/2$ the plot is tangent to the imaginary axis as it approaches the origin. The shape of the plot is drawn arbitrarily since the information about the limits and intersection with the axes does not provide information on the shape of the plot. We can improve the approximation by calculating $\text{Re}[G]$ and $\text{Im}[G]$ at a few frequencies, but since we can draw the exact plot using MATLAB, we usually do not try to improve the approximation of the sketch. For $\omega < 0$, the plot is the mirror image of the plot for $\omega > 0$. The sketch shows points where $\omega = 0^+$ and $\omega = 0^-$; $\omega = 0^+$ when it approaches zero from the positive side and $\omega = 0^-$ when ω approaches zero from the negative side.

Example 6-2: Polar Plot of a Second-Order Transfer Function

Sketch the polar plot of

$$G(s) = \frac{K}{s(\tau s + 1)}, \quad K > 0, \quad \tau > 0.$$

Solution:

$$G(j\omega) = \frac{K}{j\omega(1 + j\omega\tau)} = \frac{-jK(1 - j\omega\tau)}{\omega(1 + \omega^2\tau^2)} = \frac{-K(\omega\tau + j)}{\omega(1 + \omega^2\tau^2)},$$

$$\text{Re}[G] = -\frac{K\tau}{1 + \omega^2\tau^2}, \quad \text{Im}[G] = -\frac{K}{\omega(1 + \omega^2\tau^2)}, \quad \angle[G] = -\frac{\pi}{2} - \tan^{-1}(\omega\tau).$$

Note that $\angle[G]$ is computed from the original expression of $G(j\omega)$, not from $\text{Re}[G]$ and $\text{Im}[G]$. This is usually the easier way to compute $\angle[G]$ when G can be written as $G = G_1 G_2 \ldots G_m$ because $\angle[G] = \angle[G1] + \angle[G_2] + \cdots + \angle[G_m]$.

Compute the limits:

- As $\omega \to 0$, $\text{Re}[G] \to -K\tau$, $\text{Im}[G] \to -\infty$, $\angle[G] \to -\frac{\pi}{2}$,

- As $\omega \to \infty$, $\text{Re}[G] \to 0$, $\text{Im}[G] \to 0$, $\angle[G] \to -\pi$.

6-1. POLAR PLOT

There are no intersections with the axes except at the origin.

The sketch is shown in Fig. 6-4. As $\omega \to 0$, the plot approaches $-\infty$ asymptotic to the vertical line $\text{Re}[G] = -K\tau$. As ω increases, the plot stays in the lower half of the plane and approaches the origin, tangent to the real axis since $\angle[G]$ approaches $-\pi$. For $\omega < 0$, the plot is the mirror image of the plot for $\omega > 0$.

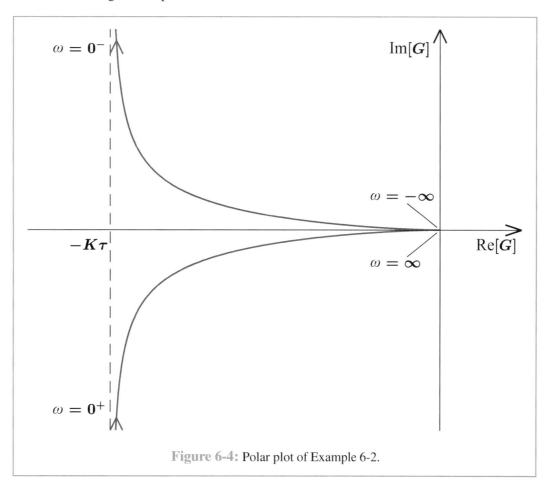

Figure 6-4: Polar plot of Example 6-2.

Example 6-3: Polar Plot of a Third-Order Transfer Function

Sketch the polar plot of

$$G(s) = \frac{1}{(s+1)(s+2)(s+3)}.$$

Solution:

$$G(j\omega) = \frac{1}{(1+j\omega)(2+j\omega)(3+j\omega)} = \frac{(1-j\omega)(2-j\omega)(3-j\omega)}{(1+\omega^2)(4+\omega^2)(9+\omega^2)},$$

$$\text{Re}[G] = \frac{6(1-\omega^2)}{(1+\omega^2)(4+\omega^2)(9+\omega^2)}, \quad \text{Im}[G] = -\frac{\omega(11-\omega^2)}{(1+\omega^2)(4+\omega^2)(9+\omega^2)},$$

$$\angle[G] = -\tan^{-1}(\omega) - \tan^{-1}\left(\frac{\omega}{2}\right) - \tan^{-1}\left(\frac{\omega}{3}\right).$$

Compute the limits:

- As $\omega \to 0$, $\text{Re}[G] \to \frac{1}{6}$, $\text{Im}[G] \to 0$, $\angle[G] \to 0$,

- As $\omega \to \infty$, $\text{Re}[G] \to 0$. $\text{Im}[G] \to 0$, $\angle[G] \to -\frac{3\pi}{2} \equiv \frac{\pi}{2}$,

Intersection with the axes:

- Real axis: $\text{Im}[G] = 0 \rightarrow \omega = 0 \rightarrow \text{Re}[G] = \frac{1}{6}$ or $\omega^2 = 11 \rightarrow \text{Re}[G] = \frac{-1}{60}$,

- Imaginary axis: $\text{Re}[G] = 0 \rightarrow \omega = 1 \rightarrow \text{Im}[G] = \frac{-1}{10}$.

The sketch is shown in Fig. 6-5. At $\omega = 0$, the plot starts at $(\frac{1}{6}, 0)$ on the real axis. As ω

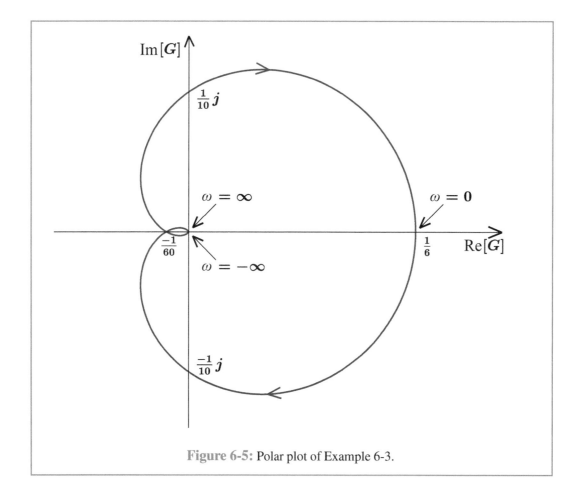

Figure 6-5: Polar plot of Example 6-3.

increases, Im$[G]$ becomes negative; hence, the plot moves into the lower half of the plane. It intersects the imaginary axis at $(0, -\frac{1}{10})$ and moves into the third quadrant. It stays in the third quadrant until it intersects the real axis at $(-\frac{1}{60}, 0)$ and crosses into the second quadrant. It stays in the second quadrant until it approaches the origin as $\omega \to \infty$. Because the limit of $\angle[G]$ is $\frac{\pi}{2}$, the plot is tangent to the imaginary axis as it approaches the origin. This is shown in the zoomed-in graph of Fig. 6-6. For $\omega < 0$, the plot is the mirror image of the plot for $\omega > 0$.

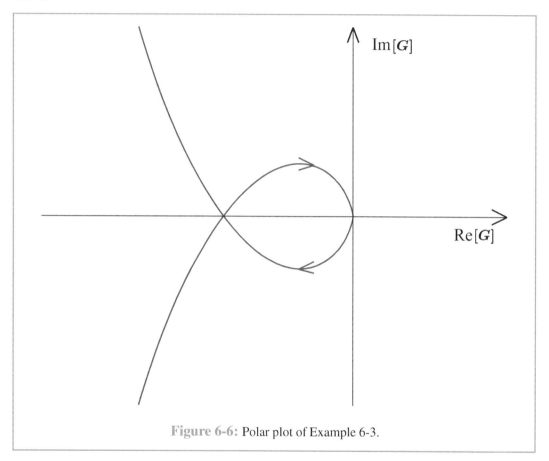

Figure 6-6: Polar plot of Example 6-3.

6-2 Bode Plots

Bode plots provide an alternative way to plot the frequency response $G(j\omega)$. Recall that $G(j\omega) = |G(j\omega)|e^{j\angle[G(j\omega)]}$, where $|G|$ is the magnitude of G and $\angle[G]$ is the angle or phase of G. In Bode plots, $G(j\omega)$ is represented by two plots: magnitude versus frequency and phase versus frequency. There are, however, some details that make these plots very useful. First, the ω-axis uses a logarithmic scale, which allows for a wide frequency range.

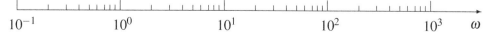

Second, the magnitude is also drawn using a logarithmic scale by plotting $20\log|G|$ versus ω. The log is taken to the base 10 and the units of $20\log|G|$ are decibels (dB).[1] When G is a product of simpler transfer functions, $G = G_1 G_2 \ldots G_m$, magnitude G is a product of magnitudes, $|G| = |G_1||G_2|\ldots|G_m|$. The logarithm converts this product into a summation,

$$\log|G| = \log|G_1| + \log|G_2| + \cdots + \log|G_m|\,.$$

Recalling that the phase $\angle[G]$ is also a summation,

$$\angle[G] = \angle[G_1] + \angle[G_2] + \cdots \angle[G_m]\,,$$

we see that the Bode plots of $G = G_1 G_2 \cdots G_m$ can be constructed by adding the Bode plots of G_1, G_2, to G_m.

Figure 6-7 shows an example of Bode plots drawn using the MATLAB command "*bode*". The grid is added using the command "*grid*".

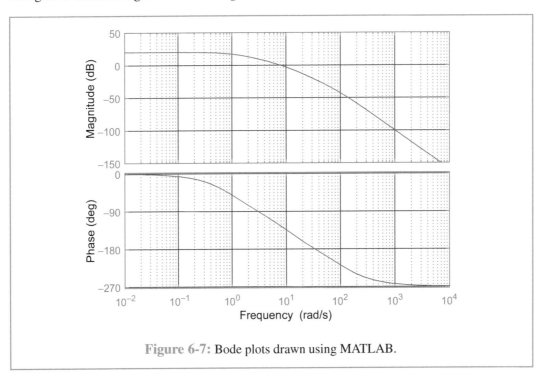

Figure 6-7: Bode plots drawn using MATLAB.

Example 6-4: Bode Plots of a Transfer Function with One Pole at the Origin

Sketch the Bode plots of

$$G(s) = \frac{K}{s}\,.$$

---

[1] In communications, the power gain is measured in decibels, $10\log(P_2/P_1)$. Because power is the square of voltage, the power gain is measured by $20\log(V_2/V_1)$.

6-2. BODE PLOTS

Solution:

$$G(j\omega) = \frac{K}{j\omega}, \qquad |G| = \frac{K}{\omega}, \qquad \angle[G] = -90°.$$

The phase plot is a horizontal line at $-90°$. For the magnitude plot,

$$20\log|G| = 20\log\left(\frac{K}{\omega}\right) = 20\log K - 20\log\omega.$$

Because the ω-scale is logarithmic, the foregoing equation represents a straight line with negative slope.

If ω_2 and ω_1 differ by a decade; that is, $\omega_2 = 10\,\omega_1$, $\dfrac{K}{\omega_2} = \dfrac{K}{10\,\omega_1}$,

$$20\log\left(\frac{K}{\omega_2}\right) = 20\log\left(\frac{K}{\omega_1}\right) - 20\log 10 = 20\log\left(\frac{K}{\omega_1}\right) - 20.$$

$20\log|G|$ drops by 20 dB per decade; so, the slope is -20 dB/dec. The Bode plots are shown in Fig. 6-8. The magnitude plot intersects the zero-dB line at $\omega = K$ because at $\omega = K$, $K/\omega = 1$ and $\log(1) = 0$.

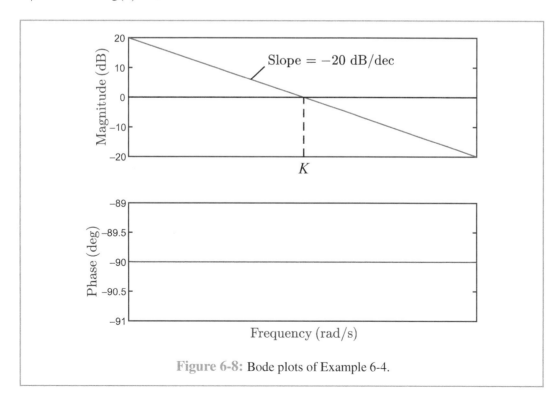

Figure 6-8: Bode plots of Example 6-4.

6-2.1 Asymptotic Bode Plots

We now examine the use of approximations to obtain asymptotic Bode plots of the transfer function.

Example 6-5: Bode Plots of a Transfer Function with One Pole

Sketch the Bode plots and asymptotic Bode plots of

$$G(s) = \frac{K}{(\tau s + 1)}, \qquad K > 1.$$

Solution:

$$G(j\omega) = \frac{K}{(1+j\omega\tau)}, \qquad |G| = \frac{K}{\sqrt{1+\omega^2\tau^2}}, \qquad \angle[G] = -\tan^{-1}(\omega\tau).$$

One particular frequency of interest is $\omega = 1/\tau$, which is called the *corner frequency* or *break frequency*. At this frequency,

$$|G| = \frac{K}{\sqrt{2}} \longrightarrow 20\log|G| \approx 20\log K - 3 \text{ dB} \quad \text{and} \quad \angle[G] = -\tan^{-1}(1) = -45°.$$

The Bode plots can be approximated by a sequence of straight lines, called asymptotes.

$$\text{For } \omega \ll \frac{1}{\tau}, \quad G(j\omega) \approx K \longrightarrow |G| = K \text{ and } \angle[G] = 0,$$

$$\text{For } \omega \gg \frac{1}{\tau}, \quad G(j\omega) \approx \frac{K}{j\omega\tau} \longrightarrow |G| = \frac{K}{\omega\tau} \text{ and } \angle[G] = -90°,$$

$$|G| = \frac{K}{\omega\tau} \longrightarrow 20\log|G| = 20\log K - 20\log(\omega\tau).$$

This is the equation of a line that starts at $20\log K$ at the break frequency $\omega = 1/\tau$ and decreases with a slope of -20 dB/dec. Thus, the asymptotic magnitude plot is formed of two lines intersecting at the break frequency. Starting at low frequency, we draw a horizontal line of amplitude $20\log K$. The line is extended until the break frequency where a line with a slope of -20 dB/dec starts, extending to high frequencies.

For the asymptotic phase plot we want to draw a line that starts at $0°$ at low frequency and goes to $-90°$ at high frequencies and passes through $-45°$ at $w = 1/\tau$. For that, we draw a line of slope $-45°$/dec that starts one decade below the break frequency, at $\omega = 0.1/\tau$, and reaches $-90°$ one decade after the break frequency, at $\omega = 10/\tau$. The asymptotic phase plot is zero for $\omega < 0.1/\tau$ and $-90°$ for $\omega > 10/\tau$.

6-2. BODE PLOTS

The asymptotic Bode plots are shown in Fig. 6-9. They are shown again in Fig. 6-10 together with the exact Bode plots. The figure shows that the asymptotic Bode plots are good approximations of the exact Bode plots.

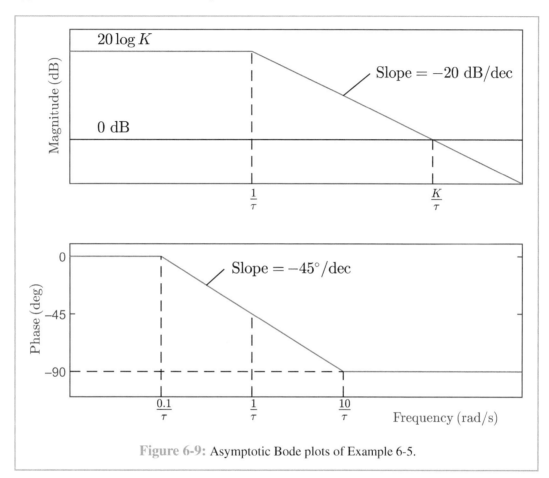

Figure 6-9: Asymptotic Bode plots of Example 6-5.

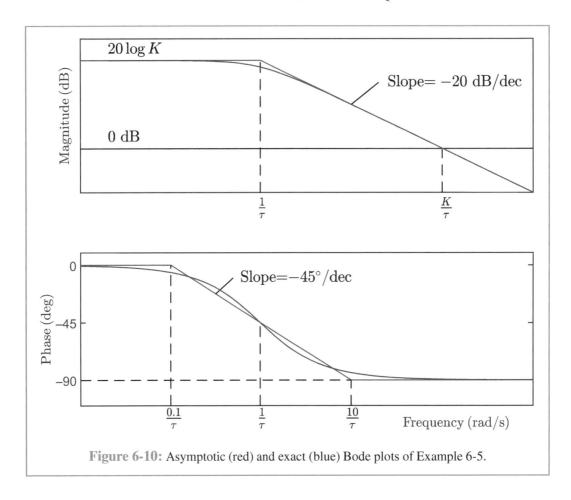

Figure 6-10: Asymptotic (red) and exact (blue) Bode plots of Example 6-5.

Example 6-6: Asymptotic Bode Plots of a Transfer Function with One Zero

Draw the asymptotic Bode plots of

$$G(s) = K(\tau s + 1) .$$

Solution:

$$G(j\omega) = K(1 + j\omega\tau), \qquad |G| = K\sqrt{1 + \omega^2 \tau^2}, \qquad \angle[G] = \tan^{-1}(\omega\tau) .$$

At the break frequency $\omega = 1/\tau$,

$$|G| = K\sqrt{2} \quad \rightarrow \quad 20\log|G| \approx 20\log K + 3 \text{ dB} \text{ and } \angle[G] = \tan^{-1}(1) = 45°.$$

$$\text{For } \omega \ll \frac{1}{\tau}, \quad G(j\omega) \approx K \quad \rightarrow \quad |G| = K \text{ and } \angle[G] = 0,$$

6-2. BODE PLOTS

For $\omega \gg \dfrac{1}{\tau}$, $G(j\omega) \approx jK\omega\tau$ → $|G| = K\omega\tau$ and $\angle[G] = 90°$,

$$|G| = K\omega\tau \quad \rightarrow \quad 20\log|G| = 20\log K + 20\log(\omega\tau).$$

This is the equation of a line that starts at $20\log K$ at the break frequency $\omega = 1/\tau$ and increases with slope 20 dB/dec. Thus, the asymptotic magnitude plot is formed of two lines intersecting at the break frequency. The asymptotic Bode plots are shown in Fig. 6-11. Starting at low frequency, we draw a horizontal line of amplitude $20\log K$. The line is extended until the break frequency where a line of slope 20 dB/dec starts, extending to high frequency. For the asymptotic phase plot, there are two frequencies of interest: one decade below the break frequency at $0.1/\tau$ and one decade after the break frequency at $10/\tau$. The asymptotic phase plot is formed of three lines. For $\omega < 0.1/\tau$, we draw a horizontal line with $\angle[G] = 0°$. Between $0.1/\tau$ and $10/\tau$ we draw a line that starts at $0°$ and increases with slope $45°$/dec to reach $90°$ at $\omega = 10/\tau$. For $\omega > 10/\tau$, the asymptotic phase plot is constant at $90°$.

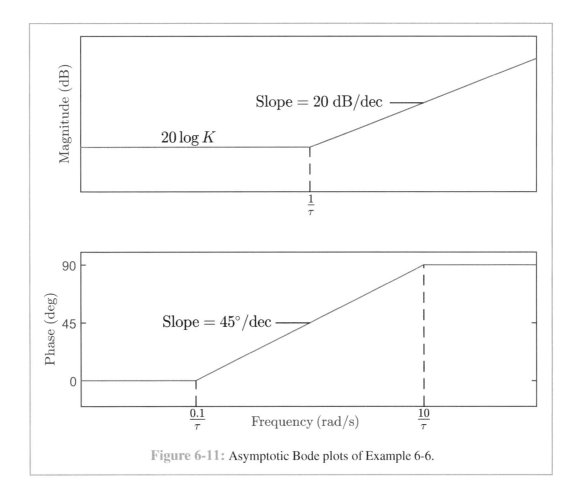

Figure 6-11: Asymptotic Bode plots of Example 6-6.

> Example 6-7: Asymptotic Bode Plots of a Transfer Function with Three Poles

Draw the asymptotic Bode plots of

$$G(s) = \frac{1000}{(s+1)(s+5)(s+10)} = \frac{20}{(s+1)(\frac{1}{5}s+1)(\frac{1}{10}s+1)}.$$

Solution: There are two ways to draw the asymptotic Bode plots of G. The first one is to represent G as a product of three transfer functions of the form $K/(\tau s + 1)$, draw the asymptotic Bode plots of each transfer function, and then add the plots together.

The second way, which we follow here, is to identify the frequencies at which there are changes in the slopes of the asymptotic plots and change the slopes accordingly. The asymptotic Bode plots are shown in Fig. 6-12. For the magnitude plot, the frequencies of interest are the break frequencies of the three first-order terms; that is, 1, 5 and 10; all frequencies are in rad/s. At each frequency a slope of 20 dB/dec is subtracted from

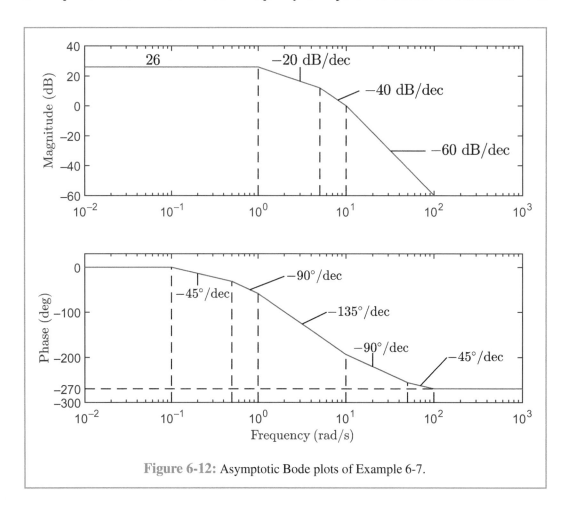

Figure 6-12: Asymptotic Bode plots of Example 6-7.

the previous slope. At low frequency, the magnitude plot starts with zero slope from $20\log 20 \approx 26$ dB since $G(0) = 20$.

For the phase plot, the frequencies of interest are one decade below and one decade after each of the break frequencies. At one decade below the break frequency a slope $-45°$/dec starts and ends at one decade after the break frequency. For the first term, a slope $-45°$/dec starts at 0.1 and ends at 10. For the second term, a slope $-45°$/dec starts at 0.5 and ends at 50. For the third term, a slope $-45°$/dec starts at 1 and ends at 100. At low frequency the phase plot starts from $0°$ because $G(0) = 20$. For comparison, both the asymptotic and exact Bode plots are shown in Fig. 6-13.

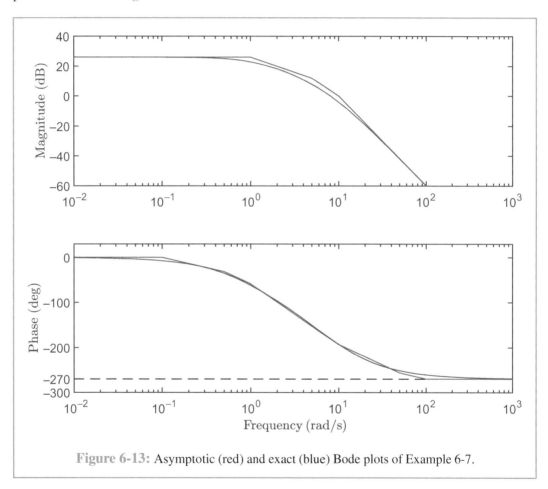

Figure 6-13: Asymptotic (red) and exact (blue) Bode plots of Example 6-7.

> Example 6-8: Asymptotic Bode Plots of a Transfer Function with Three Poles and One Zero

Draw the asymptotic Bode plots of

$$G(s) = \frac{50(s+10)}{s(s+1)(s+50)} = \frac{10(\frac{1}{10}s+1)}{s(s+1)(\frac{1}{50}s+1)}.$$

Solution: The procedure is similar to the previous example but there are two important differences. First, at low frequency $G(j\omega) \approx \frac{10}{j\omega}$. Therefore $|G| = 10/\omega$ and $\angle[G] = -90°$. Thus, the asymptotic phase plot starts from $-90°$. To draw the asymptotic magnitude plot at low frequency, start at $\omega = 10$ rad/s on the zero-dB line and draw a line of slope -20 dB/dec. The asymptotic magnitude plot starts on this line at low frequency but it may not follow it all the way because the slope will change at the first break frequency.

The second difference from the previous example is the presence of the numerator term $(\frac{1}{10}s+1)$. As we saw in Example 6-6, the magnitude plot of this term will add a positive slope of 20 dB/dec starting at the break frequency and its phase adds a positive slope of $45°$/dec starting one decade below the break frequency and ending one decade after it.

The asymptotic Bode plots are shown in **Fig. 6-14**. The frequencies of interest for the magnitude plot are 1, 10, and 50 rad/s (the break frequencies of the first-order terms) and the frequencies of interest for the phase plot are 0.1, 1, 5, 10, 100, and 500 rad/s (one decade below and one decade after the break frequencies of the first-order terms). The magnitude plot starts with a slope -20 dB/dec, which changes to -40 dB/dec at $\omega = 1$ rad/s. The numerator term starts to be effective at $\omega = 10$ rad/s with a slope 20 dB/dec and that brings the net slope to -20 dB/dec. This continues until $\omega = 50$ rad/s where the denominator term with 50 rad/s break frequency becomes effective, bringing the net slope back to -40 dB/dec.

In the phase plot, a negative slope $-45°$/dec starts at $\omega = 0.1$ rad/s and ends at $\omega = 10$ rad/s. A positive slope $45°$/dec starts at $\omega = 1$ rad/s and ends at $\omega = 100$ rad/s. Finally, a negative slope $-45°$/dec starts at $\omega = 5$ rad/s and ends at $\omega = 500$ rad/s. The resulting net slope is shown in the asymptotic phase plot. **Figure 6-15** shows both the asymptotic and exact Bode plots.

6-2. BODE PLOTS

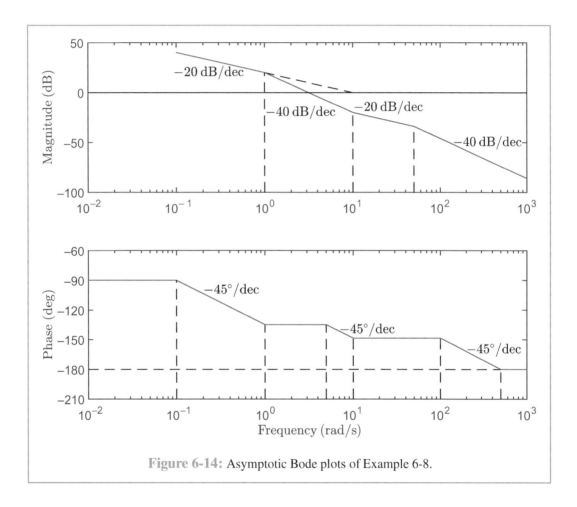

Figure 6-14: Asymptotic Bode plots of Example 6-8.

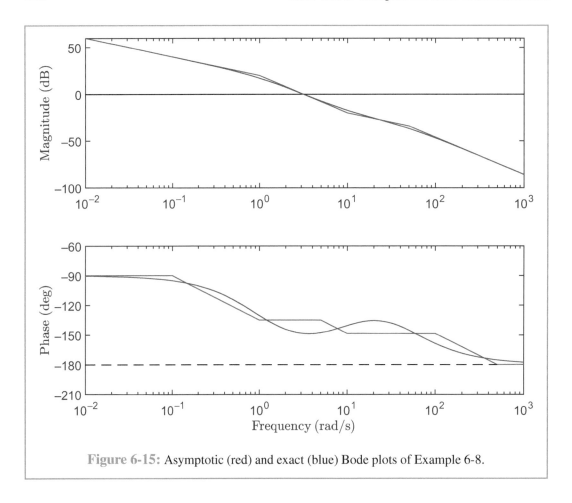

Figure 6-15: Asymptotic (red) and exact (blue) Bode plots of Example 6-8.

6-2.2 Underdamped Second-Order Transfer Functions

Consider

$$G(s) = \frac{\omega_n^2}{s^2 + 2\zeta\omega_n s + \omega_n^2}, \quad \zeta < 1, \qquad G(j\omega) = \frac{\omega_n^2}{\omega_n^2 - \omega^2 + j2\zeta\omega_n\omega}.$$

We start by investigating the magnitude plot.

$$|G(j\omega)| = \frac{\omega_n^2}{\sqrt{(\omega_n^2 - \omega^2)^2 + 4\zeta^2\omega_n^2\omega^2}},$$

$$\omega \ll \omega_n \;\longrightarrow\; |G(j\omega)| \approx \frac{\omega_n^2}{\sqrt{\omega_n^4 + 4\zeta^2\omega_n^2\omega^2}} \approx 1 \;\longrightarrow\; 20\log|G| \approx 0,$$

$$\omega \gg \omega_n \;\longrightarrow\; |G(j\omega)| \approx \frac{\omega_n^2}{\sqrt{\omega^4 + 4\zeta^2\omega_n^2\omega^2}} \approx \left(\frac{\omega_n}{\omega}\right)^2 \;\longrightarrow\; 20\log|G| \approx -40\log\left(\frac{\omega}{\omega_n}\right).$$

6-2. BODE PLOTS

The asymptotic magnitude plot is formed of the intersection of two lines. Starting from low frequency a horizontal line at 0 dB intersects at $\omega = \omega_n$ with a line of slope -40 dB/dec extending to high frequency. For small ζ the exact magnitude plot can deviate significantly from the asymptotic plot. In fact, the magnitude plot may have a maximum value for small ζ. To find the frequency of the maximum point, differentiate $|G(j\omega)|^2$ with respect to ω and equate to zero, to obtain

$$\omega^2 = \omega_n^2(1 - 2\zeta^2) \, .$$

Therefore, a maximum exists for $\zeta < 1/\sqrt{2}$. Denote the frequency at which the maximum is achieved by ω_p; that is, $\omega_p = \omega_n \sqrt{1 - 2\zeta^2}$. Substituting $\omega = \omega_p$ in $|G(j\omega)|$ results in the maximum

$$M_{p\omega} = \frac{1}{2\zeta\sqrt{1-\zeta^2}} \, .$$

For very small ζ the peak of the exact magnitude plot deviates significantly from the asymptotic plot. Figure 6-16 shows the asymptotic and exact magnitude plots of

$$G(s) = \frac{1}{s^2 + 2\zeta s + 1}, \qquad \text{for } \zeta = 0.5 \text{ and } \zeta = 0.1 \, .$$

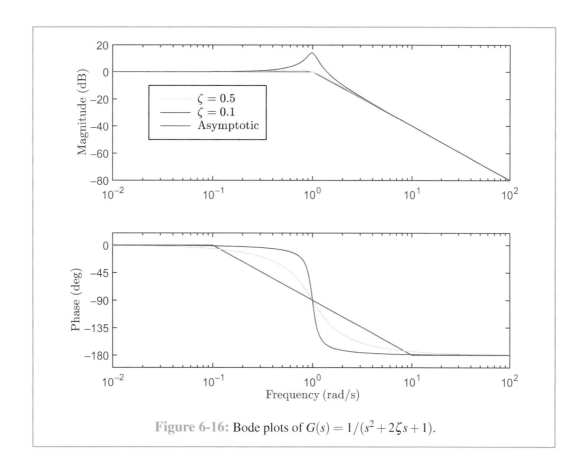

Figure 6-16: Bode plots of $G(s) = 1/(s^2 + 2\zeta s + 1)$.

We turn now to the phase plot.

$$\text{For } \omega < \omega_n, \ \angle[G(j\omega)] = -\tan^{-1}\left(\frac{2\zeta\omega\omega_n}{\omega_n^2 - \omega^2}\right),$$

$$\angle[G(0)] = 0; \quad \text{as } \omega \to \omega_n^-, \ \angle[G(j\omega)] \to -90°,$$

$$\text{For } \omega > \omega_n, \ \angle[G(j\omega)] = -180° + \tan^{-1}\left(\frac{2\zeta\omega\omega_n}{\omega^2 - \omega_n^2}\right),$$

$$\text{As } \omega \to \omega_n^+, \ \angle[G(j\omega)] \to -90°; \quad \text{as } \omega \to \infty, \ \angle[G(j\omega)] \to -180°.$$

The asymptotic phase plot is formed of three lines. A line of slope $-90°$/dec extends from 0 at $\omega = 0.1\omega_n$ to $-180°$ at $\omega = 10\omega_n$. For $\omega < 0.1\omega_n$ there is a horizontal line at $0°$ and for $\omega > 10\omega_n$, there is a horizontal line at $-180°$. The asymptotic phase plot coincides with the exact plot at $\omega = \omega_n$ where $\angle[G] = -90°$. Similar to the magnitude plot, the exact phase plot can deviate significantly from the asymptotic phase plot for small ζ. This is shown in **Fig. 6-16**.

Example 6-9: Bode Plots of a Third-Order Transfer Function with a Pair of Complex Poles

---

Draw the asymptotic and exact Bode plots of

$$G(s) = \frac{5000}{(s+5)(s^2+10s+100)}.$$

Solution: For the second-order term, $\omega_n = 10$ rad/s and $\zeta = 0.5$.

$$G(0) = 10 \quad \to \quad 20\log G(0) = 20 \text{ and } \angle[G(0)] = 0°,$$

$$\text{As } \omega \to \infty, \ G(j\omega) \to \frac{5000}{(j\omega)^3}.$$

Thus, as $\omega \to \infty$, the slope of the magnitude plot approaches -60 dB/dec and the phase plot approaches $-270°$.

The frequencies of interest for the asymptotic magnitude plot are 5 and 10 rad/s. The plot starts from 20 dB at low frequency. A negative slope of -20 dB/dec starts at $\omega = 5$ rad/s, due to the first-order term. A negative slope of -40 dB/dec starts at $\omega = 10$ rad/s, due to the second-order term,, bringing the net slope to -60 dB/dec, which extends to high frequencies.

The frequencies of interest for the asymptotic phase plot are 0.5, 1, 50, and 100 rad/s. The plot starts from $0°$ at low frequency. A negative slope of $-45°$/dec starts at $\omega = 0.5$ rad/s, due to the first-order term. A negative slope of $-90°$/dec starts at $\omega = 1$ rad/s, due to the second-order term, bringing the net slope to $-135°$/dec. At $\omega = 50$, the negative slope of $-45°$/dec ends, resulting in a net slope of $-90°$/dec. At $\omega = 100$ rad/s, the negative slope of $-90°$/dec ends. At this frequency, $\angle[G] = -270°$ and the asymptotic phase plot extends horizontally to high frequency. The asymptotic Bode plots are shown in **Fig. 6-17**, together with the exact plots, which are calculated using MATLAB.

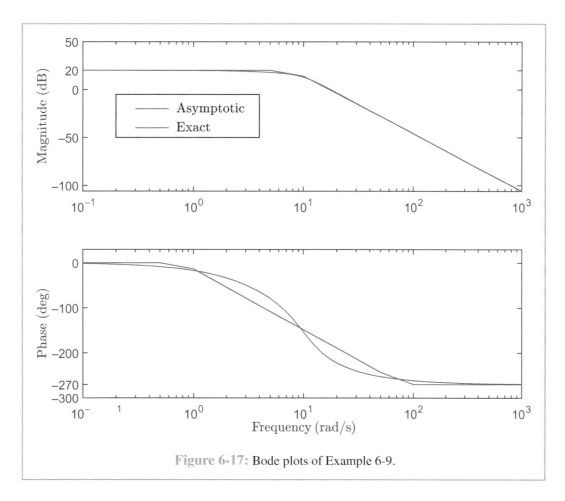

Figure 6-17: Bode plots of Example 6-9.

6-2.3 Error Constants from Bode Plots

For a unity feedback control system with a forward transfer function $G(s)$, the error constants K_p, K_v, or K_a can be determined from the Bode magnitude plot of $G(j\omega)$. For Type 0 systems, $K_p = G(0)$. The magnitude plot starts from $20\log|G(0)|$ at low frequency. Therefore, K_p can be determined from the low-frequency level of the magnitude plot. In particular, if the magnitude plot starts from b dB at low frequency, then $K_p = 10^{(b/20)}$.

For Type 1 systems, $G(s)$ has a pole at $s = 0$ and $K_v = \lim_{s \to 0} sG(s)$. Hence, the low-frequency approximation of $|G(j\omega)|$ is K_v/ω. The magnitude plot starts at low frequency with slope -20 dB/dec. Extend the low-frequency line until it intersects the zero-dB line at frequency ω_1; hence, $K_v = \omega_1$.

For Type 2 systems, $G(s)$ has two poles at $s = 0$ and $K_a = \lim_{s \to 0} s^2 G(s)$. Hence, the low-frequency approximation of $|G(j\omega)|$ is K_a/ω^2. The magnitude plot starts at low frequency with slope -40 dB/dec. Extend the low-frequency line until it intersects the zero-dB line at frequency ω_2; hence, $K_a = \omega_2^2$.

Example 6-10: Finding the Type of Feedback System and the Error Constant from the Bode Magnitude Plot

Figure 6-18 shows the Bode magnitude plots of the forward transfer functions $G(s)$ for three unity feedback systems. Determine the type of each system and the corresponding error constant.

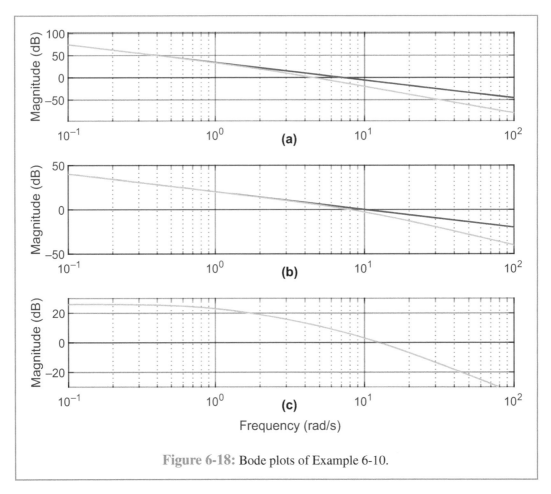

Figure 6-18: Bode plots of Example 6-10.

Solution: For case (a), the slope of the magnitude plot at low frequency is -40 dB/dec. Hence, the system is Type 2. The extension of the low frequency line intersects the zero-dB line at $\omega \approx 7$ rad/s. Hence, $K_a \approx 49$.

For case (b), the slope of the magnitude plot at low frequency is -20 dB/dec. Hence, the system is Type 1. The extension of the low frequency line intersects the zero-dB line at $\omega \approx 10$ rad/s. Hence, $K_v \approx 10$.

For case (c), the magnitude plot starts flat at 26 dB. Hence, the system is Type 0 and $K_p = 10^{(26/20)} \approx 20$.

6-3 Time-Delay Systems

Signals may be delayed, for example, due to communication time. A time-delay unit is an element whose output $y(t)$ is delayed from its input $u(t)$ by a delay time τ; that is,

$$y(t) = u(t - \tau).$$

Taking the Laplace transform of both sides yields

$$\mathcal{L}\{y(t)\} = \mathcal{L}\{u(t-\tau)\} \longrightarrow Y(s) = e^{-\tau s} U(s).$$

Therefore, the transfer function of the time-delay unit is $e^{-\tau s}$.

A time-delay system has its input or output delayed by time τ. The transfer function of such a system takes the form

$$G(s) = e^{-\tau s} \hat{G}(s),$$

where the undelayed function $\hat{G}(s)$ is a proper rational function of s. The frequency response of $G(s)$ is given by

$$G(j\omega) = e^{-j\omega\tau} \hat{G}(j\omega),$$
$$|G(j\omega)| = |e^{-j\omega\tau}| |\hat{G}(j\omega)| = |\hat{G}(j\omega)| \quad \text{because } |e^{-j\omega\tau}| = 1,$$
$$\angle[G(j\omega)] = \angle[e^{-j\omega\tau}] + \angle[\hat{G}(j\omega)] = -\omega\tau + \angle[\hat{G}(j\omega)].$$

Thus, the Bode magnitude plot of $G(j\omega)$ is the same as the Bode magnitude plot of $\hat{G}(j\omega)$, while the Bode Phase plot of $G(j\omega)$ has the curve $\omega\tau$ subtracted from the Bode phase plot of $\hat{G}(j\omega)$. While $\omega\tau$ is linear in ω, the curve $\omega\tau$ in the Bode plot will not be linear because the frequency scale is logarithmic. Note that the unit of $\omega\tau$ is radian, so to plot it on a Bode phase plot it should be converted into degrees.

Example 6-11: Bode Plots of a Transfer Function with Time Delay

---

Draw the Bode plots of

$$G(s) = \frac{10 e^{-(\pi/4)s}}{s(s+1)}.$$

Solution: The Bode plots of G and $\hat{G} = 10/[s(s+1)]$ are shown in **Fig. 6-19**.

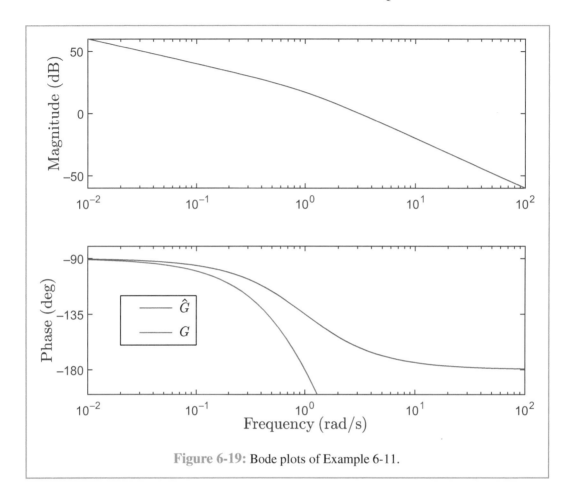

Figure 6-19: Bode plots of Example 6-11.

6-4 Nyquist Criterion

6-4.1 Contour Mapping

Let $F(s)$ be a rational function of s. If s is a complex number, $F(s)$ also is a complex number. The complex number s can be represented by a point in a complex plane, called the s-plane, where the real and imaginary axes are $\text{Re}[s]$ and $\text{Im}[s]$. Similarly, $F(s)$ can be represented by a point in a complex plane, called the F-plane, where the real and imaginary axes are $\text{Re}[F(s)]$ and $\text{Im}[F(s)]$. Hence, the point s in the s-plane is mapped into the point $F(s)$ in the F-plane.

If s varies along a closed contour in the s-plane, $F(s)$ also varies along a closed contour in the F-plane, as illustrated in Fig. 6-20.

6-4. NYQUIST CRITERION

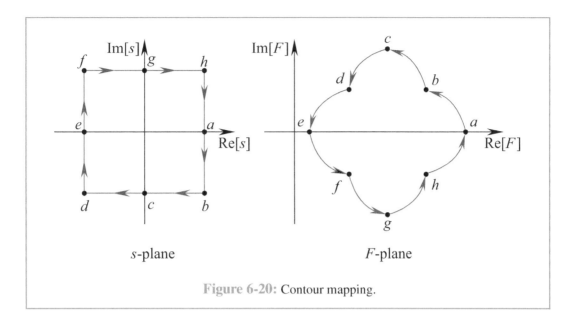

Figure 6-20: Contour mapping.

6-4.2 Cauchy's Principle of the Argument

Let
$$F(s) = K \frac{\prod_{i=1}^{m}(s+z_i)}{\prod_{i=1}^{n}(s+p_i)}.$$

How many times does the F-plane contour encircle the origin? The answer is determined by

$$\angle[F(s)] = \sum_{i=1}^{m} \angle[(s+z_i)] - \sum_{i=1}^{n} \angle[(s+p_i)].$$

If the net change in $\angle[F(s)]$, as s completes a clockwise revolution around the s-plane contour, is $2k\pi$ for an integer k, the F-plane contour encircles the origin k times. If k is positive, the encirclement is clockwise. If it is negative, the encirclement is counterclockwise. If the net change in $\angle[F(s)]$ is zero, the F-plane contour does not encircle the origin.

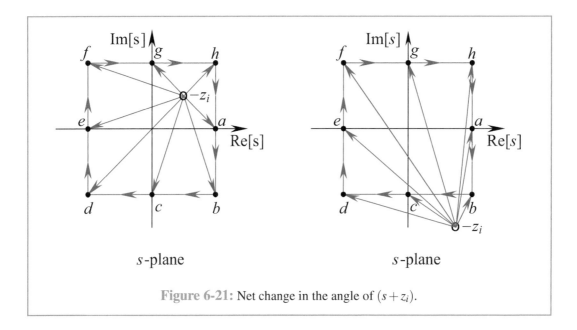

Figure 6-21: Net change in the angle of $(s+z_i)$.

The net change in $\angle[(s+z_i)]$ depends on whether the zero $-z_i$ is inside or outside the s-plane contour. Assume the zero is not on the contour itself. Figure 6-21 shows two cases, one where the zero is inside and another where the zero is outside the s-plane contour. For a point s on the contour, $(s+z_i)$ is represented by the vector connecting $-z_i$ to s. Let s start at point a and make a complete clockwise revolution until it comes back to a. The vectors $(s+z_i)$ at the representative points $a, b, \ldots, h, a$ are shown. When $-z_i$ is inside the contour, the net change in $\angle[(s+z_i)]$ is 2π, but when $-z_i$ is outside the contour, the net change in $\angle[(s+z_i)]$ is zero.

Similarly, if a pole $-p_i$ is not on the s-plane contour, the net change in $\angle[(s+p_i)]$ is 2π if the pole is inside the contour and zero if it is outside. Thus, provided no poles or zeros of $F(s)$ lie on the s-plane contour,

$$\text{Net change in } \angle[F(s)] = 2\pi(Z - P),$$

where Z and P are, respectively, the numbers of zeros and poles of $F(s)$ inside the s-plane contour. Let N be the number of *counterclockwise* encirclements of the origin by the F-plane contour. Then

$$-2\pi N = 2\pi(Z - P) \quad \longrightarrow \quad N = P - Z.$$

Example 6-12: Encirclement of the Origin by the F-Plane Contour

The function

$$F(s) = \frac{(s+2)(s+3)}{s(s+1)}$$

has two poles at 0 and -1 and two zero at -2 and -3. Figure 6-22 shows four different s-plane contours. The first contour encircles the pole at $s = 0$. Thus, $P = 1$, $Z = 0$, and $N = 1$; the F-plane contour encircles the origin once counterclockwise. The second contour encircles the poles at 0 and -1. Thus, $P = 2$, $Z = 0$, and $N = 2$; the F-plane contour encircles the origin twice counterclockwise. The third contour encircles the poles at 0 and -1 and the zero at -2. Thus, $P = 2$, $Z = 1$, and $N = 1$; the F-plane contour encircles the origin once counterclockwise. The fourth contour encircles the poles at 0 and -1 and the zeros at -2 and -3. Thus, $P = 2$, $Z = 2$, and $N = 0$; the F-plane contour does not encircle the origin.

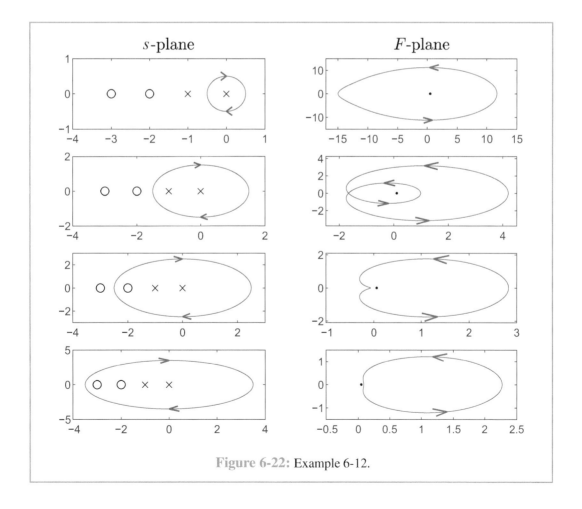

Figure 6-22: Example 6-12.

6-4.3 Stability of Feedback Systems

Consider the feedback control system of **Fig. 6-1** where $GH(s)$ is a proper rational function of s; that is,

$$GH(s) = \frac{n(s)}{d(s)},$$

in which $n(s)$ and $d(s)$ are polynomials with $\deg(n(s)) \leq \deg(d(s))$. The characteristic equation of the system is

$$1 + GH(s) = 0.$$

The system is stable if all the zeros of this equation have negative real parts; that is, all the zeros are in the Left-Half Plane (LHP). Recall that the zeros of the characteristic equation are the poles of the closed-loop transfer function

$$\frac{G(s)}{1 + GH(s)}.$$

We will apply Cauchy's principle of the argument to determine whether or not the zeros of $1 + GH(s)$ are in the RHP. Let $F(s) = 1 + GH(s)$. Then,

$$F(s) = \frac{n(s) + d(s)}{d(s)}.$$

The poles of $F(s)$ are known and the task is to determine if its zeros are in the LHP. Let P and Z be the numbers of poles and zeros of $F(s)$ in the Right-Half Plane (RHP) and recall that P is known since it is the number of poles of $GH(s)$ in the RHP. The system is stable if $Z = 0$. Cauchy's principle of the argument is applied as follows:

(1) Choose a contour in the s-plane, called the *Nyquist contour*, that includes the whole RHP in its interior.

(2) Map the Nyquist contour into the F-plane

(3) Determine the number N of counterclockwise encirclements of the origin by the F-plane contour.

Hence,

$$Z = P - N.$$

6-4. NYQUIST CRITERION

To construct the Nyquist contour, assume, for now, that $GH(s)$ has no poles on the imaginary axis; we will deal with poles on the imaginary axis later on. The contour is constructed as shown in Fig. 6-23. Take a path on the imaginary axis from $\omega = -r$ to $\omega = r$ and close the contour by a semicircle of radius r centered at the origin in the RHP. Then, let r tend to ∞ so that the Nyquist contour includes the whole RHP.

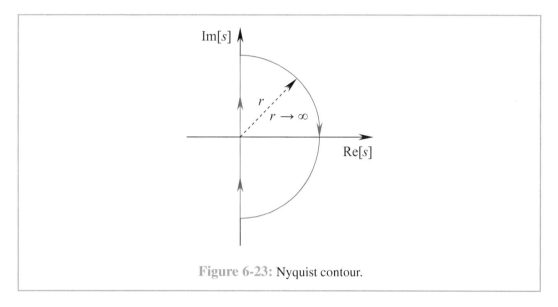

Figure 6-23: Nyquist contour.

The next step is to map the Nyquist contour into the F-plane. Here, we make a slight modification. Since $F(s) = 1 + GH(s)$, we map the Nyquist contour into the GH-plane instead of the F-plane. The number of encirclements of the origin by the F-plane contour is the same as the number of encirclements of the point $(-1+0j)$ by the GH-plane contour. The GH-plane contour is called the *Nyquist plot*.

Drawing the Nyquist plot requires mapping two segments of the Nyquist contour. The first segment is the imaginary axis of the s-plane. This segment maps into the graph of $GH(j\omega)$ in the GH-plane. This is precisely the polar plot we examined in Section 6-1. The second segment is the semicircle of a radius approaching ∞. To map this segment, take $s = re^{j\theta}$ with θ changing from $\frac{\pi}{2}$ to $-\frac{\pi}{2}$, substitute s in $GH(s)$, and compute the limit as $r \to \infty$. The limit will be zero when $GH(s)$ is strictly proper; that is, when $\deg(n(s)) < \deg(d(s))$.

In summary,

> **Nyquist Stability Criterion:** Let
>
> - P be the number of RHP poles of $GH(s)$;
>
> - N be the number of counterclockwise encirclements of $-1+0j$ by the Nyquist plot in the GH-plane;
>
> - Z be the number of closed-loop poles in RHP.
>
> Then,
> $$Z = P - N.$$
> The system is stable when $Z = 0$.

Example 6-13: Checking Stability by the Nyquist Criterion

Determine the stability of a feedback control system with

$$GH(s) = \frac{1}{(s+1)(s+2)(s+3)}.$$

Solution: $GH(s)$ has no poles in RHP; that is, $P = 0$. The polar plot is shown in Fig. 6-5. Since $GH(s)$ is strictly proper, $s = re^{j\theta}$ maps into the origin as $r \to \infty$. Hence, the Nyquist plot coincides with the polar plot. The Nyquist plot intersects the negative real axis at $-1/60 + 0j$. Thus, the plot does not encircle $-1 + 0j$; that is, $N = 0$. Therefore, $Z = P - N = 0$. The system is stable.

6-4.4 Application of the Nyquist Criterion with Gain Adjustment

Suppose $GH(s) = K\widehat{GH}(s)$, $K > 0$. The characteristic equation is

$$1 + K\widehat{GH}(s) = 0 \quad \longleftrightarrow \quad \frac{1}{K} + \widehat{GH}(s) = 0.$$

The number of encirclements of $(-1+0j)$ by the Nyquist plot of $GH(s)$ is the same as the number of encirclements of $-(1/K)+0j$ by the Nyquist plot of $\widehat{GH}(s)$. Hence, we can expand the application of the Nyquist criterion as follows:

(1) Given $GH(s) = K\widehat{GH}(s)$, $K > 0$, draw the Nyquist plot of $\widehat{GH}(s)$ at $K = 1$.

(2) By changing K, determine the number of counterclockwise encirclements of $-(1/K)+0j$ for different values of K.

(3) Apply the equation $Z = P - N$ to find Z for different values of K.

6-4. NYQUIST CRITERION

> **Example 6-14:** Checking Stability by the Nyquist Criterion for All Values of a Positive Gain

Determine the stability of a feedback control system with

$$GH(s) = \frac{K}{(s+1)(s+2)(s+3)}, \quad K > 0.$$

Solution: $GH(s)$ has no poles in RHP; that is, $P = 0$. From the previous example, the Nyquist plot for $K = 1$ is shown in Fig. 6-5. The plot intersects the negative real axis at $(-1/60 + 0j)$.

$$\text{For } -\frac{1}{K} < -\frac{1}{60}, \quad N = 0 \quad \rightarrow \quad Z = P - N = 0,$$

$$\text{For } -\frac{1}{K} > -\frac{1}{60}, \quad N = -2 \quad \rightarrow \quad Z = P - N = 2.$$

Note that $N = -2$ because the encirclements are clockwise.

$$-\frac{1}{K} < -\frac{1}{60} \quad \longleftrightarrow \quad \frac{1}{K} > \frac{1}{60} \quad \longleftrightarrow \quad K < 60.$$

Thus, the system is stable for $K < 60$ and unstable (with two RHP closed-loop poles) for $K > 60$.

> **Example 6-15:** Checking Stability by the Nyquist Criterion for All Values of a Positive Gain

Determine the stability of a feedback control system with

$$GH(s) = \frac{6K}{(s-1)(s+2)(s+3)}, \quad K > 0.$$

Solution: The poles of $GH(s)$ are $1, -2,$ and -3. Hence, $P = 1$.

Step 1: Draw the Nyquist plot for $K = 1$. Start with the polar plot

$$GH(j\omega) = \frac{-6}{(1-j\omega)(2+j\omega)(3+j\omega)} = \frac{-6(1+j\omega)(2-j\omega)(3-jw)}{(1+\omega^2)(4+\omega^2)(9+\omega^2)},$$

$$\text{Re}[GH(j\omega)] = \frac{-6(6+4\omega^2)}{(1+\omega^2)(4+\omega^2)(9+\omega^2)},$$

$$\text{Im}[GH(j\omega)] = \frac{-6\omega(1-\omega^2)}{(1+\omega^2)(4+\omega^2)(9+\omega^2)},$$

$$\angle[GH(j\omega)] = \pi + \tan^{-1}(\omega) - \tan^{-1}\left(\frac{\omega}{2}\right) - \tan^{-1}\left(\frac{\omega}{3}\right).$$

Compute the limits:

(a) At $\omega = 0$, $\text{Re}[GH] = -1$, $\text{Im}[GH] = 0$, $\angle[GH] = \pi$;

(b) As $\omega \to \infty$, $\text{Re}[GH] \to 0$, $\text{Im}[GH] \to 0$, $\angle[GH] \to \frac{\pi}{2}$.

Intersection with the axes:

(c) Real axis: $\text{Im}[GH] = 0 \longrightarrow \omega = 0$ or $\omega^2 = 1$;

$$\omega = 0 \longrightarrow \text{Re}[GH] = -1, \quad \omega^2 = 1 \longrightarrow \text{Re}[GH] = -0.6.$$

(d) No intersections with the imaginary axis.

Since $GH(s)$ is strictly proper, $s = re^{j\theta}$ maps into the origin as $r \to \infty$. Hence, the Nyquist plot coincides with the polar plot. The Nyquist plot is sketched in Fig. 6-24.

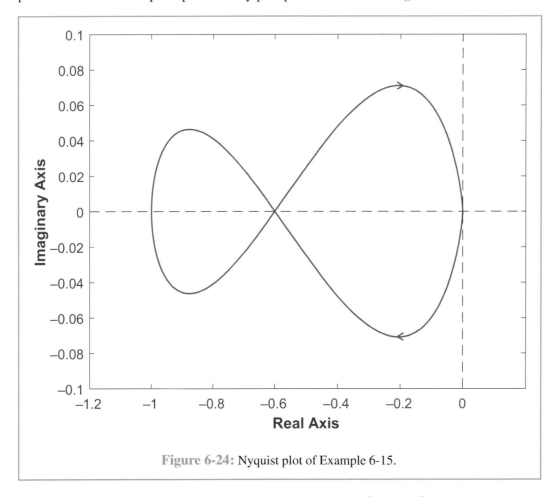

Figure 6-24: Nyquist plot of Example 6-15.

At $\omega = 0$, the plot starts at $(-1, 0)$. For small $\omega > 0$, $\text{Im}[GH(j\omega)] < 0$; hence, the plot moves into the lower half of the plane. At $\omega = 1$, the plot crosses the real axis at $(-0.6, 0)$ moving into the upper half of the plane and stays there because $\text{Im}[GH(j\omega)] > 0$ for $\omega > 1$. As $\omega \to \infty$, the plot approaches the origin tangent to the imaginary axis since the limit of $\angle[GH(j\omega)]$ is $\pi/2$. For $\omega < 0$, the plot is the mirror image of the plot for $\omega > 0$.

6-4. NYQUIST CRITERION

Step 2: Apply the Nyquist criterion with $(-1/K, 0)$ as the critical point. There are three cases:

(a) For $-\frac{1}{K} < -1$, the critical point is to the left of the Nyquist plot. Hence,

$$N = 0 \quad \Rightarrow \quad Z = P - N = 1,$$

$$-\frac{1}{K} < -1 \quad \Leftrightarrow \quad \frac{1}{K} > 1 \quad \Leftrightarrow \quad K < 1.$$

For $K < 1$, there is one closed-loop pole in the RHP.

(b) For $-1 < -\frac{1}{K} < -0.6$, the critical point is in the left lobe of the Nyquist plot. It is encircled once counterclockwise. Hence,

$$N = 1 \quad \Rightarrow \quad Z = P - N = 0,$$

$$-1 < -\frac{1}{K} < -0.6 \quad \Leftrightarrow \quad 1 > \frac{1}{K} > 0.6 \quad \Leftrightarrow \quad 1 < K < \frac{1}{0.6}.$$

For $1 < K < 1/0.6$, there are no closed-loop poles in the RHP.

(c) For $-\frac{1}{K} > -0.6$, the critical point is in the right lobe of the Nyquist plot. It is encircled once clockwise. Hence,

$$N = -1 \quad \Rightarrow \quad Z = P - N = 2,$$

$$-\frac{1}{K} > -0.6 \quad \Leftrightarrow \quad \frac{1}{K} < 0.6 \quad \Leftrightarrow \quad K > \frac{1}{0.6}.$$

For $K > 1/0.6$, there are two closed-loop poles in the RHP.

Therefore, the system is stable for $1 < K < 1/0.6$.

6-4.5 The Case When $GH(s)$ Has Poles on the Imaginary Axis

To apply Cauchy's principle of the argument, $F(s)$, and consequently $GH(s)$, cannot have poles on the imaginary axis. In this case, the Nyquist contour is modified to take infinitesimally small detours in the RHP around the imaginary-axis poles. Because the detours are in the RHP, poles on the imaginary axis are not included in the count P. As shown in Fig. 6-25, for a pole at $j\omega$, the points of the detour are given by

$$s = j\omega + \varepsilon e^{j\phi}, \quad \phi : -\frac{\pi}{2} \to 0 \to \frac{\pi}{2} \text{ and } \varepsilon \to 0.$$

ϕ is described by three angles to indicate that the detour is in the RHP.

Mapping of the detour is obtained by substituting s in $GH(s)$ and taking the limit as $\varepsilon \to 0$. Because $GH(s)$ has the term $(s - j\omega)$ in its denominator, $|GH(s)| \to \infty$ as $\varepsilon \to 0$. Therefore, the mapping will be an arc with an infinitely large radius. The mapping of the angle ϕ determines the angle that the infinite-radius arc traverses.

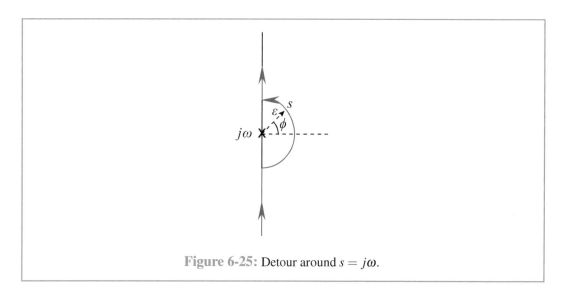

Figure 6-25: Detour around $s = j\omega$.

Example 6-16: Application of the Nyquist Criterion When $GH(s)$ Has a Pole at the Origin

---

Determine the stability of a feedback control system with

$$GH(s) = \frac{K}{s(s+1)}, \quad K > 0.$$

Solution: The Nyquist contour is shown in **Fig. 6-26**. Different segments of the contour map as follows:

(a) Segment a-b → polar plot.

(b) Segment b-c-d → origin.

(c) Segment d-e → the mirror image of the polar plot of a-b.

(d) Segment e-f-a: $s = \varepsilon e^{j\phi}$, $\phi : -\frac{\pi}{2} \to 0 \to \frac{\pi}{2}$ and $\varepsilon \to 0$. Also, at $K = 1$,

$$GH(\varepsilon e^{j\phi}) = \frac{1}{\varepsilon e^{j\phi}(\varepsilon e^{j\phi} + 1)} \approx \frac{1}{\varepsilon} e^{-j\phi},$$

$\frac{1}{\varepsilon} \to \infty$ as $\varepsilon \to 0$ and $-\phi$ changes from $\frac{\pi}{2} \to 0 \to -\frac{\pi}{2}$.

The segment e-f-a maps into an infinite-radius arc that traverses in the RHP from $\frac{\pi}{2}$ to 0 to $-\frac{\pi}{2}$.

6-4. NYQUIST CRITERION

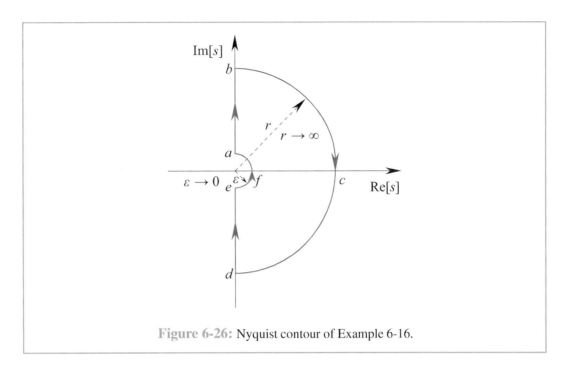

Figure 6-26: Nyquist contour of Example 6-16.

Polar plot of $GH(j\omega)$ at $K = 1$:

$$GH(j\omega) = \frac{1}{j\omega(1+j\omega)} = \frac{-j(1-j\omega)}{\omega(1+\omega^2)},$$

$$\text{Re}[GH(j\omega)] = \frac{-1}{1+\omega^2}, \quad \text{Im}[GH(j\omega)] = \frac{-1}{\omega(1+\omega^2)}, \quad \angle[GH(j\omega)] = -\frac{\pi}{2} - \tan^{-1}(\omega).$$

(1) As $\omega \to 0$, $\text{Re}[GH] \to -1$, $\text{Im}[GH] \to -\infty$ and $\angle[GH] \to -\frac{\pi}{2}$.

(2) As $\omega \to \infty$, $\text{Re}[GH] \to 0$, $\text{Im}[GH] \to 0$, and $\angle[GH] \to -\pi$.

(3) No intersection with the axes except at the origin.

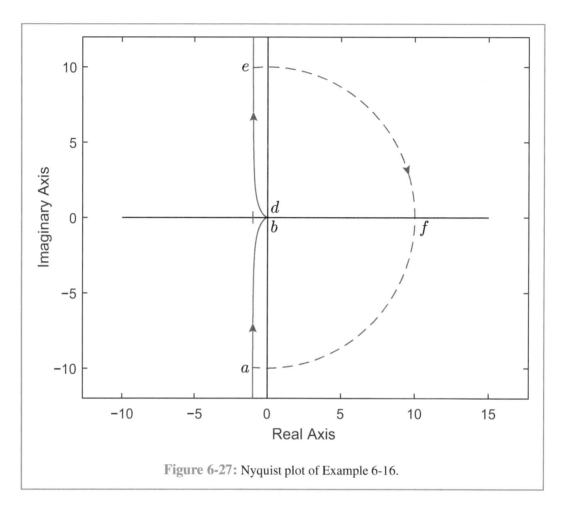

Figure 6-27: Nyquist plot of Example 6-16.

The Nyquist plot is shown in Fig. 6-27. Segment a-b starts from $-\infty$ asymptotic to the vertical line $\text{Re}[s] = -1$. As $\omega \to \infty$, it approaches the origin tangent to the real axis. Segment d-e is the mirror image of segment a-b. Segment e-f-a closes the plot with infinite-radius arc.

Apply the Nyquist criterion. The critical point at $-1/K + 0j$ is to left of the Nyquist plot. Thus, $N = 0$. With $P = 0$, we have $Z = 0$. The system is stable for all $K > 0$.

Example 6-17: Application of the Nyquist Criterion When $GH(s)$ Has a Pole at the Origin and a Pole in the RHP

Determine the stability of a feedback control system with

$$GH(s) = \frac{K(s+2)}{s(s-1)}, \quad K > 0.$$

6-4. NYQUIST CRITERION

Solution: The poles of $GH(s)$ are at 0 and 1; hence, $P = 1$. Because of the pole at the origin, the Nyquist contour, which is taken as in the previous example, is shown in Fig. 6-26. Different segments of the contour map as follows:

(a) Segment a-b $\to$ polar plot.

(b) Segment b-c-d $\to$ origin.

(c) Segment d-e $\to$ the mirror image of the polar plot of a-b.

(d) Segment e-f-a: $s = \varepsilon e^{j\phi}$, $\phi : -\frac{\pi}{2} \to 0 \to \frac{\pi}{2}$ and $\varepsilon \to 0$. Also, at $K = 1$,

$$GH(\varepsilon e^{j\phi}) = \frac{\varepsilon e^{j\phi} + 2}{\varepsilon e^{j\phi}(\varepsilon e^{j\phi} - 1)} \approx \frac{2}{-\varepsilon e^{j\phi}} = \frac{2}{\varepsilon} e^{j(\pi - \phi)},$$

$\frac{1}{\varepsilon} \to \infty$ as $\varepsilon \to 0$ and $\pi - \phi$ changes from $\frac{3\pi}{2} \to \pi \to \frac{\pi}{2}$.

The segment e-f-a maps into an infinite-radius arc that traverses in the left-half plane from $-\frac{\pi}{2}$ to π to $\frac{\pi}{2}$.

Polar plot of $GH(j\omega)$ at $K = 1$:

$$GH(j\omega) = \frac{j\omega + 2}{j\omega(j\omega - 1)} = \frac{-(2 + j\omega)}{j\omega(1 - j\omega)} = \frac{j(2 + j\omega)(1 + j\omega)}{\omega(1 + \omega^2)},$$

$$\mathrm{Re}[GH] = \frac{-3}{1 + \omega^2}, \quad \mathrm{Im}[GH] = \frac{2 - \omega^2}{\omega(1 + \omega^2)},$$

$$\angle[GH] = \pi + \tan^{-1}\left(\frac{\omega}{2}\right) - \frac{\pi}{2} + \tan^{-1}(\omega) = \frac{\pi}{2} + \tan^{-1}\left(\frac{\omega}{2}\right) + \tan^{-1}(\omega).$$

(1) As $\omega \to 0$, $\mathrm{Re}[GH] \to -3$, $\mathrm{Im}[GH] \to \infty$, and $\angle[GH] \to \frac{\pi}{2}$.

(2) As $\omega \to \infty$, $\mathrm{Re}[GH] \to 0$, $\mathrm{Im}[GH] \to 0$, and $\angle[GH] \to \frac{3\pi}{2} \equiv -\frac{\pi}{2}$;

(3) Intersection with the real axis: $\mathrm{Im}[GH] = 0 \to \omega^2 = 2 \to \mathrm{Re}[GH] = -1$.

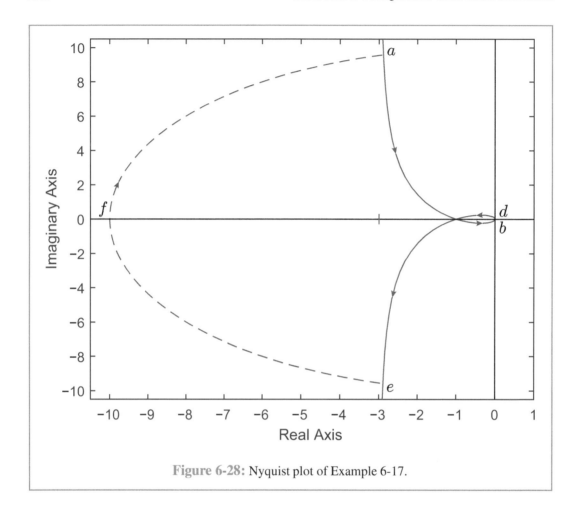

Figure 6-28: Nyquist plot of Example 6-17.

The Nyquist plot is shown in Fig. 6-28. Segment a-b starts from ∞ asymptotic to the vertical line $\text{Re}[s] = -3$. It crosses the real axis at -1 and approaches the origin as $\omega \to \infty$ tangent to the imaginary axis. Segment d-e is the mirror image of segment a-b. Segment e-f-a closes the plot with infinite-radius arc in the left-half plane.

Apply the Nyquist criterion:

$$-\frac{1}{K} < -1 \quad \rightarrow \quad N = -1 \quad \rightarrow \quad Z = P - N = 2,$$

$$-\frac{1}{K} < -1 \quad \leftrightarrow \quad \frac{1}{K} > 1 \quad \leftrightarrow \quad K < 1,$$

$$-\frac{1}{K} > -1 \quad \rightarrow \quad N = 1 \quad \rightarrow \quad Z = P - N = 0.$$

The system is stable for $K > 1$ and unstable for $K < 1$ with two RHP closed-loop poles.

6-5 Relative Stability

One of the main concerns in control system analysis and design is the effect of parameter uncertainty. If we design a control system to be stable, will stability be robust to changes in the plant model due to parameter uncertainty? The ability of the system to maintain stability is called *relative stability*. It is a way to quantify the strength of the stability of the system.

The Nyquist stability criterion gives us a way to measure relative stability since stability is determined by the number of encirclements of the critical point $(-1+0j)$ in the GH-plane. The stability is stronger if the Nyquist plot is far from $(-1+0j)$ so that a relatively large uncertainty can be tolerated without destroying stability. On the other hand, if the Nyquist plot is close to $(-1+0j)$, a small uncertainty might change the number of encirclements of $(-1+0j)$.

Figure 6-29 shows an example of a Nyquist plot for a system with $P=1$. The nominal Nyquist plot, shown in blue color, encircles $(-1+0j)$ once counterclockwise, which shows that the system is stable. Parameter perturbations can cause changes in the Nyquist plot. Figure 6-29 shows the envelope of uncertainty, in dashed red color, for a certain range of

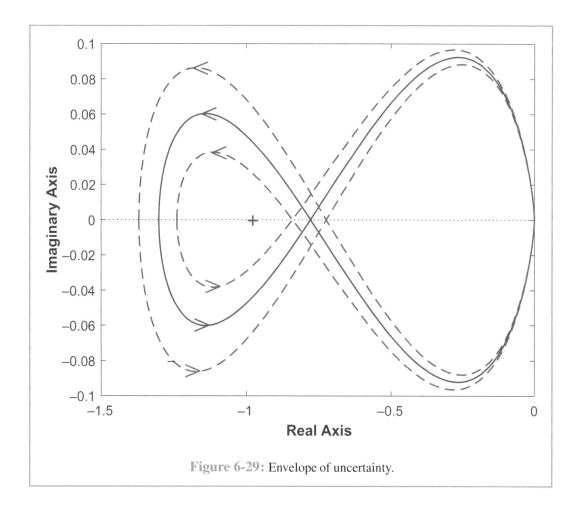

Figure 6-29: Envelope of uncertainty.

parameter perturbations that do not change the number P of RHP poles. For all parameter perturbations in this range, the perturbed Nyquist plot will be within the envelope of uncertainty. In this example the nominal Nyquist plot is far enough from $(-1+0j)$ so that stability is maintained for the prescribed parameter perturbations.

6-5.1 Stability Margins

For systems with no RHP open-loop poles; that is, $P = 0$, relative stability is measured by the gain and phase margins, which are defined in Fig. 6-30. Since $P = 0$, the closed-loop system is stable if the Nyquist plot does not encircle $(-1+0j)$, as shown in the figure. The distance from the Nyquist plot to the point $(-1+0j)$ is measured by two parameters, the ratio $1/d$ and the angle α.

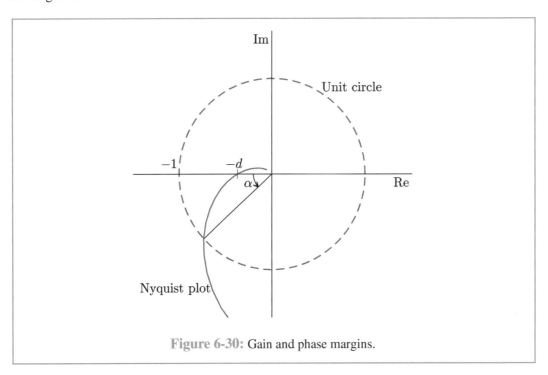

Figure 6-30: Gain and phase margins.

The Nyquist plot intersects the real axis at $-d$. At this point the phase is $180°$. Increasing the gain $|GH(j\omega)|$ by a factor $1/d$, while maintaining $\angle[GH(j\omega)] = 180°$, brings the system to the verge of instability. This factor is usually expressed in decibels: $20\log(1/d)$ dB.

▶ The *gain margin* (GM) is the increase in the open-loop gain, expressed in decibels (dB), required at $180°$ phase shift, to make the closed-loop system unstable. ◀

The magnitude of the Nyquist plot in Fig. 6-30 is unity at an angle α from the $180°$ line. Delaying the phase $\angle[GH(j\omega)]$ by an angle α while maintaining $|GH(j\omega)| = 1$, brings the system to the verge of instability.

6-5.2 Stability Margins from Bode Plots

The gain and phase margins can be easily determined from the Bode plots. **Figure 6-31** shows an example of the Bode plots for a stable system. The frequency at which the magnitude plot crosses the zero–dB line is called the *crossover frequency* and denoted by ω_c. The phase margin is determined at this frequency. It is the distance from the $-180°$ line to the phase plot at ω_c; that is,

$$PM = (\text{phase at } \omega_c) + 180°.$$

The frequency at which the phase plot crosses the $-180°$ line is called the *phase crossover frequency* and denoted by ω_{pc}. The gain margin is determined at this frequency. It is the

▶ The *phase margin* (*PM*) is the phase delay in the open-loop transfer function required, at unity gain, to make the closed-loop system unstable. ◀

The gain and phase margins are called the *stability margins*. Both margins must be positive for the closed-loop system to be stable.

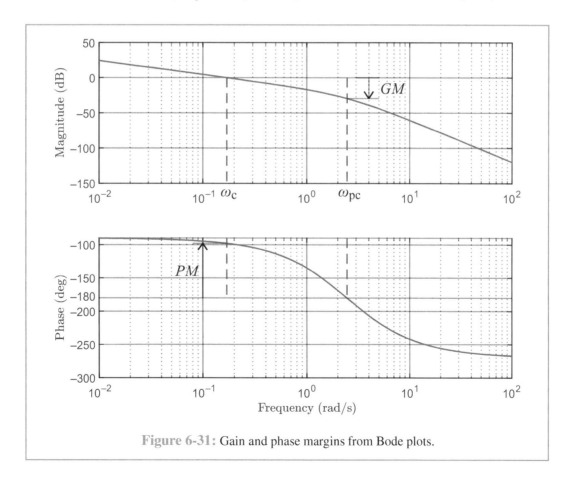

Figure 6-31: Gain and phase margins from Bode plots.

distance from the zero–dB line to the magnitude plot at ω_{pc}. For a stable system, the magnitude at ω_{pc} will be under the zero–dB line.

The Bode plots illustrate the meaning of the gain and phase margins. Suppose you start with a system that has the open-loop transfer function $GH(s)$ with positive gain and phase margins GM and PM, respectively. Suppose now that you replace $G(s)$ by $KG(s)$ for some positive gain K. What is the effect of the gain K on the Bode plots? The phase plot does not change. The magnitude plot will move up if $K > 1$ and will move down if $K < 1$. Because the phase plot does not change, the phase crossover frequency ω_{pc} does not change. When $K > 1$, increasing K will move the magnitude at ω_{pc} closer to the zero–dB line, thus reducing the gain margin. When $20\log K = GM$, the gain margin of the system will be reduced to zero. Thus, $K = 10^{GM/20}$ is the gain that makes the system unstable.

Suppose now that you replace $G(s)$ by $e^{-\tau s} G(s)$ for some time delay τ. What is the effect of the time delay on the Bode plots? The magnitude plot does not change because $|e^{-j\omega\tau}| = 1$. The phase plot will become more negative since $\omega\tau$ is subtracted from the phase of $G(j\omega)$. Since the magnitude plot does not change, the crossover frequency ω_c does not change. Because the phase at ω_c is now more negative, adding the time delay reduces the phase margin of the system. When

$$\omega\tau \times \frac{180}{\pi} = PM,$$

the system becomes unstable. Thus, the foregoing equation gives the time delay that makes the system unstable. The left-hand side of this equation is multiplied by $180/\pi$ because $\omega\tau$ is in radians while PM is in degrees.

There are special cases when the gain or phase margin can approach infinity. Figure 6-32 shows such cases. The figures are generated with the MATLAB command *margin*. The infinite-gain-margin case is for the transfer function

$$\frac{10}{s(s+1)}.$$

The phase plot does not cross the $-180°$ line. It approaches the line asymptotically as $\omega \to \infty$. Hence, $\omega_{pc} = \infty$ rad/s. The infinite-gain margin means that the system remains stable for any increase in the gain of the open-loop transfer function.

The infinite-phase-margin case is for the transfer function

$$\frac{0.1}{s(s+1)(s+2)}.$$

The magnitude plot is always under the zero–dB line (equivalently, the gain is always less than one). Therefore, no phase delay, no matter how large, would drive the system unstable.

6-5. RELATIVE STABILITY

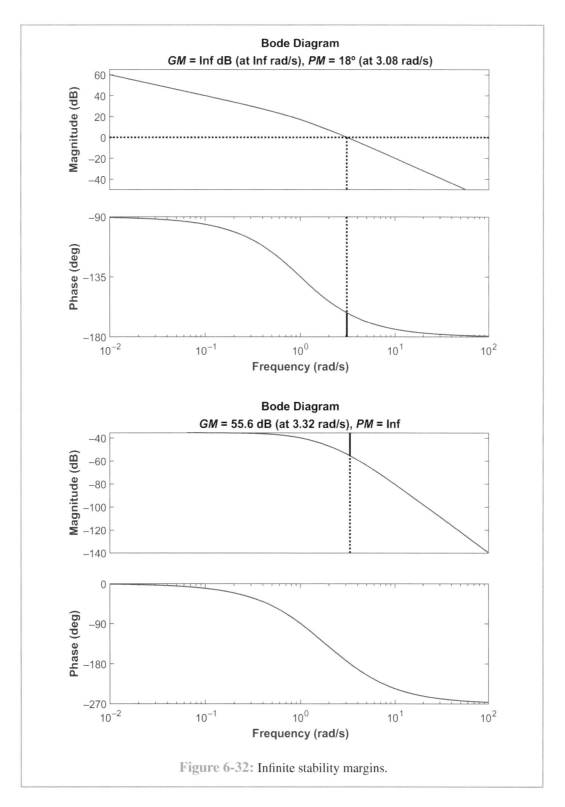

Figure 6-32: Infinite stability margins.

Example 6-18: The Gain and Phase Margins from the Bode Plots

The Bode plots of $G(s)$ in a unity-feedback control system are shown in Fig. 6-33.

(a) Find the gain and phase margins and the corresponding crossover frequencies.

(b) Replace $G(s)$ by $KG(s)$ with $K \geq 1$. Find the range of K for stability.

(c) Replace $G(s)$ by $e^{-\tau s} G(s)$ with $\tau \geq 0$. Find the range of τ for stability.

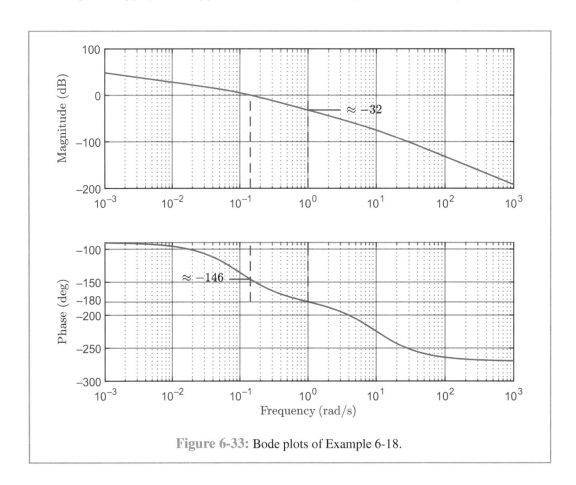

Figure 6-33: Bode plots of Example 6-18.

Solution:

(a) The locations of the crossover frequencies are shown in Fig. 6-33. $\omega_{pc} = 1$ rad/s. At this frequency, the magnitude is approximately -32 dB. Hence, $GM \approx 32$ dB. Also, ω_c is about halfway between 0.1 and 0.2 rad/s. At this frequency the phase is $\approx -146°$. Hence, $PM \approx -146° + 180° = 34°$. Because the frequency scale is logarithmic, ω_c is not 1.5 rad/s, although we may use 1.5 as an approximation. However, if we want a more accurate answer, we need to know how to read frequencies on a logarithmic scale, as follows.

Let the frequency ω be between frequencies ω_1 and ω_2 on a logarithmic scale. Let A be the linear distance from ω_1 to ω and B be the linear distance from ω_1 to ω_2.

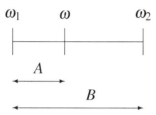

Let $r = A/B$. Then

$$\omega = \left(\frac{\omega_2}{\omega_1}\right)^r \omega_1 .$$

Applying this calculation to our example, we have $\omega_1 = 0.1$, $\omega_2 = 0.2$, and $r = 0.5$. Therefore, $\omega_c = 2^{0.5} = 1.4$ rad/s.

(b)

$$20 \log K_{max} = GM = 32 \quad \rightarrow \quad K_{max} = 10^{32/20} = 39.8.$$

The closed-loop system is stable for $1 \leq K < 39.8$.

(c)

$$\tau_{max} \, \omega_c \times \frac{180}{\pi} = PM \quad \rightarrow \quad \tau_{max} = \frac{34 \times \pi}{1.4 \times 180} = 0.424 \text{ s}.$$

The closed-loop system is stable for $0 \leq \tau < 0.424$.

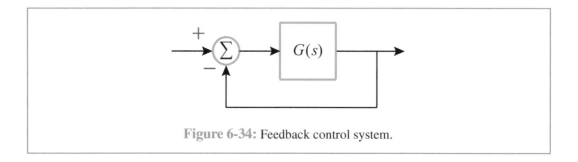

Figure 6-34: Feedback control system.

6-6 Relations between Frequency-Domain and Time-Domain Responses

Consider the unity feedback control system of Fig. 6-34. The closed-loop transfer function is given by

$$T(s) = \frac{G(s)}{1+G(s)}.$$

When we investigate the behavior of this system we can look at it from two different perspectives. We may look at the time response of $T(s)$, represented by the step response, or at the frequency response. It is important to understand how these two responses are related to each other. The step response is characterized by the percent overshoot (PO) and the speed of the response, measured by the rise time (T_r), the peak time (T_p), or the settling time (T_s).

The frequency response is represented by the magnitude plot of $T(j\omega)$ as a function of ω. A typical response is shown in Fig. 6-35, where the frequency is in a logarithmic scale and the magnitude is in decibels. There are three parameters that characterize this response: $M_{p\omega}$, ω_p, and ω_{BW}. Frequency ω_p is called the *peak frequency*, as it is the frequency at which where $|T(j\omega)|$ has its *peak*. The magnitude $M_{p\omega}$ is the ratio of the peak to the dc gain, $|T(j\omega_p)|/|T(0)|$, and ω_{BW} is the *bandwidth* of the system. It is the frequency at which the magnitude of $T(j\omega)$ is 3 dB under its dc value; that is, $|T(j\omega_{BW})/T(0)| = 1/\sqrt{2}$.

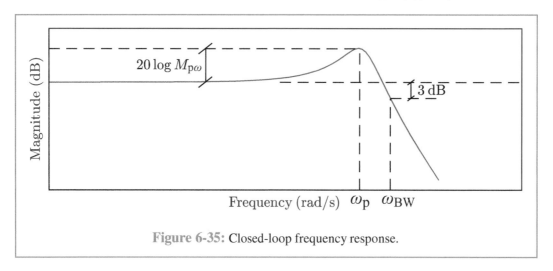

Figure 6-35: Closed-loop frequency response.

6-6. RELATIONS BETWEEN FREQUENCY-DOMAIN AND TIME-DOMAIN

Another set of relations that are important to understand are those that govern how the closed-loop frequency response $T(j\omega)$ is related to the open-loop frequency response $G(j\omega)$. For this relationship, the open-loop frequency response is characterized by the crossover frequency (ω_c) and the phase margin (*PM*).

The aforementioned relationships can be calculated exactly for the underdamped second-order transfer function

$$T(s) = \frac{\omega_n^2}{s^2 + 2\zeta\omega_n s + \omega_n^2} = \frac{G(s)}{1 + G(s)}, \quad \zeta < 1,$$

for which

$$G(s) = \frac{\omega_n^2}{s(s + 2\zeta\omega_n s)}.$$

We saw in Chapter 3 that the step response parameters are determined by the damping ratio ζ and the natural frequency ω_n. In particular,

$$PO = 100 \times e^{-\zeta\pi/\sqrt{1-\zeta^2}},$$

$$T_r \approx \frac{2.16\zeta + 0.6}{\omega_n}, \text{ for } 0.3 \leq \zeta \leq 0.75, \quad T_p = \frac{\pi}{\omega_n\sqrt{1-\zeta^2}}, \quad T_s = \frac{4}{\zeta\omega_n}.$$

For the frequency response, we saw in Section 6-2 that

$$|T(j\omega)| = |\omega_n^2/(\omega_n^2 - \omega^2 + j2\zeta\omega\omega_n)|$$

does not peak if $\zeta > 1/\sqrt{2}$, and peaks for $\zeta < 1/\sqrt{2}$, where

$$\omega_p = \omega_n\sqrt{1 - 2\zeta^2}, \quad M_{p\omega} = \frac{1}{2\zeta\sqrt{1-\zeta^2}}.$$

The bandwidth frequency ω_{BW} can be determined by equating $|T(j\omega)|^2$ to $\frac{1}{2}$.

$$\frac{\omega_n^4}{(\omega_n^2 - \omega^2)^2 + 4\zeta^2\omega^2\omega_n^2} = \frac{1}{2} :$$

$$\omega^4 - 2\omega_n^2(1 - 2\zeta^2)\omega^2 - \omega_n^4 = 0.$$

By the quadratic formula, the bandwidth is

$$\omega_{BW} = \omega_n\sqrt{(1-2\zeta^2) + \sqrt{1 + (1-2\zeta^2)^2}}.$$

For the open-loop transfer function $G(s) = \omega_n^2/[s(s + 2\zeta\omega_n)]$,

$$|G(j\omega)| = \frac{\omega_n^2}{\omega\sqrt{\omega^2 + 4\zeta^2\omega_n^2}}, \quad \angle[G(j\omega)] = -\frac{\pi}{2} - \tan^{-1}\left(\frac{\omega}{2\zeta\omega_n}\right).$$

The crossover frequency ω_c can be determined by equating $|G(j\omega)|$ to 1:

$$\frac{\omega_n^4}{\omega^2(\omega^2 + 4\zeta^2\omega_n^2)} = 1 \, ,$$

$$\omega^4 + 4\zeta^2\omega_n^2\omega^2 - \omega_n^4 = 0 \, .$$

By the quadratic formula,

$$\omega_c = \omega_n\sqrt{-2\zeta^2 + \sqrt{1+4\zeta^4}} \, .$$

Substitution of $\omega = \omega_c$ in $\angle[G(j\omega)]$ yields the phase margin

$$PM = 180° + \left[-90° - \tan^{-1}\left(\frac{\omega_c}{2\zeta\omega_n}\right)\right] = 90° - \tan^{-1}\left(\frac{\omega_c}{2\zeta\omega_n}\right) \, .$$

Using the fact that if $\tan(\theta) = b/a$, $\tan(90° - \theta) = a/b$, we express the phase margin as

$$PM = \tan^{-1}\left(\frac{2\zeta\omega_n}{\omega_c}\right) \, .$$

All quantities ω_c/ω_n, ω_p/ω_n, ω_{BW}/ω_n, PM, and $M_{p\omega}$ depend on the damping ratio ζ. The expressions show the general trend of how these quantities change with ζ. Some of these expressions can be approximated by simpler formulas for a certain range of ζ.

Starting with ω_c, Fig. 6-36(a) shows the ratio ω_c/ω_n for $\zeta \in [0.1, 0.9]$. For $\zeta \in [0.3, 0.75]$, least-squares parameter fitting yields the straight-line approximation

$$\omega_c \approx \frac{(3.35 - 2\zeta)}{3}\omega_n \, .$$

Figure 6-36(b) shows that the relative approximation error is less than 1%.

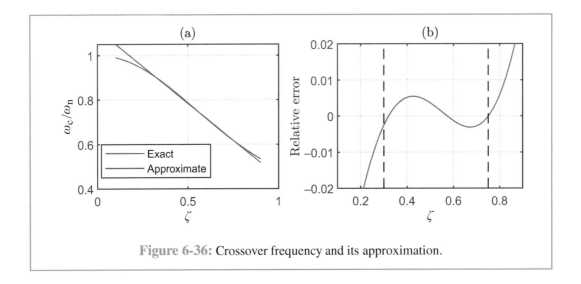

Figure 6-36: Crossover frequency and its approximation.

6-6. RELATIONS BETWEEN FREQUENCY-DOMAIN AND TIME-DOMAIN

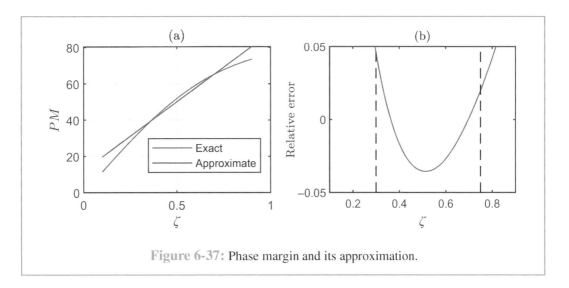

Figure 6-37: Phase margin and its approximation.

Figure 6-37(a) shows the phase margin for $\zeta \in [0.1, 0.9]$. For $\zeta \in [0.3, 0.75]$, least-squares parameter fitting yields the straight line approximation

$$PM \approx 12 + 76\zeta \;.$$

Figure 6-37(b) shows that the relative approximation error is less than 5%.

Figure 6-38(a) shows the ratio $\omega_{\text{BW}}/\omega_{\text{n}}$ for $\zeta \in [0.1, 0.9]$. For $\zeta \in [0.3, 0.75]$, least-squares parameter fitting yields the straight line approximation

$$\omega_{\text{BW}} \approx (1.83 - 1.16\zeta)\omega_{\text{n}} \;.$$

Figure 6-38(b) shows that the relative approximation error is less than 2%.

Sometimes it is convenient to relate ω_{BW} to ω_{c}. Figure 6-38(c) shows the ratio $\omega_{\text{BW}}/\omega_{\text{c}}$ for $\zeta \in [0.1, 0.9]$. A simple approximation of this relationship is

$$\omega_{\text{BW}} \approx 1.6\omega_{\text{c}} \;.$$

Figure 6-38(d) shows that, for $\zeta \in [0.3, 0.75]$, the relative approximation error is less than 5%.

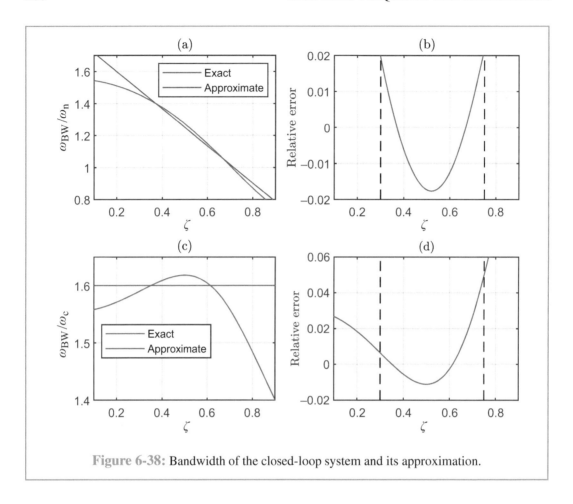

Figure 6-38: Bandwidth of the closed-loop system and its approximation.

All the relations for the second-order transfer function are summarized in Fig. 6-39.

Can the relations of Fig. 6-39 be used for other systems? They can be used as approximations for systems whose closed-loop transfer function can be approximated by $T(s)$ of Fig. 6-39. We have discussed these cases in Subsection 3-4.4.

The more general message we get from the relations of Fig. 6-39 is the general trend of the change in the response characteristics with the change of the damping ratio ζ and natural frequency ω_n of a pair of dominant closed-loop poles. That trend is shown in Table 6-1. Increase in ζ results in increase in PM and decrease in $M_{p\omega}$ and PO. The opposite is also true; that is, a decrease in ζ results in a decrease in PM and an increase in $M_{p\omega}$ and PO. The effect of changing ω_n for a fixed value of ζ is read similarly.

6-6. RELATIONS BETWEEN FREQUENCY-DOMAIN AND TIME-DOMAIN

$$G(s) = \frac{\omega_n^2}{s(s+2\zeta\omega_n)}, \quad T(s) = \frac{G(s)}{1+G(s)} = \frac{\omega_n^2}{s^2+2\zeta\omega_n s + \omega_n^2}$$

Step response of $T(s)$:

$$PO = 100 \times e^{-\zeta\pi/\sqrt{1-\zeta^2}}$$

$$T_r \approx \frac{2.16\zeta + 0.6}{\omega_n}, \quad T_p = \frac{\pi}{\omega_n\sqrt{1-\zeta^2}}, \quad T_s = \frac{4}{\zeta\omega_n}$$

Frequency response of $T(s)$:

$$\text{For } \zeta < \frac{1}{\sqrt{2}}, \quad \omega_p = \omega_n\sqrt{1-2\zeta^2}, \quad M_{p\omega} = \frac{1}{2\zeta\sqrt{1-\zeta^2}}$$

$$\omega_{BW} = \omega_n\sqrt{(1-2\zeta^2) + \sqrt{1+(1-2\zeta^2)^2}} \approx (1.83 - 1.16\zeta)\omega_n \approx 1.6\omega_c$$

Frequency response of $G(s)$:

$$\omega_c = \omega_n\sqrt{-2\zeta^2 + \sqrt{1+4\zeta^4}} \approx \frac{(3.35 - 2\zeta)}{3}\omega_n$$

$$PM = \tan^{-1}\left(\frac{2\zeta\omega_n}{\omega_c}\right) \approx 12 + 76\zeta$$

Approximations are valid for $0.3 \leq \zeta \leq 0.75$.

Figure 6-39: Time-domain–frequency-domain relations.

Table 6-1: **The effect of changing ζ and ω_n on the frequency and time responses.**

Dominant poles ζ, ω_n	Open-loop frequency response $G(s)$	Closed-loop frequency response $T(s)$	Closed-loop step response $T(s)$
$\zeta \uparrow$	$PM \uparrow$	$M_{p\omega} \downarrow$	$PO \downarrow$
Fix ζ, $\omega_n \uparrow$	$\omega_c \uparrow$	$\omega_p, \omega_{BW} \uparrow$	$T_r, T_p, T_s \downarrow$

Example 6-19: Estimation of the Closed-Loop-Response Parameters from the Open-Loop Bode Plots

The Bode plots of $G(s)$ in a unity-feedback control system are shown in Fig. 6-40.

(a) Find ω_c and PM.

(b) Estimate ζ and ω_n.

(c) Estimate PO, T_r, T_p, and T_s of the closed-loop step response.

(d) Estimate $M_{p\omega}$, ω_p, ω_{BW} of the closed-loop frequency response.

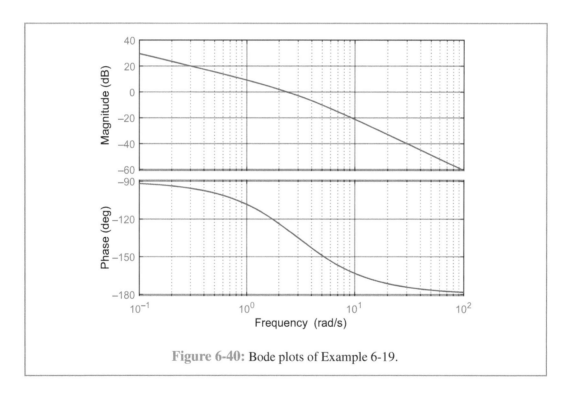

Figure 6-40: Bode plots of Example 6-19.

Solution: The phase plot starts from $-90°$ and approaches $-180°$ asymptotically. Therefore, the transfer function has a pole at the origin and $\deg(\text{denominator}) - \deg(\text{numerator}) = 2$.

(a) From the Bode plots, $\omega_c \approx 2.36$ rad/s and $\angle[G(2.36j)] \approx -128°$. Hence $PM \approx 180 - 128 = 52°$.

(b)
$$PM = 12 + 76\zeta \quad \rightarrow \quad 52 = 12 + 76\zeta \quad \rightarrow \quad \zeta = 0.53,$$

$$\omega_c = \frac{3.35 - 2\zeta}{3}\omega_n = \frac{3.35 - 2 \times 0.53}{3}\omega_n = 0.76\omega_n \quad \rightarrow \quad \omega_n = \frac{2.36}{0.76} = 3.1 \text{ rad/s}.$$

6-6. RELATIONS BETWEEN FREQUENCY-DOMAIN AND TIME-DOMAIN

(c)
$$PO = 14\%,\ T_r = 0.56\text{ s},\ T_p = 1.195\text{ s},\ T_s = 2.43\text{ s}.$$

(d)
$$\omega_p = 2.05\text{ rad/s},\ M_{p\omega} = 1.11,\ \omega_{BW} = 1.6\omega_c = 3.78\text{ rad/s}.$$

Now that we have estimated all these quantities using the relations of Fig. 6-39, it is useful to get a sense of how good the estimates are. To that end, let us reveal that the Bode plots of Fig. 6-40 are for the transfer function

$$G(s) = \frac{9}{s(s+3)}.$$

Table 6-2 compares the estimates with the exact values.

Table 6-2: **Example 6-19.**

	ω_c	PM	ζ	ω_n	PO	T_r	T_p	T_s	$M_{p\omega}$	ω_p	ω_{BW}
Estimate	2.36	52	0.53	3.1	14	0.56	1.195	2.43	1.11	2.05	3.78
Exact	2.36	51.8	0.5	3	16.3	0.546	1.197	2.69	1.15	2.15	3.81

| Example 6-20: | Estimation of the Closed-Loop-Response Parameters from the Open-Loop Bode Plots |

Repeat Example 6-19 for the Bode plots of Fig. 6-41.

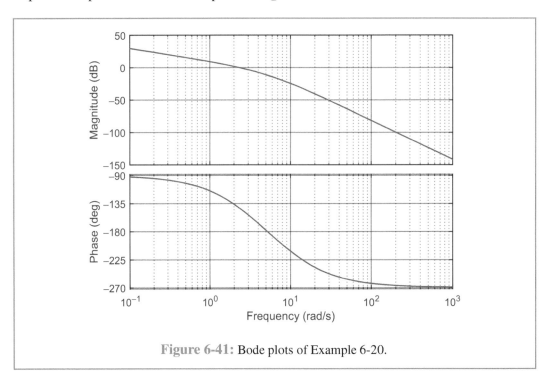

Figure 6-41: Bode plots of Example 6-20.

Solution: The phase plot starts from $-90°$ and approaches $-270°$ asymptotically. Therefore, the transfer function has a pole at the origin and $\deg(\text{denominator}) - \deg(\text{numerator}) = 3$.

(a) From the Bode plots, $\omega_c \approx 2.5$ rad/s and $\angle[G(2.5j)] \approx -145°$. Hence, $PM \approx 180 - 145 = 35°$.

(b)
$$PM = 12 + 76\zeta \quad \rightarrow \quad 35 = 12 + 76\zeta \quad \rightarrow \quad \zeta = 0.3,$$

$$\omega_c = \frac{3.35 - 2\zeta}{3}\omega_n = \frac{3.35 - 2 \times 0.3}{3}\omega_n = 0.917\omega_n \quad \rightarrow \quad \omega_n = \frac{2.5}{0.917} = 2.73 \text{ rad/s}.$$

(c)
$$PO = 37.2\%, \quad T_r = 0.46 \text{ s}, \quad T_p = 1.21 \text{ s}, \quad T_s = 4.88 \text{ s}.$$

(d)
$$\omega_p = 2.47 \text{ rad/s}, \quad M_{p\omega} = 1.75, \quad \omega_{BW} = 1.6\omega_c = 4 \text{ rad/s}.$$

6-6. RELATIONS BETWEEN FREQUENCY-DOMAIN AND TIME-DOMAIN

As done in the previous example, we reveal that the Bode plots of Fig. 6-41 are for the transfer function

$$G(s) = \frac{81}{s(s+3)(s+9)}.$$

The closed-loop transfer function

$$T(s) = \frac{G(s)}{1+G(s)} = \frac{81}{(s+10.12)(s^2+1.877s+8)}$$

has a pair of dominant poles with $\zeta = 0.33$ and $\omega_n = 2.83$ rad/s. Table 6-3 compares the estimates with the exact values.

Table 6-3: **Example 6-20.**

	ω_c	PM	ζ	ω_n	PO	T_r	T_p	T_s	$M_{p\omega}$	ω_p	ω_{BW}
Estimate	2.5	35	0.3	2.73	37.2	0.46	1.21	4.88	1.75	2.47	4
Exact	2.3	38.1	0.33	2.83	38.1	0.51	1.28	4.02	1.55	2.5	3.94

Summary

Concepts

- The frequency response of a transfer function can be represented graphically by the polar plot or by the Bode plots.
- A sketch of the polar plot can be drawn from the knowledge of the asymptotic behavior of the frequency response as $\omega \to 0$ and $\omega \to \infty$, and the intersection points with the real and imaginary axes.
- Bode plots consist of two plots: magnitude versus frequency and phase versus frequency, with the frequency in a log scale. Usually the magnitude is in decibels, the phase in degrees, and the frequency in rad/s.
- Bode plots of $G = G_1 G_2 \ldots G_m$ are the sums of the Bode plots of G_1, G_2, to G_m if the Bode magnitude plots are in decibels.
- Bode plots can be approximated by the asymptotic Bode plots, which are drawn from the knowledge of the break frequencies of first and second-order transfer functions.
- Multiplication of a transfer function by a constant gain does not change the phase plot, and multiplication by a time-delay transfer function does not change the magnitude plot.
- The Nyquist stability criterion determines the number of closed-loop pole in the RHP (right-half plane) by counting the number of open-loop poles in the RHP and the number of encirclement of a critical point by the Nyquist plot.
- Relative stability is a way to quantify the robustness of the stability of the feedback system. It is measured by the gain and phase margins, which are the gain increase and phase delay, respectively, that bring the system to the verge of instability.
- The gain and phase margins and their corresponding crossover frequencies can be easily determined form the Bode plots.
- The open-loop frequency response, the closed-loop frequency response, and the closed-loop step response are related to each other. For the underdamped second-order transfer function, the parameters of these three responses can be expressed or approximated in terms of the damping ratio ζ and the undamped natural frequency ω_n.

Mathematical Models

Nyquist criterion: $Z = P - N$; the system is stable for $Z = 0$.

Time-domain–frequency-domain relations: See Table 6.1 and Figure 6-39.

Important Terms Provide definitions or explain the meaning of the following terms:

asymptotic Bode plots	crossover frequency	peak frequency
bandwidth	gain margin	phase crossover frequency
Bode plots	Nyquist contour	phase margin
Cauchy's principle of the argument	Nyquist plot	polar plot
	Nyquist stability criterion	relative stability
contour mapping	peak	

PROBLEMS

6.1 Let
$$G(s) = \frac{10}{s(s+1)(s+4)}.$$

(a) Find the expressions of $\angle G(j\omega)$, $\text{Re}[G(j\omega)]$, and $\text{Im}[G(j\omega)]$.

(b) Sketch the polar plot. Using MATLAB, draw the polar plot and compare with your sketch.

(c) Draw the asymptotic Bode plots. Using MATLAB, draw the Bode plots and compare with yours.

6.2 Repeat Problem 6.1 for
$$G(s) = \frac{10(s+4)}{(s+2)(s+5)}.$$

6.3 Repeat Problem 6.1 for
$$G(s) = \frac{10(s+3)(s+4)}{s(s+2)(s+5)}.$$

6.4 Repeat Problem 6.1 for
$$G(s) = \frac{10}{s(s+1)(s+10)}.$$

6.5 Repeat Problem 6.1 for
$$G(s) = \frac{10(s+5)}{(s+1)(s+10)(s+50)}.$$

6.6 Repeat Problem 6.1 for
$$G(s) = \frac{32}{s(s+2)(s+8)}.$$

6.7 Repeat Problem 6.1 for

$$G(s) = \frac{60}{(s+1)(s+3)(s+5)}.$$

6.8 Repeat Problem 6.1 for

$$G(s) = \frac{10^4}{(s+1)(s+10)(s+100)}.$$

6.9 Repeat Problem 6.1 for

$$G(s) = \frac{100(s+5)}{s(s+10)(s+100)}.$$

6.10 Repeat Problem 6.1 for

$$G(s) = \frac{20(s+5)(s+50)}{(s+1)(s+10)(s+100)}.$$

6.11 Consider a unity feedback control system with

$$G(s) = \frac{K}{s(s-1)(s+2)}, \quad K > 0.$$

Using the Nyquist criterion find the range of K for stability. Verify your answer using the Routh-Hurwitz criterion.

6.12 Repeat Problem 6.11 for

$$G(s) = \frac{64K(s+1)}{(s+2)(s+4)(s+8)}, \quad K > 0.$$

6.13 Repeat Problem 6.11 for

$$G(s) = \frac{15K}{s(s+3)(s+5)}, \quad K > 0.$$

6.14 Repeat Problem 6.11 for

$$G(s) = \frac{10K(s+2)}{(s-1)(s+1)(s+10)}, \quad K > 0.$$

6.15 Repeat Problem 6.11 for

$$G(s) = \frac{K}{s(s+1)(s-2)}, \quad K > 0.$$

PROBLEMS

6.16 Repeat Problem 6.11 for

$$G(s) = \frac{K(s+1)}{s(s-1)(s+5)}, \quad K > 0.$$

6.17 Repeat Problem 6.11 for

$$G(s) = \frac{K}{s^2(s+1)}, \quad K > 0.$$

6.18 Repeat Problem 6.11 for

$$G(s) = \frac{5K(s+1)}{s(s-1)(s+3)}, \quad K > 0.$$

6.19 Repeat Problem 6.11 for

$$G(s) = \frac{K}{s(s-2)(s+8)}, \quad K > 0.$$

6.20 Repeat Problem 6.11 for

$$G(s) = \frac{K}{(s+1)(s+3)(s+5)}, \quad K > 0.$$

6.21 Repeat Problem 6.11 for

$$G(s) = \frac{K(s+5)}{s(s+10)(s+100)}, \quad K > 0.$$

6.22 Repeat Problem 6.11 for

$$G(s) = \frac{K(s-5)(s+50)}{(s+1)(s+10)(s+100)}, \quad K > 0.$$

6.23 The Bode plots of $G(s)$ in a unity feedback control system are shown in Fig. P6.23.

(a) Find the gain and phase margins and the corresponding crossover frequencies.

(b) Replace $G(s)$ by $KG(s)$ with $K \geq 1$. Find the range of K for stability.

(c) Replace $G(s)$ by $e^{-\tau s}G(s)$ with $\tau \geq 0$. Find the range of τ for stability.

(d) Find the type of the feedback system and the corresponding error constant.

(e) Find the steady-state error to a unit step input.

(f) Estimate the percent overshoot, rise time, and settling time of the step response.

(g) Estimate the bandwidth of the closed-loop frequency response.

(h) Estimate the peak amplitude and peak frequency of the closed-loop frequency response.

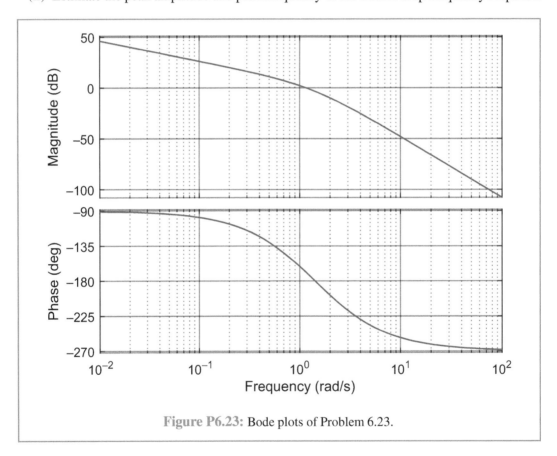

Figure P6.23: Bode plots of Problem 6.23.

6.24 Repeat Problem 6.23 for the Bode plots shown in Figure P6.24.

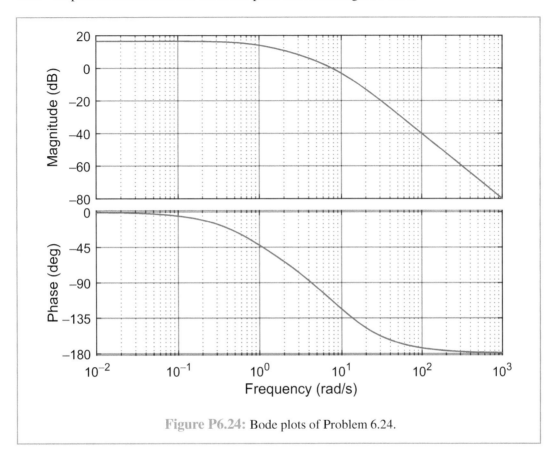

Figure P6.24: Bode plots of Problem 6.24.

6.25 Repeat Problem 6.23 for the Bode plots shown in Figure P6.25.

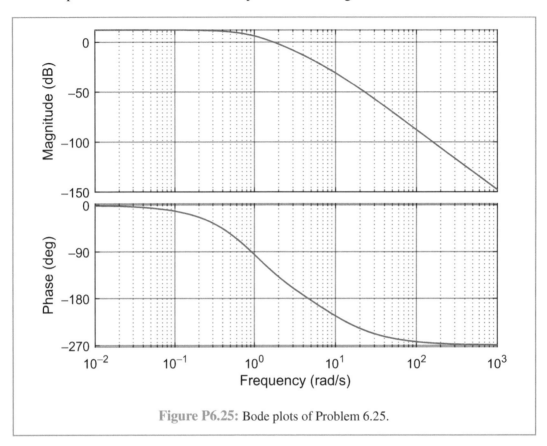

Figure P6.25: Bode plots of Problem 6.25.

6.26 Repeat Problem 6.23 for the Bode plots shown in Fig. P6.26.

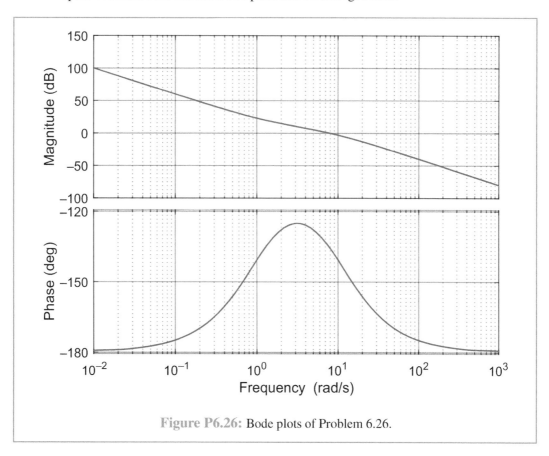

Figure P6.26: Bode plots of Problem 6.26.

6.27 Repeat Problem 6.23 for the Bode plots shown in Fig. P6.27.

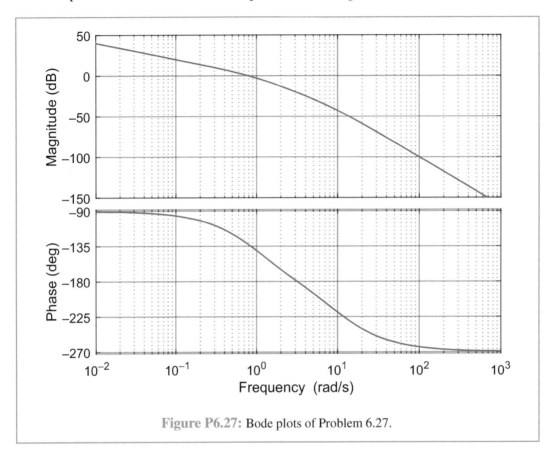

Figure P6.27: Bode plots of Problem 6.27.

6.28 The Bode plots of $G(s)$ in a unity feedback control system are shown in Fig. P6.28.

(a) Find the gain and phase margins and the corresponding crossover frequencies.
(b) What is the range of $K \geq 1$ for stability if $G(s)$ is replaced by $KG(s)$?
(c) What is the gain margin if $G(s)$ is replaced by $2G(s)$?
(d) What is the range of $\tau \geq 0$ for stability if $G(s)$ is replaced by $e^{-\tau s}G(s)$?
(e) What is the phase margin if $G(s)$ is replaced by $e^{-0.2s}G(s)$?

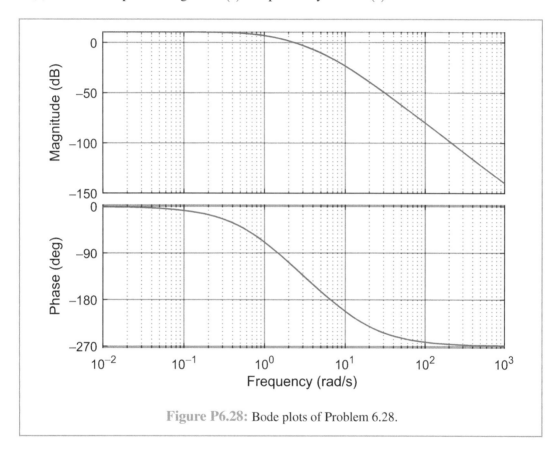

Figure P6.28: Bode plots of Problem 6.28.

6.29 A riderless bicycle can be stabilized by equipping it with a steering mechanism and feedback control from the measured lean angle. In [1], the dynamics of the bicycle, in constant forward velocity, from the steering angle to the lean angle is modeled by

$$G_1(s) = 3.373 \left(\frac{s+8}{s^2 - 18.9} \right),$$

and the steering mechanism is modeled by

$$G_2(s) = e^{-0.015s} \frac{(33.9)^2}{s^2 + 40.68s + (33.9)^2}.$$

The transfer function from the input of the steering mechanism to the lean angle is $G_p = G_1 G_2$. A PID controller is designed in [1] as

$$G_c(s) = 2.5177\left(1 + \frac{1.5431}{s} + 0.075s\right).$$

(a) Verify the stability of the system using Nyquist criterion.

(b) Plot $|G_p(j\omega)\,G_c(j\omega)|$. Determine the crossover frequency ω_c. Find the minimum of $|G_p(j\omega)\,G_c(j\omega)|$ for $\omega \leq 0.1\omega_c$ and the maximum of $|G_p(j\omega)\,G_c(j\omega)|$ for $\omega \geq 10\omega_c$.

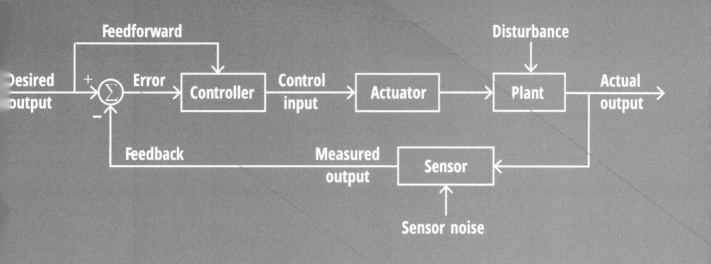

Chapter 7

Frequency-Domain Design

Chapter Contents

　　　　Overview, 278
7-1　Design Specifications, 278
7-2　Design by Gain Adjustment, 280
7-3　Lead Compensation, 287
7-4　Lag Compensation, 295
7-5　PI Compensation, 299
7-6　Lead-Lag Compensation, 302
7-7　Advantages of the Frequency-Domain
　　　 Approach, 306
　　　Chapter Summary, 307
　　　Problems, 308

Objectives

Upon learning the material presented in this chapter, you should be able to:

1. Know the frequency-domain design specifications.
2. Design by adjustment of a constant gain.
3. Design lead, lag, and lead-lag compensators.
4. Design a PI controller.
5. Appreciate the advantages of the frequency-domain approach in analysis and design.

Overview

There are two approaches to the design of feedback control systems: the time-domain approach and the frequency-domain approach. Time-domain design is covered in Chapter 5 and will be covered again in Chapter 8 using state-space models.

In this chapter we present the frequency-domain design approach. The design uses Bode plots of the open-loop transfer function. Section 7-1 states the design specifications in the frequency domain. In Section 7-2 the controller is a gain, which is adjusted to meet one of the design specifications. If other design specifications are not met by this gain choice, compensation is used. Section 7-3 uses lead compensation to improve the dynamic response of the system as measured by the phase margin and the crossover frequency. Sections 7-4 and 7-5 show how to reduce the steady-state error by increasing the low-frequency gain. It is realized by lag compensation in Section 7-4 and PI (proportional-integral) compensation in Section 7-5. To improve the dynamic response and increase the low-frequency gain, lead and lag compensation are combined in the lead-lag compensator of Section 7-6. Section 7-7 concludes our discussion in this chapter, and the previous one, by summarizing the advantages of the frequency-domain approach in analysis and design.

7-1 Design Specifications

Given the plant transfer function $G_p(s)$ in the feedback control system of Fig. 7-1, the objective is to design the controller transfer function $G_c(s)$ to meet frequency-domain design specifications.

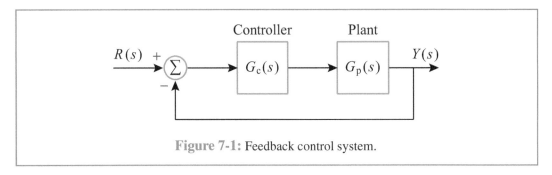

Figure 7-1: Feedback control system.

Frequency-domain design specifications are classified into three groups:

Group 1: Stability requirements on the phase margin and crossover frequency. From the relations presented in Section 6-5, we know that the phase margin and crossover frequency shape the dynamic response of the closed-loop system, whether looked at in the frequency domain through the closed-loop frequency response or in the time domain through the closed-loop step response.

Group 2: Requirements on the error constants K_p, K_v, or K_a in order to meet requirements on the steady-state tracking errors.

Group 3: Requirements on the open-loop frequency response $G_p(j\omega) G_c(j\omega)$ in order to track the reference signal and attenuate the effects of disturbance and measurement noise.

7-1. DESIGN SPECIFICATIONS

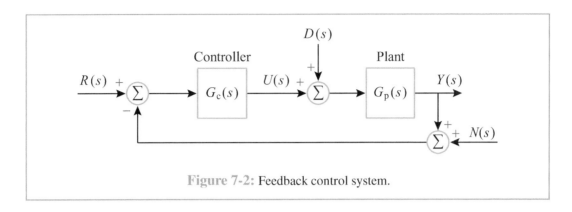

Figure 7-2: Feedback control system.

To elaborate on the third group of requirements, let us recall Fig. 4-16 from Section 4-6, repeated here as Fig. 7-2. The figure shows the feedback control system subject to three inputs: the reference R, the disturbance D, and the measurement noise N. The tracking error is given by

$$E(s) = R(s) - Y(s) = \frac{1}{1+G(s)} R(s) - \frac{G_p(s)}{1+G(s)} D(s) + \frac{G(s)}{1+G(s)} N(s)$$
$$= S(s) R(s) - S(s) G_p(s) D(s) + [1 - S(s)] N(s) , \qquad (7.1)$$

where $G(s) = G_p(s) G_c(s)$ is the *loop gain* and $S(s) = 1/[1+G(s)]$ is the *sensitivity transfer function*. To reduce the error due to R and D, we want $|G| \gg 1$ so that $|S| \ll 1$. This design, however, makes $|1 - S| \approx 1$, which allows the noise N to affect the error without attenuation. Attenuating the effect of the measurement noise requires $|G| \ll 1$ so that $|S| \approx 1$ and $|1 - S| \approx 0$. It is clear that we cannot achieve both requirements at the same time.

The frequency-domain approach allows us to trade off these two conflicting requirements. The frequency spectrum of the reference R and disturbance D is usually in a low frequency band $[0, \omega_1]$. The frequency spectrum of the measurement noise, on the other hand, is usually flat on a wide frequency band, but the noise amplitude will be small if we have good sensors. One way to trade off these requirements is to design the loop gain $|G(j\omega)|$ to be high over a frequency band that covers the reference and disturbance frequency bands, and then let it roll off to a low gain for higher frequencies. Figure 7-3 shows typical requirements on the loop gain.

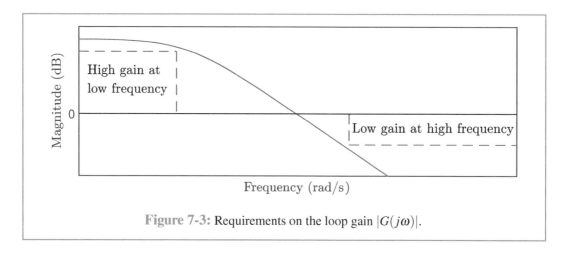

Figure 7-3: Requirements on the loop gain $|G(j\omega)|$.

7-2 Design by Gain Adjustment

Let $G_c(s) = K > 0$ and choose K to meet one of the design specifications. We shall see shortly how to do it for different specifications. Once K is chosen, we have to check the stability of the system for the chosen value of K. This can be done analytically by calculating the roots of the characteristic equation

$$1 + KG_p(s) = 0, \tag{7.2}$$

or it can be done using the Bode plots:

(1) Draw the magnitude and phase Bode plots for $K = 1$.

(2) If the magnitude plot needs to be raised by A dB to meet the design specification, determine the new crossover frequency ω_c. It is the frequency at which the magnitude plot equals $-A$ dB.

(3) Project the crossover frequency down to the phase plot and make sure that $\angle[G(j\omega_c)]$ is above the $-180°$ line, for then the phase margin $PM = \angle[G(j\omega_c)] + 180°$ will be positive.

This procedure is illustrated in Fig. 7-4. Note that if the magnitude plot needs to be lowered by A dB to meet the design specification, then the new crossover frequency is the frequency at which the magnitude plot is A dB.

7-2.1 Gain Adjustment to Achieve a Desired Crossover Frequency

Choose K such that $\omega_c = \omega_{c\ des}$, where $\omega_{c\ des}$ is the desired crossover frequency.

Step 1: Check that the system will have a positive phase margin at $\omega_{c\ des}$. This can be done by finding $\angle[G(j\omega_{c\ des})]$ and verifying that $PM = \angle[G(j\omega_{c\ des})] + 180° > 0$. The angle $\angle[G(j\omega_{c\ des})]$ can be calculated from the analytic expression of $G(s)$ or from the phase plot

7-2. DESIGN BY GAIN ADJUSTMENT

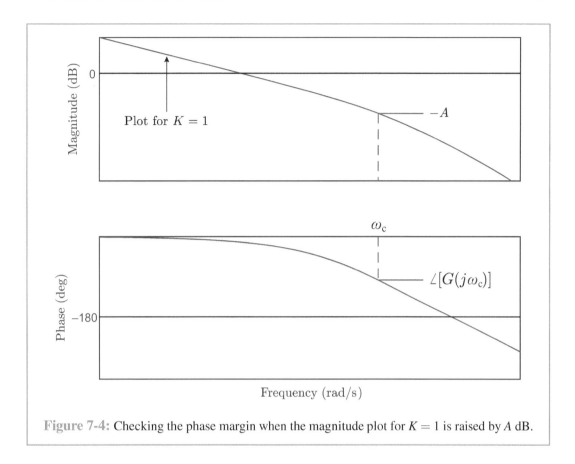

Figure 7-4: Checking the phase margin when the magnitude plot for $K = 1$ is raised by A dB.

of $G(s)$ at $K = 1$. If the phase margin at $\omega_{c\ des}$ is not positive, stop. This desired crossover frequency cannot be achieved.

Step 2: Choose K such that $|G(j\omega_{c\ des})| = 1$.

This can be done analytically. Let $G(s) = K\hat{G}(s)$, where $\hat{G}(s)$ is a given transfer function. Set

$$K = \frac{1}{|\hat{G}(j\omega_{c\ des})|}, \tag{7.3}$$

or from the Bode plot, draw the magnitude plot of G at $K = 1$. Mark the desired crossover frequency $\omega_{c\ des}$ on the zero–dB line and determine whether or not the magnitude plot needs to move up or down so that it intersects the zero–dB line at $\omega_{c\ des}$. Figure 7-5 shows a case where the magnitude plot needs to be raised by A dB; $K = 10^{A/20}$.

7-2.2 Gain Adjustment to Achieve a Desired Phase Margin

Choose K such that $PM = PM_{\text{desired}}$, where PM_{desired} is the desired phase margin. Let $G(s) = K\hat{G}(s)$, where $\hat{G}(s)$ is a given transfer function..

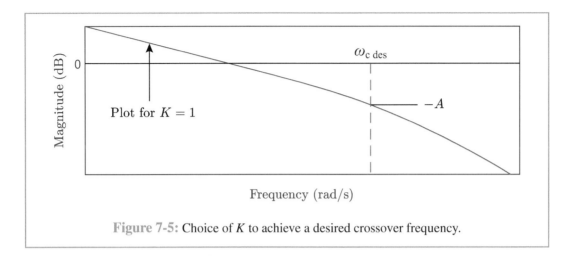

Figure 7-5: Choice of K to achieve a desired crossover frequency.

Step 1: From the phase plot of $\hat{G}(s)$, find ω_c at which $\angle[\hat{G}(j\omega_c)] = PM_{\text{desired}} - 180°$; see Fig. 7-6.

Step 2: Choose K such that $|G(j\omega_c)| = 1$.

This can be determined analytically: $K = 1/|\hat{G}(j\omega_c)|$,

or from the Bode plot by drawing the magnitude plot of G at $K = 1$. Determine whether the magnitude plot needs to move up or down so that it intersects the zero-dB line at ω_c. Figure 7-6 shows a case where the magnitude plot needs to be raised by A dB; $K = 10^{A/20}$.

7-2.3 Gain Adjustment to Achieve a Desired K_p, K_v, or K_a

Calculate the error constant in terms of K, as was done in Section 4-3, and choose K to achieve the desired error constant, or find K from the Bode plots. Figure 7-7 shows how to compute K for a Type 0 system and Fig. 7-8 shows how to do it for a Type 1 system.

- For the case of K_p, draw the Bode magnitude plots for $K = 1$. At the low frequency end of the magnitude plot, where the plot is flat, mark the desired K_p in decibels. Assuming that the desired K_p is higher than the one for $K = 1$, you need to raise the plot by A dB. Set $K = 10^{A/20}$.

- For the case of K_v, draw the Bode magnitude plot for $K = 1$. On the zero–dB line mark the frequency that is equal to the desired K_v. From this point draw a line with a slope -20 dB/dec, extending to the low-frequency end of the graph. At the low frequency where the magnitude plot for $K = 1$ has a slope -20 dB/dec, mark the required change in the magnitude. Assuming that the desired K_v is higher than the one for $K = 1$, you need to raise the plot by A dB. Set $K = 10^{A/20}$.

7-2. DESIGN BY GAIN ADJUSTMENT

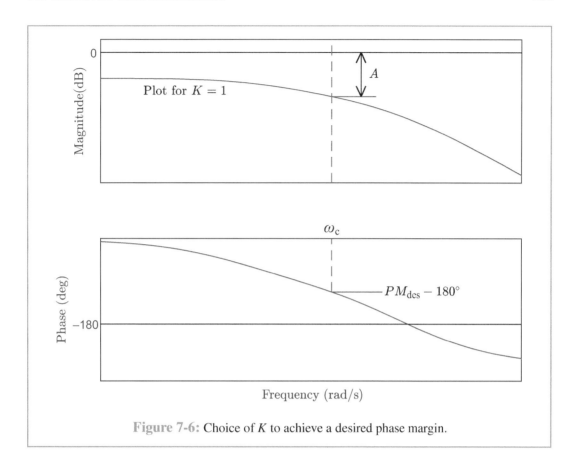

Figure 7-6: Choice of K to achieve a desired phase margin.

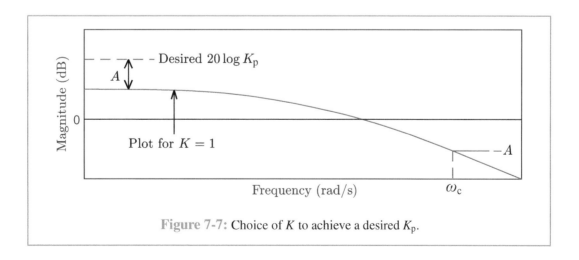

Figure 7-7: Choice of K to achieve a desired K_p.

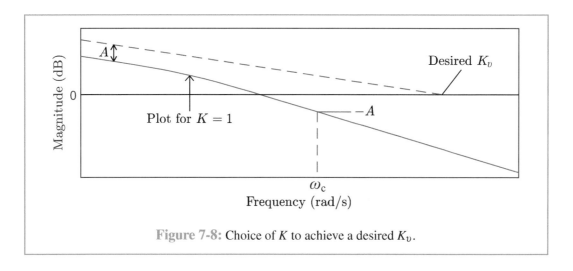

Figure 7-8: Choice of K to achieve a desired K_v.

7-2.4 Gain Adjustment to Shape the Low Frequency Gain

Choose K such that for a specified gain a, $|G(j\omega)| \geq a$, for all $\omega \in [0, \omega_1]$. Typically, $|G(j\omega)|$ has low-pass filtering characteristics; that is, $|G(j\omega)| \geq |G(j\omega_1)|$ for all $\omega \in [0, \omega_1]$. Therefore it is sufficient to choose K such that $|G(j\omega_1)| \geq a$.

This can be done analytically. Let $G(s) = K\hat{G}(s)$, where $\hat{G}(s)$ is a given transfer function. Set

$$K \geq \frac{a}{|\hat{G}(j\omega_1)|},$$

or from the Bode plot, draw the magnitude plot of G at $K = 1$. At $\omega = \omega_1$ determine the increase A (dB) in the magnitude plot needed to bring $|G(j\omega_1)|$ to the desired level. Set $K = 10^{A/20}$.

Example 7-1: Gain Choice to Meet One Design Specification

Let

$$G(s) = \frac{K}{(s+1)(5s+1)}.$$

(1) Find K such that $\omega_c = 1$ rad/s.

(2) Find K such that $PM = 60°$.

(3) Find K such that $K_p \geq 100$.

(4) Find K such that

$$20\log|G(j\omega)| \geq 20 \text{ dB} \quad \text{for } \omega \in [0, 1] \text{ rad/s}.$$

7-2. DESIGN BY GAIN ADJUSTMENT

Solution: The Bode phase plot is shown in Fig. 7-9. The system has an infinite gain margin. Hence, it is stable for all $K > 0$.

(1) Choose K such that $|G(j\omega)| = 1$ at $\omega = 1$ rad/s:

$$|G(j)| = \frac{K}{\sqrt{(1+1)(25+1)}} = \frac{K}{7.21} = 1 \quad \rightarrow \quad K = 7.21.$$

(2) For $PM = 60°$, $\angle[G] = -120°$. From the phase plot, $\omega_c \approx 0.9$ rad/s. Choose K such that $|G(0.9j)| = 1$:

$$|G(0.9j)| = \frac{K}{\sqrt{(0.81+1)(20.25+1)}} = \frac{K}{6.2} = 1 \quad \rightarrow \quad K = 6.2.$$

(3) $K_p = G(0) = K$. Set $K \geq 100$.

(4) Choose K such that $20\log|G(j)| \geq 20$:

$$20\log|G(j)| \geq 20 \quad \leftrightarrow \quad |G(j)| \geq 10 \quad \leftrightarrow \quad \frac{K}{7.21} \geq 10 \quad \leftrightarrow \quad K \geq 72.1.$$

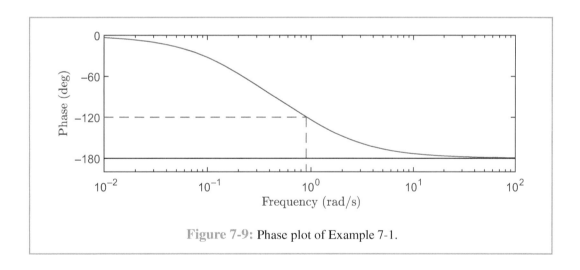

Figure 7-9: Phase plot of Example 7-1.

Example 7-2: Gain Choice Using Bode Plots without Analytic Expression of the Transfer Function

Let $G(s) = K\hat{G}(s)$. The Bode plots of $\hat{G}(s)$ are shown in Fig. 7-10.

(1) Find K such that $\omega_c = 1$ rad/s.

(2) Find K such that $PM = 60°$.

(3) Find K such that $K_v = 1$.

(4) Find K such that

$$20\log|G(j\omega)| \geq 30 \text{ dB} \quad \text{for } \omega \in [0, 0.1] \text{ rad/s}.$$

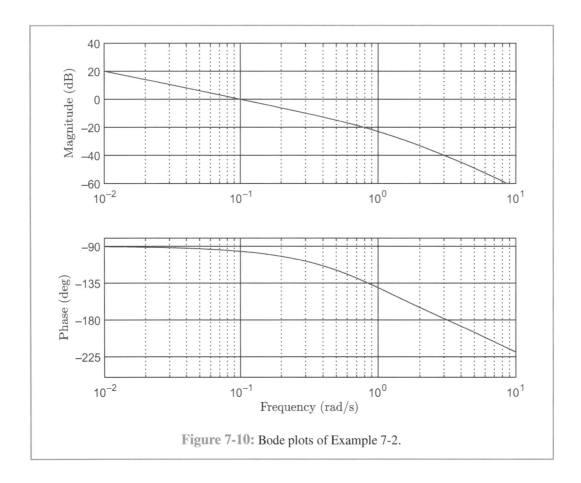

Figure 7-10: Bode plots of Example 7-2.

Solution: The calculations are illustrated in Fig. 7-11.

(1) First we need to check if the system has a positive phase margin at $\omega_c = 1$ rad/s. From the phase plot, it can be seen that at 1 rad/s the phase plot is above the $-180°$ line. Hence, the phase margin is positive. At 1 rad/s, the magnitude plot is about -22.5 dB. We need to raise the magnitude plot by 22.5 dB. Hence, $K = 10^{22.5/20} = 13.3$.

(2) For $PM = 60°$, the phase should be $-120°$ at ω_c. From the phase plot, this occurs at $\omega \approx 0.5$ rad/s. This frequency is the desired ω_c. The magnitude plot at 0.5 rad/s is about -15 dB. We need to raise the magnitude plot by 15 dB. Hence, $K = 10^{15/20} = 5.6$.

7-3. LEAD COMPENSATION

(3) Starting at $\omega = 1$ rad/s on the zero–dB line, draw a line with a slope -20 dB/dec extending to the low-frequency end of the graph. From the graph we see that we need to raise the magnitude plot by 20 dB. Since we saw earlier that the system will be stable when the magnitude plot is raised by 22.5 dB, it will be stable when it is raised by 20 dB. Hence, $K = 10^{20/20} = 10$.

(4) We need to lift the magnitude plot up so that it is equal to 30 dB at 0.1 rad/s. The magnitude is almost 0 dB at 0.1 rad/s. Hence, we need to lift it by 30 dB. To check stability of the system we need to find the phase margin at the new ω_c. The new ω_c is the frequency at which the magnitude plot for $K = 1$ is -30 dB. At this frequency the phase plot is above the $-180°$ line. Hence, the system is stable and $K = 10^{30/20} = 31.6$.

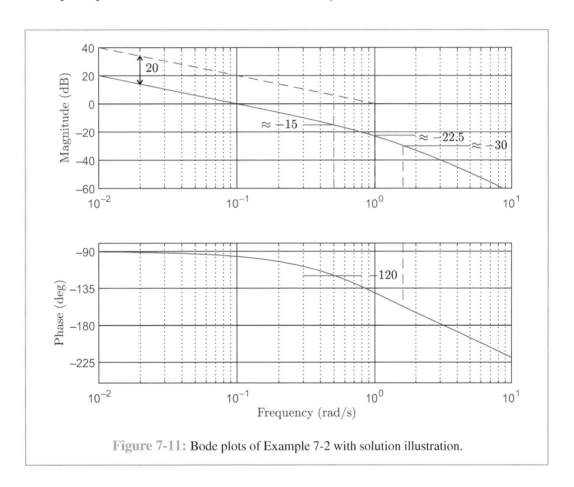

Figure 7-11: Bode plots of Example 7-2 with solution illustration.

7-3 Lead Compensation

We saw in Chapter 5 that the lead compensator improves stability in the time-domain design because it shifts the root locus to the left. It does the same function in the frequency-domain design because it can increase the phase margin.

The transfer function of the lead compensator, written in a form suitable for frequency-domain calculations, is

$$G_c(s) = K\left(\frac{Ts+1}{\beta Ts+1}\right), \quad K>0, \quad T>0, \quad 0<\beta<1. \tag{7.4}$$

It has a zero at $-1/T$ and a pole at $-1/(\beta T)$. Because $\beta < 1$, the zero is to the right of the pole. The key property of the lead compensator that enables increasing the phase margin is that its phase is positive. The phase of the lead compensator is given by

$$\angle[G_c(j\omega)] = \tan^{-1}(T\omega) - \tan^{-1}(\beta T\omega). \tag{7.5}$$

The break frequency $1/T$ is lower than the break frequency $1/(\beta T)$. Therefore, the angle $\tan^{-1}(T\omega)$ starts to increase positively at an earlier frequency than that for the angle $\tan^{-1}(\beta T\omega)$. Eventually, the angle $\tan^{-1}(\beta T\omega)$ starts to increase positively and as $\omega \to \infty$, $\angle[G_c(j\omega)] \to 0°$. The general shape of $\angle[G_c(j\omega)]$ is shown in **Fig. 7-12**. The phase has a

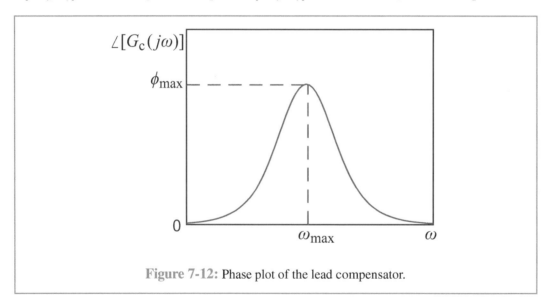

Figure 7-12: Phase plot of the lead compensator.

maximum ϕ_{max} at frequency ω_{max}. To find ω_{max}, differentiate $\angle[G_c(j\omega)]$ with respect to ω and equate to zero:

$$\frac{d}{d\omega}\angle[G_c(j\omega)] = \frac{T}{1+T^2\omega^2} - \frac{\beta T}{1+\beta^2 T^2\omega^2} = \frac{T(1-\beta)(1-\beta T^2\omega^2)}{(1+T^2\omega^2)(1+\beta^2 T^2\omega^2)}.$$

Equating the derivative to zero yields

$$\omega_{max} = \frac{1}{T\sqrt{\beta}}. \tag{7.6}$$

Frequency ω_{max} is the geometric mean of the break frequencies $1/T$ and $1/(\beta T)$. On a logarithmic scale, the geometric mean is halfway between the two break frequencies.

7-3. LEAD COMPENSATION

The maximum phase is given by

$$\phi_{max} = \angle[G_c(j\omega_{max})] = \tan^{-1}\left(\frac{1}{\sqrt{\beta}}\right) - \tan^{-1}\left(\sqrt{\beta}\right). \quad (7.7)$$

Using the trigonometric identity

$$\tan(\phi_1 - \phi_2) = \frac{\tan\phi_1 - \tan\phi_2}{1 + \tan\phi_1 \tan\phi_2},$$

we obtain

$$\tan\phi_{max} = \frac{\frac{1}{\sqrt{\beta}} - \sqrt{\beta}}{1 + \frac{1}{\sqrt{\beta}} \cdot \sqrt{\beta}} = \frac{1-\beta}{2\sqrt{\beta}}.$$

Therefore,

$$\sin\phi_{max} = \frac{1-\beta}{1+\beta} \quad \rightarrow \quad \beta = \frac{1-\sin\phi_{max}}{1+\sin\phi_{max}}. \quad (7.8)$$

In summary,

Phase Relations of the Lead Compensator

$$\omega_{max} = \frac{1}{T\sqrt{\beta}}, \quad \sin\phi_{max} = \frac{1-\beta}{1+\beta}, \quad \beta = \frac{1-\sin\phi_{max}}{1+\sin\phi_{max}}$$

Table 7-1 shows how ϕ_{max} changes with β. For large ϕ_{max}, β is very small. It is hard to implement the lead compensator with a very small β. Therefore we usually do not design the lead compensator for $\phi_{max} > 70°$. If a large positive angle is needed, we can cascade multiple lead compensators. For example, if 90° is needed we can cascade two lead compensators providing 45° each.

Table 7-1: ϕ_{max} as function of β.

ϕ_{max} (degrees)	20	30	40	50	60	70	80
β	0.49	0.33	0.22	0.13	0.07	0.03	0.008

The Bode plots of the lead compensator, both exact and asymptotic, are shown in Fig. 7-13. The magnitude plot starts from $20\log K$ and increases by $20\log(1/\beta)$ to approach $20\log(K/\beta)$ as $\omega \to \infty$. At $\omega_{max} = 1/(T\sqrt{\beta})$, the magnitude plot is halfway between its two limits; that is, at $10\log(K/\beta)$.

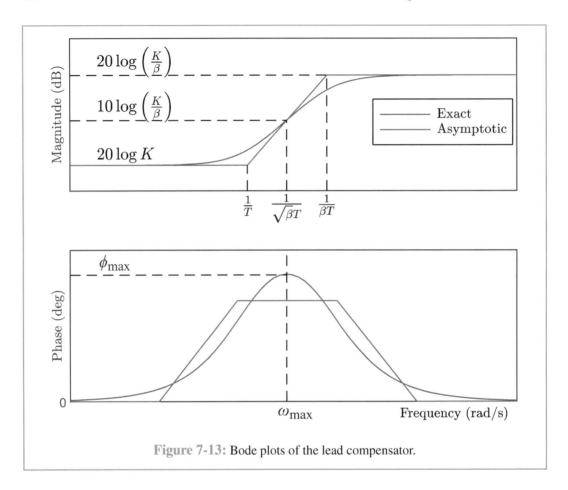

Figure 7-13: Bode plots of the lead compensator.

7-3.1 Design of a Lead Compensator to Increase the Phase Margin at a Desired Crossover Frequency

Given a specified value for ω_c, design $G_c(s)$ such that $PM = PM_{\text{desired}}$.

Step 1: Calculate the uncompensated phase margin PM_{unc}. This is the phase margin if $G_c(s) = K$. That is,

$$PM_{\text{unc}} = \angle[G_p(j\omega_c)] + 180°.$$

Step 2: Choose $\phi_{\max}$ to increase the phase at ω_c to the desired level:

$$\phi_{\max} = PM_{\text{desired}} - PM_{\text{unc}}.$$

Step 3: Calculate β:

$$\beta = \frac{1 - \sin \phi_{\max}}{1 + \sin \phi_{\max}}.$$

7-3. LEAD COMPENSATION

Step 4: Choose T so as to locate $\omega_{\max}$ at ω_c.

$$\omega_c = \frac{1}{T\sqrt{\beta}} \quad \rightarrow \quad T = \frac{1}{\omega_c \sqrt{\beta}} . \tag{7.9}$$

Step 5: Choose K such that $|G_c(j\omega_c)G_p(j\omega_c)| = 1$.

Example 7-3: Increasing the Phase Margin at a Desired Crossover Frequency

Consider the plant transfer function

$$G_p(s) = \frac{0.5}{s(s+2.5)} .$$

Design a compensator $G_c(s)$ such that $\omega_c = 6$ rad/s and $PM = 50°$.

Solution:

$$\angle[G_p(6j)] = -90° - \tan^{-1}\left(\frac{6}{2.5}\right) = -157.38°$$

and

$$PM_{\text{unc}} = \angle[G_p(6j)] + 180° = -157.38° + 180° = 22.62° .$$

The uncompensated phase margin is less than the desired phase margin. We use a lead compensator to increase the phase at the desired ω_c:

$$\phi_{\max} = PM_{\text{desired}} - PM_{\text{unc}} = 50° - 22.62° = 27.38°,$$

$$\beta = \frac{1 - \sin\phi_{\max}}{1 + \sin\phi_{\max}} = \frac{1 - \sin 27.38°}{1 + \sin 27.38°} = 0.37,$$

$$T = \frac{1}{\omega_c \sqrt{\beta}} = \frac{1}{6\sqrt{0.37}} = 0.274,$$

$$\left| K\left(\frac{0.274s+1}{0.274 \times 0.37 s + 1}\right)\left(\frac{0.5}{s(s+2.5)}\right)\right|_{s=6j} = 1,$$

$$K = \left|\frac{(1+0.274 \times 0.37 \times 6j)6j(2.5+6j)}{(1+0.274 \times 6j) \times 0.5}\right| = 47.43,$$

and

$$G_c(s) = \frac{47.43(0.274s+1)}{(0.274 \times 0.37s + 1)} = \frac{128.2(s+3.65)}{(s+9.86)} .$$

MATLAB Check: margin($Gp*Gc$) yields $PM = 50°$, $\omega_c = 6$ rad/s.

To get a sense of how the lead compensator is working, Fig. 7-14 shows the Bode plots for the uncompensated system, the compensator, and the compensated system. The Bode plots for the uncompensated system are for $KG_p(j\omega)$, where $K = 78$ was chosen so as to place the crossover frequency at 6 rad/s; that is, $|KG_p(j6)| = 1$.

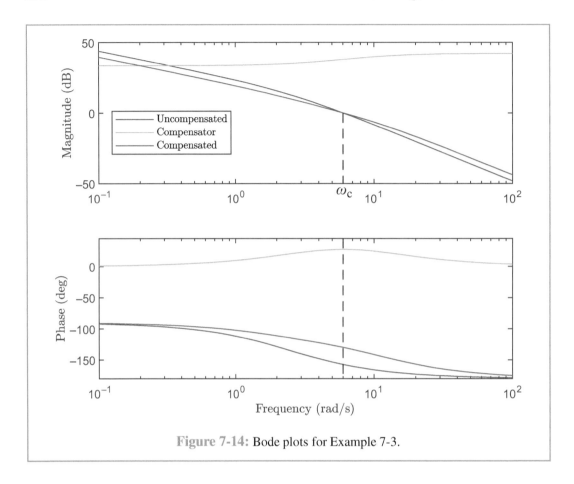

Figure 7-14: Bode plots for Example 7-3.

7-3.2 Design of a Lead Compensator to Increase the Phase Margin without Changing the Low-Frequency Gain

This is a more challenging task than the previous one. Keeping the low-frequency gain means that the gain K in the compensator transfer function

$$G_c(s) = K\left(\frac{Ts+1}{\beta Ts+1}\right)$$

is fixed and we only have the freedom to choose β and T. This situation arises when the design specifications require K_p or K_v to have a certain value, but doing it with $G_c(s) = K$ yields a low value for the phase margin. When we add a lead compensator we need to decide where to locate ω_{max}. It cannot be located at the uncompensated crossover frequency because the magnitude of the compensator is positive, which when added to the uncompensated magnitude plot will shift the crossover frequency to the right; see Fig. 7-15. Because the increase in the magnitude of the compensator at ω_{max} is $10\log(1/\beta)$, we can estimate the compensated crossover frequency as the frequency at which the uncompensated magnitude plot is $-10\log(1/\beta)$. But at this frequency, the angle $G_p(j\omega)$ will typically be more negative

7-3. LEAD COMPENSATION

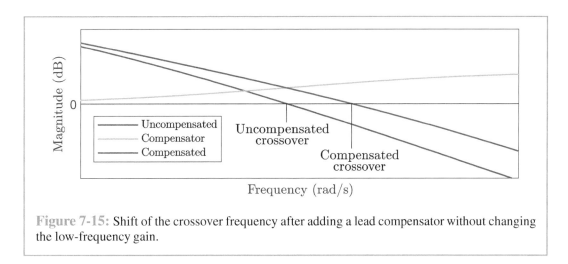

Figure 7-15: Shift of the crossover frequency after adding a lead compensator without changing the low-frequency gain.

than its value at the uncompensated crossover frequency. We can anticipate this reduction in our choice of ϕ_{max} by adding a correction factor.

The design steps for this case are:

Step 1: Choose ϕ_{max}:

$$\phi_{max} = PM_{desired} - PM_{unc} + \text{correction factor}.$$

The correction factor is usually 5 to 10 degrees.

Step 2: Calculate β:

$$\beta = \frac{1 - \sin\phi_{max}}{1 + \sin\phi_{max}}.$$

Step 3: Estimate the new crossover frequency ω_c as the frequency at which the uncompensated magnitude plot is $-10\log(1/\beta)$.

Step 4: Choose T:

$$T = \frac{1}{\omega_c \sqrt{\beta}}.$$

Step 5: Keep K as in the uncompensated design.

We may have to redo the design by changing the correction factor.

Example 7-4: Increasing the Phase Margin without Changing the Low-Frequency Gain

---

Consider the plant transfer function

$$G_p(s) = \frac{0.5}{s(s+2.5)}.$$

(1) Design a compensator $G_c(s) = K$ so that $K_v = 10$. Find the phase margin and the crossover frequency.

(2) Design a lead compensator to increase the phase margin by 20° while keeping $K_v = 10$.

Solution:

(1)

$$K_v = \lim_{s \to 0} s \frac{0.5K}{s(s+2.5)} = \frac{K}{5} = 10 \quad \rightarrow \quad K = 50.$$

MATLAB Check: margin($50 * G_p$) yields $PM = 28°$, $\omega_c = 4.7$ rad/s.

(2) Set 10° as the correction factor

$$\phi_{max} = PM_{desired} - PM_{unc} + \text{correction factor} = 20° + 10° = 30°,$$

$$\beta = \frac{1 - \sin 30°}{1 + \sin 30°} = 0.33,$$

$$10 \log\left(\frac{1}{\beta}\right) = 4.8 \text{ dB}.$$

Figure 7-16 shows the Bode plots of $50G_p(s)$, with the range of axes chosen to focus on the area of interest. To estimate the compensated crossover frequency, we determine the frequency at which the magnitude plot of $50G_p$ is at -4.8 dB. From Fig. 7-16, this frequency is 6.4 rad/s and $\angle[G_p(6.4j)] = -158.6$. This angle is 6.6° lower than the angle at the uncompensated crossover frequency. The correction factor of 10 is expected to work. We continue to work with $\beta = 0.33$ and $\omega_c = 6.4$ rad/s. Hence,

$$T = \frac{1}{\omega_c \sqrt{\beta}} = 0.272 \text{ s}.$$

Keep $K = 50$, which leads to

$$G_c(s) = \frac{50(0.272s + 1)}{(0.272 \times 0.33s + 1)} = \frac{151.5(s + 3.676)}{(s + 11.14)}.$$

MATLAB Check: margin($G_p * G_c$) yields $PM = 51.7°$, $\omega_c = 6.35$ rad/s.

The phase margin has been improved by 23.7°, which is more than the required 20°. It is acceptable to achieve a phase margin slightly higher than what is required because it offers stronger relative stability.

7-4 Lag Compensation

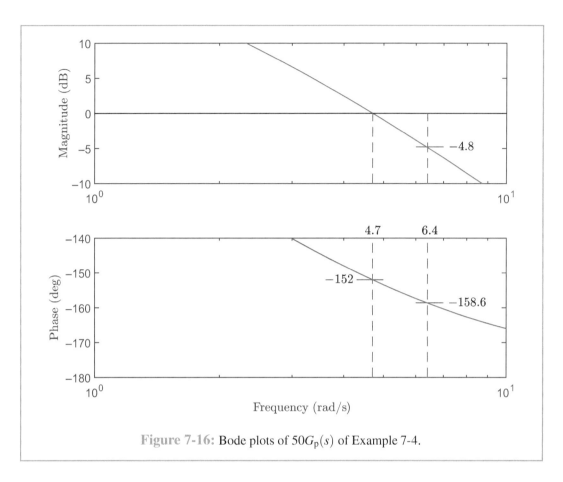

Figure 7-16: Bode plots of $50G_p(s)$ of Example 7-4.

The lag compensator is used to increase the low-frequency gain. Its transfer function is

$$G_c(s) = K\left(\frac{\alpha(Ts+1)}{\alpha Ts+1}\right), \qquad K > 0, \quad T > 0, \quad \alpha > 1.$$

The transfer function has a zero at $-1/T$ and a pole at $-1/(\alpha T)$. Because $\alpha > 1$, the pole is to the right of the zero.

The Bode plots and asymptotic Bode plots of the lag compensator are shown in Fig. 7-17 for $K = 1$. The magnitude plot starts at low frequency from a high gain of $20\log\alpha$ dB and approaches 0 dB as $\omega \to \infty$. Therefore, adding a lag compensator allows us to increase the low-frequency gain over the gain of the uncompensated system. However, the phase of the compensator is negative and reaches a minimum at the geometric mean of the break frequencies $1/T$ and $1/(\alpha T)$, which is $1/(T\sqrt{\alpha})$. This negative phase can compromise the stability of the system by decreasing the phase margin of the uncompensated system.

To reduce this effect, the break frequencies of the lag compensator are chosen at a low frequency relative to the crossover frequency of the uncompensated system. In particular, if

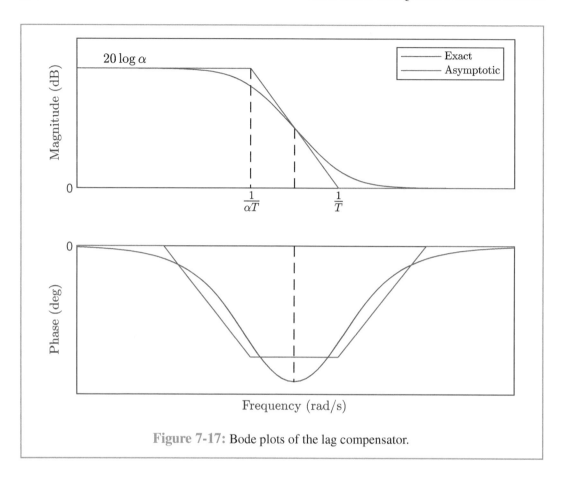

Figure 7-17: Bode plots of the lag compensator.

ω_c is the crossover frequency of the uncompensated system, T is chosen such that

$$0.1\omega_c \leq \frac{1}{T} \leq 0.2\omega_c.$$

Even with this choice there will be a residual negative phase at ω_c. Table 7-2 shows the residual angle at ω_c for different values of α. The table shows the cases when $1/T = 0.1\omega_c$ and $1/T = 0.2\omega_c$. This residual negative angle can be compensated for at the uncompensated design stage. In particular, if during the uncompensated design we anticipate the addition of

Table 7-2: **Residual negative angle of the lag compensator at the uncompensated crossover frequency.**

α	10	20	30	40	50
$\angle[G_c(j\omega_c)]$ (degree) for $1/T = 0.1\omega_c$	-5.1	-5.4	-5.5	-5.57	-5.6
$\angle[G_c(j\omega_c)]$ (degree) for $1/T = 0.2\omega_c$	-10.1	-10.7	-10.9	-11.02	-11.08

7-4. LAG COMPENSATION

a lag compensator, we design for

$$PM = PM_{\text{desired}} + \text{correction factor},$$

and the correction factor is usually 5 to 10 degrees.

7-4.1 Design of a Lag Compensator to Increase the Low-Frequency Gain with Minimal Effect on the Phase Margin of the Uncompensated System

Given an uncompensated system with desired phase margin at crossover frequency ω_c, it is required to increase the low-frequency gain by a factor F. The process involves the following steps:

Step 1: Choose $\alpha \geq F$.

Step 2: Choose T such that $0.1\omega_c \leq \frac{1}{T} \leq 0.2\omega_c$.

Step 3: Keep K as in the uncompensated design.

Example 7-5: Increasing K_p While Maintaining the Phase Margin

Given the plant transfer function

$$G_p(s) = \frac{0.2}{(s+0.2)(s+1)},$$

design a compensator $G_c(s)$ such that $K_p \geq 100$ and $PM \approx 50°$.

Solution: We start by designing $G_c(s) = K$ to achieve the required phase margin. Noting that $G_p(0) = 1$, we anticipate that we will need to use a lag compensator to increase the dc gain to meet the requirement on K_p. With that anticipation, we design for $PM = 55°$. From the phase plot of $G_p(j\omega)$, shown in Fig. 7-18, 55° phase margin is achieved at $\omega_c \approx 1.04$ rad/s.

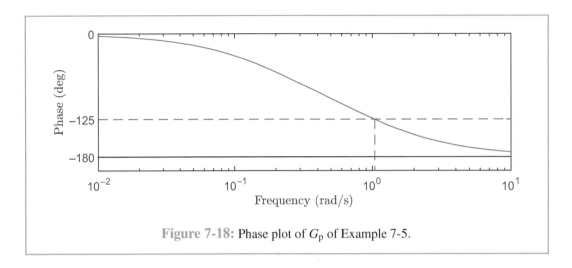

Figure 7-18: Phase plot of G_p of Example 7-5.

$$|KG_p(1.04j)| = 1 \quad \rightarrow \quad \frac{0.2K}{\sqrt{(0.04+1.0816)(1+1.0816)}} = 1 \quad \rightarrow \quad K = 7.64.$$

Uncompensated $K_p = K = 7.64$. Hence, we need to increase K_p by $\frac{100}{7.64} = 13.09$.

Set $\alpha = 14$, $\quad \frac{1}{T} = 0.1\omega_c = 0.104 \quad \rightarrow \quad T = 9.615 \approx 10.$

Keep $K = 7.64$. Hence,

$$G_c(s) = \frac{7.64 \times 14(10s+1)}{140s+1} = \frac{7.64(s+0.1)}{s + \frac{1}{140}},$$

$$K_p = G_p(0)G_c(0) = 7.64 \times 14 = 103.6.$$

MATLAB check: margin($G_p * G_c$) yields $PM = 49.6°$, $\omega_c = 1.04$ rad/s.

Figure 7-19 shows the Bode plots of the uncompensated system ($7.64G_p$), the compensator (G_c), and the compensated system ($G_p G_c$). It is clear how the lag compensator increases the low-frequency gain with little effect on the phase margin of the uncompensated system.

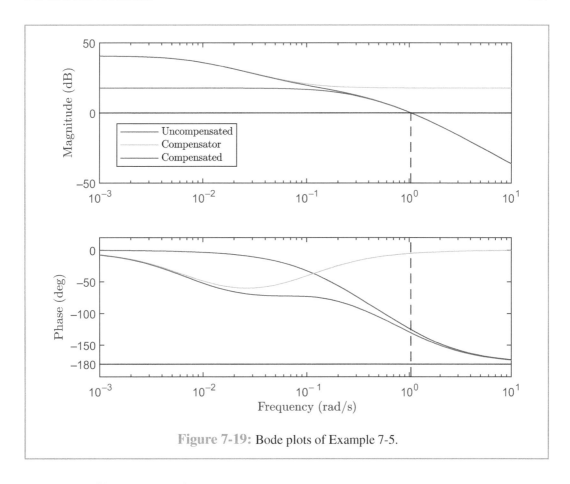

Figure 7-19: Bode plots of Example 7-5.

7-5 PI Compensation

The PI controller increases the low-frequency gain because it increases the type of the feedback system. Its transfer function is

$$G_c(s) = K\left(\frac{Ts+1}{Ts}\right), \qquad K > 0, \quad T > 0.$$

Figure 7-20 shows the exact and asymptotic Bode plots of $G_c(s)$ for $K = 1$.

The magnitude plot starts from infinity at zero frequency and rolls off with -20 dB/dec slope towards the break frequency $1/T$. It approaches 0 dB asymptotically as $\omega \to \infty$. The phase plot starts from $-90°$ and increases to $0°$, passing through $-45°$ at the break frequency.

7-5.1 Design of a PI Controller to Increase the Type of the Feedback System and Achieve a Desired Phase Margin

The design includes the following steps:

Step 1: Choose T.

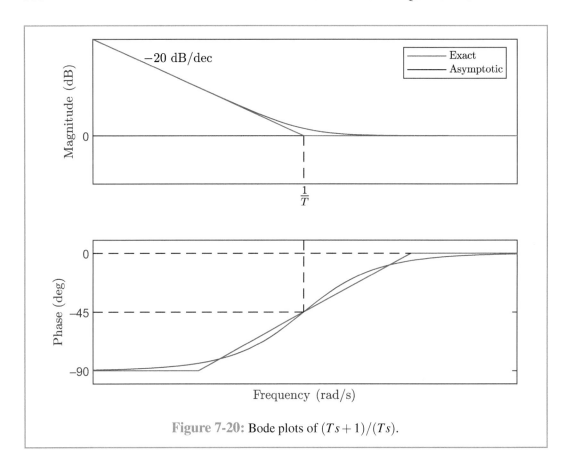

Figure 7-20: Bode plots of $(Ts+1)/(Ts)$.

Step 2: Draw the Bode phase plot of

$$\hat{G}(s) = \left(\frac{Ts+1}{Ts}\right) G_p(s)$$

and find the frequency ω_c at which

$$\angle[\hat{G}(j\omega_c)] = PM_{\text{desired}} - 180°,$$

where PM_{desired} is the desired phase margin.

Step 3: Choose K such that $|K\hat{G}(j\omega_c)| = 1$.

Example 7-6: Design of a PI Controller

Given the plant transfer function

$$G_p(s) = \frac{0.2}{(s+0.2)(s+1)},$$

7-5. PI COMPENSATION

design a compensator $G_c(s)$ such that $K_p \geq 100$ and $PM \approx 50°$.

Solution: We use a PI controller, which yields $K_p = \infty$; so the requirement on K_p is met. To meet the requirement on the phase margin, we choose T, draw the Bode phase plot for

$$\hat{G}(s) = \left(\frac{Ts+1}{Ts}\right) G_p(s)$$

and find the frequency ω_c at which $\angle[\hat{G}(j\omega_c)] = -130°$. Then we calculate K such that $|K\hat{G}(j\omega_c)| = 1$. This process was repeated for different values of T, and the results are shown in Table 7-3. The table shows that the crossover frequency ω_c and the gain K increase as T increases.

Table 7-3: **The effect of the choice of T of the PI controller of Example 7-6.**

T (s)	2	4	6	8	10
ω_c (rad/s)	0.268	0.728	0.914	0.977	1.02
K	0.82	4.42	6.24	6.91	7.39

We pick the case $T = 10$ to compare the PI controller with the lag compensator of Example 7-5. Recall that in Example 7-5, a lag compensator was designed to meet the same design specifications for the same plant as in the current example. The case $T = 10$ results in $\omega_c = 1.02$ rad/s, which is comparable to $\omega_c = 1.04$ rad/s of Example 7-5. The transfer functions are

$$\text{Lag Compensator:} \quad G_c(s) = \frac{7.64(s+0.1)}{s+0.0071},$$

$$\text{PI controller:} \quad G_c(s) = \frac{7.39(s+0.1)}{s}.$$

The main difference is that the pole at -0.0071 in the lag compensator is replaced by a pole at the origin in the PI controller.

Figure 7-21 shows the Bode plots of $G_p G_c$ for the two controllers. The plots are indistinguishable near 1 rad/s, where the crossover frequencies are located, but they are quite different at low frequencies. For the PI controller, the magnitude plot rolls off from infinity and the phase starts from $-90°$, while for the lag compensator the magnitude plot starts flat from a finite value and the phase starts from zero degrees. Implementing the lag compensator with the small coefficient 0.0071 requires high accuracy; an issue that does not arise in the PI controller. On the other hand, the PI controller will likely be implemented with an anti-windup scheme, as discussed in Section 5-7.

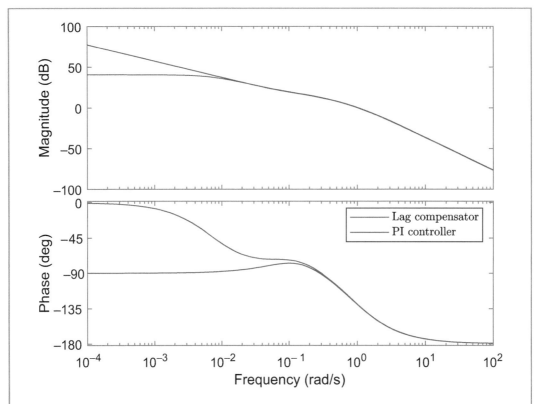

Figure 7-21: Comparison of the Bode plots of the loop gain for the lag compensator and the PI controller of Example 7-6.

7-6 Lead-Lag Compensation

We can improve the phase margin with a lead compensator. We can increase the low-frequency gain with a lag compensator. We can achieve both with a lead-lag compensator. The process is illustrated by the following two examples.

Example 7-7: Design of a Lead-Lag Compensator

Given the plant transfer function

$$G_p(s) = \frac{20}{(s+0.2)(s+1)(s+100)},$$

design a compensator $G_c(s)$ such that

(i) $K_p \approx 100$; (ii) $PO \approx 20\%$; (iii) $T_p \approx 1$ s.

7-6. LEAD-LAG COMPENSATION

Solution: To pursue the design in the frequency domain, we need to convert the requirements (ii) and (iii) into equivalent frequency-domain requirements. Although the transfer function is third-order, the pole at -100 is far to the left in the complex plane relative to the poles at -0.2 and -1. It is reasonable to expect that the closed-loop transfer function will have a pair of dominant complex poles. Therefore, we use the relations of Fig. 6-39 as approximations. For 20% overshoot we set $\zeta = 0.5$. Hence,

$$T_p = \frac{\pi}{\omega_n \sqrt{1-\zeta^2}} = 1 \;\;\longrightarrow\;\; \omega_n = \frac{\pi}{\sqrt{1-0.25}} = 3.63 \text{ rad/s},$$

$$PM = 12 + 76\zeta = 12 + 76 \times 0.5 = 50°,$$

$$\omega_c = \frac{3.35 - 2\zeta}{3}\omega_n = \frac{3.63(3.35 - 2 \times 0.5)}{3} = 2.84 \text{ rad/s}.$$

Hence, the frequency-domain specifications are

(i) $K_p \approx 100$; (ii) $PM \approx 50°$; (iii) $\omega_c \approx 2.84$ rad/s.

We use a lead compensator to meet requirements (ii) and (iii) and, if needed, a lag compensator will be used to meet requirement (i). Anticipating the probable use of a lag compensator, we design the lead compensator for $PM = 55°$. We start by investigating the uncompensated system when $G_c = K$:

$$\angle[G_p(2.84j)] = -\tan^{-1}\left(\frac{2.84}{0.2}\right) - \tan^{-1}(2.84) - \tan^{-1}\left(\frac{2.84}{100}\right)$$

$$= -85.97° - 70.6° - 1.63° = -158.2°.$$

The uncompensated phase margin is

$$PM_{unc} = -158.2° + 180° = 21.8°.$$

A lead compensator is used to increase the phase margin at the desired ω_c. Its transfer function is

$$G_c(s) = K\left(\frac{Ts+1}{\beta Ts+1}\right).$$

The maximum angle of the lead compensator is chosen as

$$\phi_{max} = 55° - 21.8° = 33.2°,$$

$$\beta = \frac{1 - \sin 33.2°}{1 + \sin 33.2°} = 0.29.$$

The maximum angle is located at the desired crossover frequency by the choice:

$$T = \frac{1}{\omega_c \sqrt{\beta}} = \frac{1}{2.84\sqrt{0.29}} = 0.65 \text{ s}.$$

The gain K of the lead compensator is chosen such that $|G_P(2.84j)G_c(2.84j)| = 1$:

$$\left| K\left(\frac{0.65s+1}{0.29 \times 0.65s+1}\right)\left(\frac{20}{(s+0.2)(s+1)(s+100)}\right)\right|_{s=2.84j} = 1 \quad \rightarrow \quad K = 23.17.$$

Thus, the lead compensator is

$$G_c(s) = \frac{23.17(0.65s+1)}{0.65 \times 0.29s+1},$$

$$K_p = G_p(0)\, G_c(0) = 23.17.$$

We need to increase K_p by the factor $100/23.17 = 4.32$. Add the lag compensator

$$\frac{\alpha(T_1s+1)}{\alpha T_1 s+1}.$$

Because the gain of the lag compensator is kept the same as in the previous step, we didn't include a gain K in the above expression. Set $\alpha = 4.5$. To reduce the effect of the negative phase of the lag compensator on the phase margin, set

$$\frac{1}{T_1} = 0.1\omega_c = 0.284 \quad \rightarrow \quad T_1 = 3.52 \text{ s}.$$

Thus, the final compensator is

$$G_c(s) = \left(\frac{23.17(0.65s+1)}{0.65 \times 0.29s+1}\right)\left(\frac{4.5(3.52s+1)}{4.5 \times 3.52s+1}\right) = \frac{79.9(s+1.54)(s+0.28)}{(s+5.31)(s+0.063)},$$

$$K_p = G_p(0)\, G_c(0) = 104.27.$$

MATLAB check:

$$PM = 50.8°, \quad \omega_c = 2.85 \text{ rad/s}, \quad PO = 20.45\%, \quad T_p = 1.19 \text{ s}.$$

$K_p = 104.27$ is slightly higher than the required 100, which is acceptable since the higher K_p is, the smaller is the steady-state error. The overshoot of 20.45% is fairly close to the required 20%. The peak time of 1.19 s is about 20% higher than the required 1 s, which is reasonable in view of the approximations used in the design calculations. If it is necessary to reduce T_p, we can repeat the design with a higher value of ω_c.

Example 7-8: Design of a Lead-Lag Compensator

Given the plant transfer function

$$G_p(s) = \frac{0.2}{(s+0.2)(s+1)},$$

7-6. LEAD-LAG COMPENSATION

design a compensator $G_c(s)$ such that

(i) Loop gain ≥ 100 (i.e., 40 dB) over the frequency band $[0, 0.1]$ rad/s;

(ii) $PM \approx 50°$;   (iii) $\omega_c \approx 3$ rad/s.

Solution: Anticipating that the Bode magnitude plot of $G_p(j\omega) G_c(j\omega)$ will be non-increasing at low frequency, requirement (i) is equivalent to

$$|G_p(0.1j) G_c(0.1j)| \geq 100.$$

We start by designing a lead compensator with

$$G_c(s) = \frac{K(Ts+1)}{\beta Ts+1}$$

to meet requirements (ii) and (iii). Anticipating the use of a lag compensator to increase the low-frequency gain, we design the lead compensator for $PM = 60°$:

$$\angle[G_p(3j)] = -\tan^{-1}(3/0.2) - \tan^{-1}(3) = -157.75°,$$
$$PM_{\text{unc}} = -157.75° + 180° = 22.25°.$$

The maximum angle of the lead compensator is chosen as

$$\phi_{\max} = 60° - 22.25° = 37.75°,$$
$$\beta = \frac{1 - \sin 37.75°}{1 + \sin 37.75°} = 0.24.$$

T is chosen to locate the maximum angle at the desired crossover frequency:

$$T = \frac{1}{\omega_c \sqrt{\beta}} = \frac{1}{3\sqrt{0.24}} = 0.68 \text{ s}.$$

The gain K is chosen such that $|G_p(3j) G_c(3j)| = 1$:

$$\left|\left(\frac{K(0.68s+1)}{0.68 \times 0.24s+1}\right)\left(\frac{0.2}{(s+0.2)(s+1)}\right)\right|_{s=3j} = 1 \;\rightarrow\; K = 23.2,$$

$$|G_p(0.1j) G_c(0.1j)| = 20.69.$$

We use a lag compensator to increase the loop gain at $\omega = 0.1$ rad/s by the factor $100/20.69 = 4.833$. The transfer function of the lag compensator is

$$\frac{\alpha(T_1 s + 1)}{\alpha T_1 s + 1}.$$

We did not include a gain K in the lag compensator's transfer function because the gain is kept the same as in the lead compensator:

$$\left|\frac{\alpha(T_1 s+1)}{\alpha T_1 s+1}\right|_{s=0.1j} = \alpha \sqrt{\frac{1+(0.1 T_1)^2}{1+(0.1 \alpha T_1)^2}}.$$

To reduce the effect of the lag compensator's negative angle on the phase margin, set

$$\frac{1}{T_1} = 0.1\omega_c = 0.3 \quad \rightarrow \quad T_1 = \frac{10}{3} \text{ s}.$$

We need to choose α such that

$$\alpha \sqrt{\frac{1+(\frac{1}{3})^2}{1+(\frac{\alpha}{3})^2}} > 4.833,$$

but this choice is not feasible because the left-hand side of the foregoing inequality is a monotonically increasing function of α that reaches a maximum value of 3.16 as $\alpha \to \infty$. We bring the break frequency $1/T_1$ closer to ω_c by setting

$$\frac{1}{T_1} = 0.5 \quad \rightarrow \quad T_1 = 2 \text{ s}.$$

We need to choose α such that

$$\alpha \sqrt{\frac{1+(0.2)^2}{1+(0.2\alpha)^2}} > 4.833.$$

The inequality is satisfied with $\alpha = 16$ for which the left-hand side is 4.867. Thus, the final compensator is

$$G_c(s) = \left(\frac{23.2(0.68s+1)}{(0.24 \times 0.68s+1)}\right)\left(\frac{16(2s+1)}{16 \times 2s+1}\right) = \frac{96.67(s+1.47)(s+0.5)}{(s+6.13)(s+0.031)}.$$

MATLAB check:

$$PM = 51.1°; \quad \omega_c = 3.02 \text{ rad/s}; \quad 20\log[G_pG_c] = 40 \text{ dB at } \omega = 0.1 \text{ rad/s}.$$

7-7 Advantages of the Frequency-Domain Approach

Controllers can be designed in the time domain, as in Chapter 5, or in the frequency domain as in this chapter. Time-domain analysis and design have the advantage that the performance can be checked by performing experiments on the system; for example, the step response can be determined experimentally by applying a step input to the closed-loop system and recording the output. What are the advantages of the frequency-domain analysis and design approach? There are four advantages:

- The frequency response can be obtained directly from experimental data. For a wide class of systems that do not have poles in the right-half plane, the frequency response at a frequency ω can be obtained by applying a sinusoidal input of frequency ω, measuring the sinusoidal output at the same frequency, and using the change in magnitude and phase to find the frequency response at that frequency. By sweeping the frequency of the input over the frequency range of interest, the frequency response is determined. We can then fit the response with a transfer function $G_p(s)$. This fitting process involves an approximation. Consequently, the frequency response obtained from $G_p(s)$ will not be exactly the same as the one determined experimentally. In time-domain analysis and design, our only option is to use the approximate transfer function $G_p(s)$. In frequency-domain analysis and design, we can perform the calculations using the experimentally determined frequency response, instead of the approximate frequency response calculated from $G_p(s)$.

- Time delay can be easily handled using Bode plots. In time-domain analysis and design, transfer functions have to be rational functions of s. Therefore, the only way to include the time-delay transfer function $e^{-\tau s}$ is to approximate it by a rational function of s, as discussed in Section 2-1. In addition to being an approximation, the rational transfer function increases the order of the system, which increases the complexity of the analysis and design.

- Robustness to model uncertainty is quantified by the gain and phase margins.

- Reference tracking, disturbance attenuation, and measurement of noise attenuation can be improved by shaping the frequency response of the loop transfer function $G = G_p G_c$.

Summary

Concepts
- Gain adjustment can meet only one design specification.
- The lead compensator can stabilize the system and improve the dynamic response because it adds a positive phase to the loop transfer function.
- The lag compensator can reduce the steady-state error because it increases the low-frequency gain.
- The lead-lag compensator combines the functions of the lead and lag compensators.
- The PI controller reduces the steady-state error because it increases the type of the feedback system.
- Meeting the conflicting requirements of increasing the loop gain to reduce the tracking error due to reference and disturbance and decreasing the sensitivity function, while reducing the loop gain to attenuate the effect of measurement noise, can be handled in the frequency domain by having high loop gain in a low frequency band that covers the frequencies of the reference and disturbance and then letting the gain roll off to low values at high frequency.

Mathematical Models

Lead compensator: $G_c(s) = K \dfrac{Ts+1}{\beta Ts+1}$, $\quad K>0, \quad T>0, \quad 0<\beta<1$

Lead phase relations: $\omega_{max} = \dfrac{1}{T\sqrt{\beta}}$

$$\sin\phi_{max} = \dfrac{1-\beta}{1+\beta}$$

$$\beta = \dfrac{1-\sin\phi_{max}}{1+\sin\phi_{max}}$$

Lag compensator: $G_c(s) = K\alpha \dfrac{Ts+1}{\alpha Ts+1}$, $\quad K>0, \quad T>0, \quad \alpha>1$

PI controller: $G_c(s) = K \dfrac{Ts+1}{Ts}$, $\quad K>0, \quad T>0$

Important Terms Provide definitions or explain the meaning of the following terms:

gain adjustment   lead compensator   phase relations of the lead compensator
lag compensator   lead-lag compensator   PI controller

PROBLEMS

> In all problems where the analytic expression of G_p is given and you are asked to design a compensator to meet certain design specifications, use MATLAB to check if you achieved the design objective.

7.1 Let
$$G(s) = \dfrac{K}{s(s+1)}.$$

(a) Find K such that $\omega_c = 2$ rad/s.
(b) Find K such that $PM = 50°$.
(c) Find K such that $K_v = 10$.
(d) Find K such that
$$20\log|G(j\omega)| \geq 20 \text{ dB} \quad \text{for } \omega \in [0,1] \text{ rad/s}.$$

7.2 Let
$$G(s) = \dfrac{6K}{(s+1)(s+2)(s+3)}.$$

(a) Find K such that $\omega_c = 2$ rad/s.
(b) Find K such that $PM = 60°$.
(c) Find K such that $K_p = 5$.
(d) Find K such that

$$20\log|G(j\omega)| \geq 10 \text{ dB} \quad \text{for } \omega \in [0, 0.3] \text{ rad/s}.$$

7.3 Let $G(s) = K\hat{G}(s)$. The Bode plots of $\hat{G}(s)$ are shown in Fig. P7.3.
 (a) Find K such that $\omega_c = 10$ rad/s.
 (b) Find K such that $PM = 50°$.
 (c) Find K such that $K_v = 20$.
 (d) Find K such that

$$20\log|G(j\omega)| \geq 30 \text{ dB} \quad \text{for } \omega \in [0, 1] \text{ rad/s}.$$

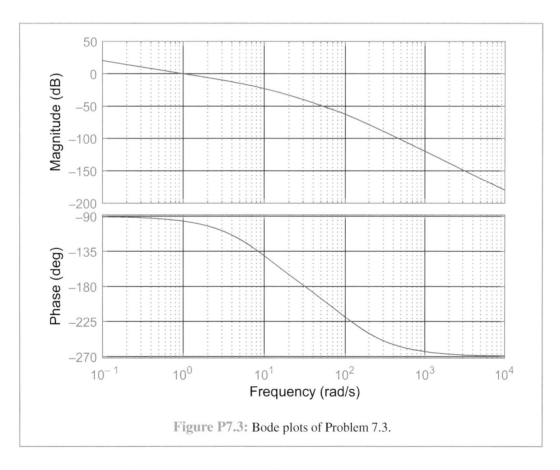

Figure P7.3: Bode plots of Problem 7.3.

7.4 Let $G(s) = K\hat{G}(s)$. The Bode plots of $\hat{G}(s)$ are shown in Fig. P7.4.

(a) Find K such that $\omega_c = 3$ rad/s.

(b) Find K such that $PM = 45°$.

(c) Find K such that $K_p = 10$.

(d) Find K such that

$$20\log|G(j\omega)| \geq 14 \text{ dB} \quad \text{for } \omega \in [0,1] \text{ rad/s}.$$

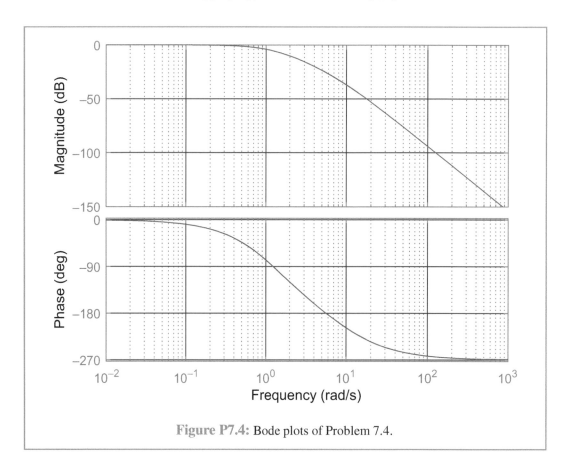

Figure P7.4: Bode plots of Problem 7.4.

7.5 Given the plant transfer function

$$G_p(s) = \frac{1}{s(s+1)(s+2)},$$

design a lead compensator such that $\omega_c = 1$ rad/s and $PM = 50°$.

7.6 Given the plant transfer function

$$G_p(s) = \frac{1}{(s+1)(s+2)},$$

design a lead compensator such that $\omega_c = 10$ rad/s and $PM = 50°$.

PROBLEMS

7.7 Given the plant transfer function

$$G_p(s) = \frac{100}{(s+1)(s+5)(s+10)},$$

design a compensator such that $\omega_c = 5$ rad/s and $PM = 45°$.

7.8 Given the plant transfer function

$$G_p(s) = \frac{100}{s(s+5)(s+10)},$$

design a compensator such that $\omega_c = 5$ rad/s and $PM = 45°$.

7.9 Consider the plant transfer function

$$G_p(s) = \frac{1}{s(s+1)(s+2)}.$$

(a) Design a compensator $G_c(s) = K$ so that $K_v = 2$. Find the phase margin and the crossover frequency.

(b) Design a lead compensator to increase the phase margin by $10°$ while keeping $K_v = 2$.

7.10 Consider the plant transfer function

$$G_p(s) = \frac{1}{(s+1)(s+2)}.$$

(a) Design a compensator $G_c(s) = K$ so that $K_p = 20$. Find the phase margin and the crossover frequency.

(b) Design a lead compensator to increase the phase margin by $30°$ while keeping $K_p = 20$.

7.11 Consider the plant transfer function

$$G_p(s) = \frac{1}{(s+1)(s+2)}.$$

(a) Design a compensator $G_c(s) = K$ so that $PM = 50°$. What is K_p?

(b) Design a lag compensator to increase K_p by a factor of 3 with little effect on the phase margin.

7.12 Consider the plant transfer function

$$G_p(s) = \frac{1}{s(s+1)(s+2)}.$$

(a) Design a compensator $G_c(s) = K$ so that $PM = 50°$. What is K_v?

(b) Design a lag compensator to increase K_v by a factor of 2 with little effect on the phase margin.

7.13 Given the plant transfer function

$$G_p(s) = \frac{100}{s(s+5)(s+10)},$$

design a lag compensator such that $K_v = 10$ and $PM \approx 45°$.

7.14 Given the plant transfer function

$$G_p(s) = \frac{100}{(s+1)(s+5)(s+10)},$$

design a lag compensator such that $K_p = 10$ and $PM \approx 45°$.

7.15 Consider the plant transfer function

$$G_p(s) = \frac{2}{s(s+1)(s+2)}.$$

(a) Find the gain and phase margins and the corresponding crossover frequencies when $G_c(s) = 1$.

(b) Let $G_c(s) = K$, where $K > 0$. Find the range of K for stability.

(c) Let $G_c(s) = 1$ and replace $G(s)$ by $e^{-Ts}G(s)$, where $T > 0$. Find the range of T for stability.

(d) Design $G_c(s) = K > 0$ to operate at 45° phase margin.

(e) Design $G_c(s)$ to operate at 40° phase margin and $K_v = 3$.

(f) Design $G_c(s)$ to operate at 45° phase margin and 1 rad/s crossover frequency.

7.16 Consider the plant transfer function

$$G_p(s) = \frac{1}{(2s+1)^2}.$$

Design a compensator to achieve 50° phase margin at 4 rad/s.

7.17 Consider the plant transfer function

$$G_p(s) = \frac{1}{s(s+1)(s+10)}.$$

(a) Design a compensator $G_c = K$ to operate at $K_v = 10$. What is the phase margin?

(b) Design a compensator $G_c = K$ to operate at 45° phase margin. What is K_v?

(c) Design a compensator $G_c(s)$ to operate at 45° phase margin and 4 rad/s crossover frequency. What is K_v?

(d) Design a compensator $G_c(s)$ to operate at 45° phase margin and $K_v = 10$.

PROBLEMS

7.18 Consider the plant transfer function

$$G_p(s) = \frac{50(s+4)}{(s+1)(s+10)(s+20)}.$$

(a) Design a lag compensator such that the system operates with a 45° phase margin and $K_p = 100$.

(b) Design a compensator to meet the following specifications: Phase margin = 45°, crossover frequency = 40 rad/s, and $K_p = 100$.

7.19 Consider the plant transfer function

$$G_p(s) = \frac{10}{s(s+1)(s+10)}.$$

(a) Find the gain and phase margins and the corresponding crossover frequencies when $G_c = 1$.

(b) Design a compensator to operate at crossover frequency $\omega_c = 2$ rad/s. What is the phase margin PM?

(c) Design a compensator to operate at $PM = 60°$ and $\omega_c = 2$ rad/s. What is K_v?

(d) Design a compensator to operate at $PM = 55°$, $\omega_c = 2$ rad/s, and $K_v = 10$.

7.20 Consider the plant transfer function

$$G_p(s) = \frac{1}{s(s+1)(s+5)}.$$

(a) Design a compensator $G_c = K$ to operate at $K_v = 3$. What is the phase margin?

(b) Design a compensator $G_c = K$ to operate at 50° phase margin. What is K_v?

(c) Design a compensator $G_c(s)$ to operate at 45° phase margin and 2 rad/s crossover frequency. What is K_v?

(d) Design a compensator $G_c(s)$ to operate at 45° phase margin and $K_v = 3$.

7.21 Repeat Problem 7.20 for

$$G_p(s) = \frac{1}{s(s+2)(s+6)}.$$

7.22 Consider the plant transfer function

$$G_p(s) = \frac{10}{s(s+10)}.$$

Design a lead-lag compensator to have phase margin $\approx 60°$, crossover frequency ≈ 20 rad/s, and loop gain ≥ 100 (40 dB) over the frequency band $[0, 0.5]$ rad/s.

7.23 Repeat Problem 7.22 for

$$G_p(s) = \frac{500}{s(s+10)(s+50)}.$$

7.24 Given the plant transfer function

$$G_p(s) = \frac{1}{(s+1)(s+2)},$$

design a PI controller such that $PM = 60°$ and $\omega_c \geq 1$ rad/s.

7.25 Given the plant transfer function

$$G_p(s) = \frac{1}{(s+1)(s+10)},$$

design a PI controller such that $PM = 50°$ and $\omega_c \geq 2$ rad/s.

7.26 The track-following control in optical disc drives locates the laser spot at the center of a data track [25]. Figure P7.26(a) shows a schematic diagram of the mechanism. A dual-stage actuator consists of a fine actuator and a coarse actuator. The coarse actuator moves the fine actuator slowly over the operation range, while the fine actuator's function is to achieve the desired tracking. In this problem we consider the design of the fine actuator. Figure P7.26(b) shows a block diagram of the control system. The transfer function of the fine actuator is given by

$$P(s) = \frac{234}{s^2 + 194.18s + 147119} \quad \text{(meters/volt)},$$

and $K_{PD} = 16.98 \times 10^6$ (volts/meter) is the photodetector gain. The disturbance d is the center position of the target track, which varies periodically with the disc rotational frequency. To meet the stability and robustness requirements and attenuate the effect of the disturbance, the design specifications are taken as:

- Gain margin ≥ 20 dB;

- Phase margin $\geq 60°$;

- Loop gain ≥ 50 dB for $\omega \leq 100$ rad/s;

- Loop gain ≤ -20 dB for $\omega \geq 10^5$ rad/s.

Design the compensator $G_c(s)$ to meet these specifications.

7.27 Consider the idle speed control of Problem 3.16 together with the linearized model of Fig. P3.16. The time delay is modeled by the transfer function $H(s) = e^{-0.04s}$. Consider the PID controller

$$\Delta U = \left(K_P + \frac{K_I}{s} + K_D s\right)(\Delta N_r - \Delta N),$$

where N_r is the desired engine speed and ΔN_r is the change from the nominal value. Design the PID controller to achieve $60°$ phase margin at 12 rad/s crossover frequency. (*Hint:* Design a PD controller first to achieve $65°$ phase margin at the desired crossover frequency. Then add integration such that the phase margin drop is less than five degrees.)

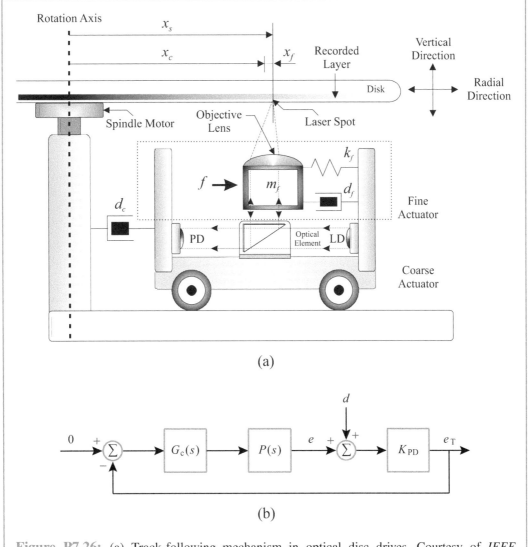

Figure P7.26: (a) Track-following mechanism in optical disc drives. Courtesy of *IEEE Transactions on Systems Technology*. (b) Feedback control loop for Problem 7.20.

7.28 A process in the pulp and paper industry is modeled by the transfer function [5]

$$G_p(s) = \frac{K_1 e^{-\tau s}}{\tau_1 s + 1},$$

where nominal values of the parameters are $K_1 = 1$, $\tau_1 = 3$ s, and $\tau = 0.2$ s. It is common in this industry to design the controller using the Internal Model Principle, where the controller cancels the process dynamics, process poles with controller zeros, and process zeros with controller poles. For the given process, a PI controller is taken as

$$G_c(s) = \frac{K(\tau_1 s + 1)}{s}.$$

(a) Choose K to achieve 60° phase margin. What is the gain margin?

(b) Study the effect of ±20% change in each of K_1, τ_1, and τ on the phase and gain margins and the corresponding crossover frequencies. Take the changes in the parameters one at a time. The controller parameters are kept at the nominal values.

(c) Comment on the results.

7.29 A process in the pulp and paper industry is modeled by the transfer function [5]

$$G_p(s) = \frac{K_1(\beta s + 1)}{s(\tau_1 s + 1)},$$

where nominal values of the parameters are $K_1 = 0.005$, $\tau_1 = 20$ s, and $\beta = 300$ s. It is common in this industry to design the controller using the Internal Model Principle, where the controller cancels the process dynamics, process poles with controller zeros, and process zeros with controller poles. For the given process, a PID controller plus filter is taken as

$$G_c(s) = \left(\frac{K(\tau_1 s + 1)(s + z)}{s}\right)\left(\frac{1}{\beta s + 1}\right).$$

(a) Choose z and K to have a 60° phase margin at $\omega_c = 3$ rad/s. What is the gain margin?

(b) Study the effect of ±20% change in each of K_1, τ_1 and β on the phase and gain margins and the corresponding crossover frequencies. Take the changes in the parameters one at a time. The controller parameters are kept at the nominal values.

(c) Comment on the results.

7.30 A closed-loop insulin pump system is described in Problem 4.30 and Fig. P4.30. The process is modeled by

$$G_p(s) = \frac{K_o}{(\tau_1 s + 1)(\tau_2 s + 1)^2},$$

where $\tau_1 = 247$ (min), $\tau_2 = 17$ (min), and $K_o = -12000/\text{TDI}$. A PID controller is taken as

$$G_c(s) = K\frac{(\tau_1 s + 1)(\tau_2 s + 1)}{s}$$

where the zeros of $G_c(s)$ are chosen to cancel two poles of $G_p(s)$.

(a) Choose K in terms of K_o to achieve 60° phase margin. Determine the crossover frequency (rad/s).

(b) Study the effect of ±20% change in K_o and ±10% change in each of τ_1 and τ_2 on the phase margin and crossover frequency. Take the changes in the parameters one at a time. The controller parameters are kept at the nominal values.

(c) Comment on the results.

7.31 State three advantages of the frequency-domain approach over the time-domain approach.

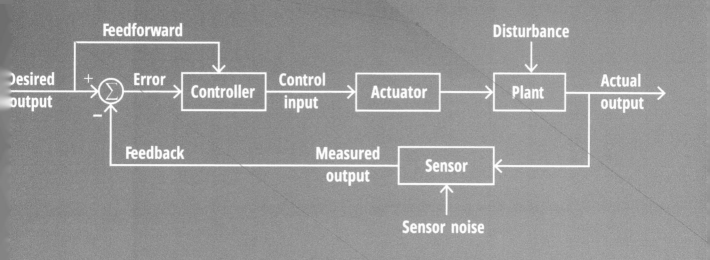

Chapter 8
State-Space Methods

Chapter Contents

    Overview, 318
8-1 State-Space Models, 318
8-2 Solution of the State Equation, 327
8-3 Stability, 330
8-4 Controllability and Observability, 333
8-5 Transfer Function of a State-Space Model, 340
8-6 State-Space Model of a Transfer Function, 344
8-7 State-Space Design, 352
    Chapter Summary, 367
    Problems, 368

Objectives

Upon learning the material presented in this chapter, you should be able to:

1. Represent linear systems by state-space models.
2. Derive state-space models of electric circuits, mechanical systems, and electromechanical systems.
3. Derive a linear state-space model by linearization of a nonlinear model.
4. Solve the linear state equation.
5. Define and test stability.
6. Define and test controllability.
7. Define and test observability.
8. Understand the duality between controllability and observability.
9. Find the transfer function of a state-space model.
10. Find a state-space model of a transfer function.
11. Design state feedback control.
12. Design observer and output feedback control.
13. Design a set-point regulator using feedforward or integral control.

Overview

State-space models present an alternative approach to modeling dynamical systems. They are well suited to deal with multi-input–multi-output systems as well as nonlinear systems. Even though we do not cover multi-input–multi-output cases in this book, state-space methods still provide useful tools for the analysis and design of single-input–single-output linear systems. They emphasize both the internal behavior of the system, due to non-zero initial conditions, and the external behavior due to inputs. They also provide new techniques for the design of feedback control.

This chapter starts by defining *state-space models* in Section 8-1, together with examples of how to derive such models for physical systems. The *solution of the state equation* is presented in Section 8-2, which is the basis for studying *stability* in Section 8-3. *Controllability* and *observability* properties of state-space models play important roles in the design of feedback control. They are presented in Section 8-4.

Because state-space models and transfer functions are two different ways to model dynamical systems, it is important to understand how they are related. This relationship is examined in Sections 8-5 and 8-6. Section 8-5 shows how to calculate the transfer function of a state-space model. Section 8-6 shows how to find a state-space model of a transfer function. For the same transfer function, several different state-space models can be created. The section presents three such models: the *controllable-canonical-form realization*, the *observable-canonical-form realization*, and the *parallel realization*. Section 8-7 presents the design of feedback control using state-space models.

8-1 State-Space Models

A state-space model of an *n*th order dynamical system is defined by *n* coupled first-order differential equations:

$$\dot{x}_1 = f_1(x_1, x_2, \ldots, x_n, u) ,$$
$$\dot{x}_2 = f_2(x_1, x_2, \ldots, x_n, u) ,$$
$$\vdots = \vdots$$
$$\dot{x}_n = f_n(x_1, x_2, \ldots, x_n, u) ,$$

where x_1 to x_n are the state variables, u is the input variable, $\dot{x}_i = dx_i/dt$, t is the time variable, and f_1 to f_n are continuous functions of their arguments.[1]

It is common practice to write this model in a more compact form by using vector notation. Define the vectors

$$x = \begin{bmatrix} x_1 \\ x_2 \\ \vdots \\ x_n \end{bmatrix}, \quad \dot{x} = \begin{bmatrix} \dot{x}_1 \\ \dot{x}_2 \\ \vdots \\ \dot{x}_n \end{bmatrix}, \quad f(x,u) = \begin{bmatrix} f_1(x_1, x_2, \ldots, x_n, u) \\ f_2(x_1, x_2, \ldots, x_n, u) \\ \vdots \\ f_n(x_1, x_2, \ldots, x_n, u) \end{bmatrix},$$

---

[1] In nonlinear analysis it is usual to require the functions f_1 to f_n to have stronger properties than continuity. For example, it is common to require their partial derivatives with respect to their arguments to be continuous. See, for example, [21].

8-1. STATE-SPACE MODELS

and rewrite the state-space model as the vector equation

$$\dot{x} = f(x,u) \,. \tag{8.1}$$

Equation (8.1) is called the *state equation*. Together, with the state equation, there is usually an *output equation* that defines the output variable y as a function of the state and input variables:

$$y = h(x,u) \,, \tag{8.2}$$

where h is a continuous function of its arguments. Together, Eqs. (8.1) and (8.2) constitute the *state-space model* of the system.

For linear systems, the functions f and h depend linearly on their arguments. This allows us to write the right-hand side of Eqs. (8.1) and (8.2) as products of matrices and vectors. We illustrate this for the case $n = 3$. Let

$$\begin{aligned} f_1(x,u) &= a_{11}x_1 + a_{12}x_2 + a_{13}x_3 + b_1 u \,, \\ f_2(x,u) &= a_{21}x_1 + a_{22}x_2 + a_{23}x_3 + b_2 u \,, \\ f_3(x,u) &= a_{31}x_1 + a_{32}x_2 + a_{33}x_3 + b_3 u \,, \end{aligned} \tag{8.3}$$

where a_{ij} and b_i are constant coefficients. Then

$$\underbrace{\begin{bmatrix} \dot{x}_1 \\ \dot{x}_2 \\ \dot{x}_3 \end{bmatrix}}_{\dot{x}} = \underbrace{\begin{bmatrix} a_{11} & a_{12} & a_{13} \\ a_{21} & a_{22} & a_{23} \\ a_{31} & a_{32} & a_{33} \end{bmatrix}}_{A} \underbrace{\begin{bmatrix} x_1 \\ x_2 \\ x_3 \end{bmatrix}}_{x} + \underbrace{\begin{bmatrix} b_1 \\ b_2 \\ b_3 \end{bmatrix}}_{B} u \,. \tag{8.4}$$

Performing the multiplications Ax and Bu and the addition of these two vectors in Eq. (8.4) produces Eq. (8.3). The matrices A and B are constructed from the coefficients of Eq. (8.3). The first row of A and B is taken from the coefficients of f_1, the second row from the coefficients of f_2, and so on. The coefficient of x_1 in f_1 is the first element of the first row of A, the coefficient of x_2 is the second element of the first row of A, the coefficient of x_3 is the third element of the first row of A, and the coefficient of u is the first row of B. If any of the variables is missing from the expression of f_1, its coefficient is set to zero. With the definition of matrices A and B, Eq. (8.1) can be written as

$$\dot{x} = Ax + Bu \,.$$

Similarly, for the output equation, if

$$h = c_1 x_1 + c_2 x_2 + c_3 x_3 + du \,,$$

the equation can be written as

$$y = \underbrace{\begin{bmatrix} c_1 & c_2 & c_3 \end{bmatrix}}_{C} x + \underbrace{d}_{D} u \,.$$

In summary, the state-space model for linear systems takes the form

$$\dot{x} = Ax + Bu , \qquad (8.5a)$$
$$y = Cx + Du . \qquad (8.5b)$$

For a single-input–single-output system of order n, the dimensions of the matrices A, B, C and D are $n \times n$, $n \times 1$, $1 \times n$, and 1×1, respectively.[2]

State-space models can be obtained from transfer functions, as we shall see in Section 8-6, or they can be written directly from the physical laws that govern the system. In the latter approach, the first step for finding a state-space model is to choose the state variables of the system. For different physical systems there are common choices of state variables. For electric circuits, the number of state variables is the number of the energy-storing elements: the capacitors and inductors.[3] A typical choice of state variables is the voltage across the capacitor and the current through the inductor. This choice is convenient because the terminal relations for the capacitor and inductor are

$$i_C = C \frac{dv_C}{dt} \quad \text{and} \quad v_L = L \frac{di_L}{dt} .$$

Choosing the state variable as $x_i = v_C$, the derivative $\dot{x}_i = i_C/C$, and i_C can be obtained from the application of Kirchoff's voltage or current laws. Similarly, for the inductor, taking $x_i = i_L$, $\dot{x}_i = v_L/L$.

For mechanical systems, the equations of motion are obtained by application of Newton's second law to each mass or rotating mass. For translational motion, the equation takes the form

$$M\ddot{y} = \text{net force}.$$

Setting the state variables as $x_1 = y$ and $x_2 = \dot{y}$, the state equations become readily available as

$$\dot{x}_1 = x_2 , \qquad \dot{x}_2 = \frac{\text{net force}}{M} .$$

For rotational motion, the equation takes the form

$$J\ddot{\theta} = \text{net torque}.$$

Setting $x_1 = \theta$ and $x_2 = \dot{\theta}$ results in

$$\dot{x}_1 = x_2 , \qquad \dot{x}_2 = \frac{\text{net torque}}{J} .$$

For mechanical systems, the number of state variables is twice the number of degrees of freedom; that is, twice the number of masses in translational motion or twice the number of rotating masses in rotational motion.

---

[2] State-space models can be easily written for multi-input–multi-output systems. When there are m inputs and q outputs, u and y are vectors of dimensions m and q, respectively, and the matrices $B, C,$ and D have the dimensions $n \times m$, $q \times n$, and $q \times m$. Our discussion in this chapter is limited to single-input–single-output systems.

[3] There are special cases where this rule does not apply. For example, if three capacitors are connected in a delta connection, the voltage of one capacitor is dependent on the voltages of the other two capacitors. In this case, the delta connection can be represented by only two state variables.

8-1.1 Examples

Example 8-1: Series RLC Circuit

Consider the electric circuit of Example 2-4. The circuit is shown again in Fig. 8-1. Find a state-space model with input v and output i.

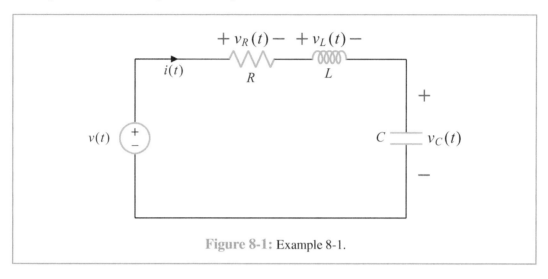

Figure 8-1: Example 8-1.

Solution: The state variables are the current i through the inductor and the voltage v_C across the capacitor; that is,

$$x_1 = i, \qquad x_2 = v_C.$$

The input and output are

$$u = v, \qquad y = i.$$

Application of Kirchoff's voltage law yields

$$v = Ri + v_L + v_C.$$

The state equations are given by

$$\dot{x}_1 = \frac{di}{dt} = \frac{1}{L} v_L = \frac{1}{L}(-Ri - v_C + v) = -\frac{R}{L} x_1 - \frac{1}{L} x_2 + \frac{1}{L} u \tag{8.6a}$$

and

$$\dot{x}_2 = \frac{dv_C}{dt} = \frac{1}{C} i = \frac{1}{C} x_1. \tag{8.6b}$$

The output equation is given by $y = i = x_1$. The quadruple $\{A, B, C, D\}$ that defines the state-space model is given by

$$A = \begin{bmatrix} -\frac{R}{L} & -\frac{1}{L} \\ \frac{1}{C} & 0 \end{bmatrix}, \qquad B = \begin{bmatrix} \frac{1}{L} \\ 0 \end{bmatrix}, \qquad C = \begin{bmatrix} 1 & 0 \end{bmatrix}, \qquad D = 0.$$

Example 8-2: Automobile Suspension

Consider the automobile suspension system of Example 2-11. Find a state-space model with input s and output y_s. The free-body diagram is shown again in Fig. 8-2 with the variables written in the time domain.

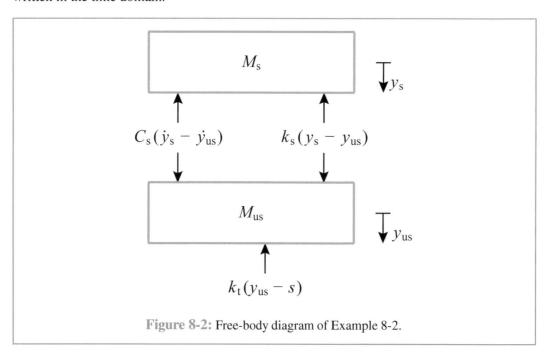

Figure 8-2: Free-body diagram of Example 8-2.

Solution: Application of Newton's second law to the two masses results in

$$M_s \ddot{y}_s = -k_s(y_s - y_{us}) - C_s(\dot{y}_s - \dot{y}_{us}) \tag{8.7a}$$

and

$$M_{us} \ddot{y}_{us} = k_s(y_s - y_{us}) + C_s(\dot{y}_s - \dot{y}_{us}) - k_t(y_{us} - s) . \tag{8.7b}$$

Assign the state variables as follows:

$$x_1 = y_s, \qquad x_2 = \dot{y}_s, \qquad x_3 = y_{us}, \qquad x_4 = \dot{y}_{us} .$$

The input and output variables are

$$u = s, \qquad y = y_s .$$

The state equations are

$$\dot{x}_1 = \dot{y}_s = x_2 ,$$

8-1. STATE-SPACE MODELS

$$\dot{x}_2 = \ddot{y}_s = \frac{1}{M_s}[-k_s(y_s - y_{us}) - C_s(\dot{y}_s - \dot{y}_{us})]$$

$$= -\frac{k_s}{M_s}(x_1 - x_3) - \frac{C_s}{M_s}(x_2 - x_4),$$

$$\dot{x}_3 = \dot{y}_{us} = x_4,$$

and

$$\dot{x}_4 = \ddot{y}_{us} = \frac{1}{M_{us}}[k_s(y_s - y_{us}) + C_s(\dot{y}_s - \dot{y}_{us}) - k_t(y_{us} - s)]$$

$$= \frac{k_s}{M_{us}}(x_1 - x_3) + \frac{C_s}{M_{us}}(x_2 - x_4) - \frac{k_t}{M_{us}}(x_3 - u).$$

The output equation is

$$y = y_s = x_1.$$

The matrices A, B, C, D are given by

$$A = \begin{bmatrix} 0 & 1 & 0 & 0 \\ -k_s/M_s & -C_s/M_s & k_s/M_s & C_s/M_s \\ 0 & 0 & 0 & 1 \\ k_s/M_{us} & C_s/M_{us} & -(k_s+k_t)/M_{us} & -C_s/M_{us} \end{bmatrix}, \quad B = \begin{bmatrix} 0 \\ 0 \\ 0 \\ k_t/M_{us} \end{bmatrix},$$

$$C = \begin{bmatrix} 1 & 0 & 0 & 0 \end{bmatrix}, \quad D = 0.$$

Example 8-3: dc Motor

Find a state-space model of the armature-controlled dc motor of Section 2-2.3 with input v_a and output ω_m. Figure 2-29 is reproduced here as Fig. 8-3.

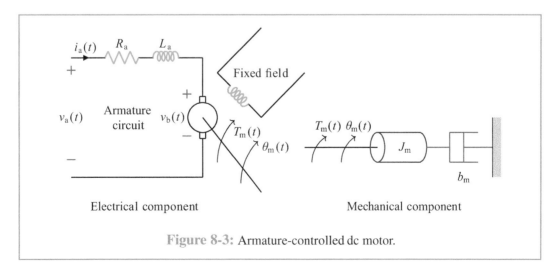

Figure 8-3: Armature-controlled dc motor.

Solution: Application of Kirchoff's voltage law to the armature circuit yields

$$v_a = R_a i_a + L_a \frac{di_a}{dt} + K_b \omega_m . \tag{8.8a}$$

Application of Newton's second law to the mechanical part yields

$$J_m \dot{\omega}_m = -b_m \omega_m + K_t i_a . \tag{8.8b}$$

Assign the state variables as follows:

$$x_1 = \theta_m , \qquad x_2 = \dot{\theta}_m = \omega_m , \qquad x_3 = i_a .$$

The input and output variables are

$$u = v_a , \qquad y = \omega_m .$$

The state equations are

$$\dot{x}_1 = \dot{\theta}_m = x_2 ,$$

$$\dot{x}_2 = \dot{\omega}_m = \frac{1}{J_m}[-b_m \omega_m + K_t i_a] = -\frac{b_m}{J_m} x_2 + \frac{K_t}{J_m} x_3 ,$$

and

$$\dot{x}_3 = \frac{di_a}{dt} = \frac{1}{L_a}[v_a - R_a i_a - K_b \omega_m] = -\frac{K_b}{L_a} x_2 - \frac{R_a}{L_a} x_3 + \frac{1}{L_a} u .$$

The output equation is

$$y = \omega_m = x_2 .$$

The quadruple $\{A, B, C, D\}$ that defines the state-space model is given by

$$A = \begin{bmatrix} 0 & 1 & 0 \\ 0 & -b_m/J_m & K_t/J_m \\ 0 & -K_b/L_a & -R_a/L_a \end{bmatrix}, \quad B = \begin{bmatrix} 0 \\ 0 \\ 1/L_a \end{bmatrix}, \quad C = \begin{bmatrix} 0 & 1 & 0 \end{bmatrix}, \quad D = 0.$$

8-1.2 Linearization

Nonlinear systems can be modeled by linear models in the neighborhood of an equilibrium point. Linearization of a nonlinear state-space model uses Jacobian matrices, which are defined as follows. For $f(x, u)$, the *Jacobian matrix* $[\partial f / \partial x]$ is defined by

$$\frac{\partial f}{\partial x}(x, u) = \begin{bmatrix} \partial f_1/\partial x_1 & \partial f_1/\partial x_2 & \cdots & \partial f_1/\partial x_n \\ \partial f_2/\partial x_1 & \partial f_2/\partial x_2 & \cdots & \partial f_2/\partial x_n \\ \vdots & & & \vdots \\ \partial f_n/\partial x_1 & \partial f_n/\partial x_2 & \cdots & \partial f_n/\partial x_n \end{bmatrix}, \tag{8.9}$$

8-1. STATE-SPACE MODELS

and the Jacobian matrix $[\partial f/\partial u](x,u)$ is defined by

$$\frac{\partial f}{\partial u}(x,u) = \begin{bmatrix} \partial f_1/\partial u \\ \partial f_2/\partial u \\ \vdots \\ \partial f_n/\partial u \end{bmatrix}.$$

For the function $h(x,u)$, the Jacobian matrix $[\partial h/\partial x]$ is defined by

$$\frac{\partial h}{\partial x}(x,u) = \begin{bmatrix} \frac{\partial h}{\partial x_1} & \frac{\partial h}{\partial x_2} & \cdots & \frac{\partial h}{\partial x_n} \end{bmatrix}.$$

Suppose that for a constant input $u^\star$, the system $\dot{x} = f(x,u)$ has an equilibrium point at $x^\star$; that is,

$$0 = f(x^\star, u^\star).$$

Then, the system can be approximated in the neighborhood of the equilibrium point by the linear state-space model

$$\dot{x}_\delta = A x_\delta + B u_\delta, \tag{8.10a}$$

$$y_\delta = C x_\delta + D y_\delta, \tag{8.10b}$$

where $x_\delta = x - x^\star$, $u_\delta = u - u^\star$, and $y_\delta = y - h(x^\star, u^\star)$ are perturbations from their equilibrium values, and the matrices A, B, C, D are given by

$$A = \frac{\partial f}{\partial x}(x^\star, u^\star), \tag{8.11a}$$

$$B = \frac{\partial f}{\partial u}(x^\star, u^\star), \tag{8.11b}$$

$$C = \frac{\partial h}{\partial x}(x^\star, u^\star), \tag{8.11c}$$

$$D = \frac{\partial h}{\partial u}(x^\star, u^\star). \tag{8.11d}$$

Example 8-4: Linearization of the Pendulum Equation

Consider the pendulum equation from Example 2-18, as related to Fig. 8-4, and suppose that a torque T is applied at the pivot point in the direction of θ. By taking moments about the pivot point, the equation of motion is given by

$$ml^2 \ddot{\theta} = -mgl \sin\theta - kl^2 \dot{\theta} + T. \tag{8.12}$$

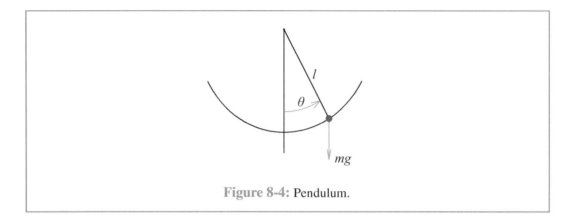

Figure 8-4: Pendulum.

Setting $x_1 = \theta$ and $x_2 = \dot{\theta}$ as the state variables, $u = T$ as the input, and $y = \theta$ as the output, the state-space model is given by

$$\dot{x} = f(x,u) = \begin{bmatrix} x_2 \\ -\frac{g}{l}\sin x_1 - \frac{k}{m}x_2 + \frac{1}{ml^2}u \end{bmatrix}, \quad y = h(x) = x_1.$$

The Jacobian matrices are

$$\frac{\partial f}{\partial x} = \begin{bmatrix} 0 & 1 \\ -\frac{g}{l}\cos x_1 & -\frac{k}{m} \end{bmatrix}, \quad \frac{\partial f}{\partial u} = \begin{bmatrix} 0 \\ \frac{1}{ml^2} \end{bmatrix}, \quad \frac{\partial h}{\partial x} = \begin{bmatrix} 1 & 0 \end{bmatrix}, \quad \frac{\partial h}{\partial u} = 0.$$

When $u = 0$, the system has equilibrium at $\sin x_1 = 0$ and $x_2 = 0$. Consider the two equilibrium points $\begin{bmatrix} 0 \\ 0 \end{bmatrix}$ and $\begin{bmatrix} \pi \\ 0 \end{bmatrix}$. Linearization at $\begin{bmatrix} 0 \\ 0 \end{bmatrix}$ results in a linear state-space model with

$$A = \begin{bmatrix} 0 & 1 \\ -\frac{g}{l} & -\frac{k}{m} \end{bmatrix}, \quad B = \begin{bmatrix} 0 \\ \frac{1}{ml^2} \end{bmatrix}, \quad C = \begin{bmatrix} 1 & 0 \end{bmatrix}, \quad D = 0,$$

while linearization at $\begin{bmatrix} \pi \\ 0 \end{bmatrix}$ results in a linear state-space model with

$$A = \begin{bmatrix} 0 & 1 \\ \frac{g}{l} & -\frac{k}{m} \end{bmatrix}, \quad B = \begin{bmatrix} 0 \\ \frac{1}{ml^2} \end{bmatrix}, \quad C = \begin{bmatrix} 1 & 0 \end{bmatrix}, \quad D = 0.$$

8-2. SOLUTION OF THE STATE EQUATION

> Example 8-5: Linearization of a dc-to-dc Power Converter

A dc-to-dc power converter can be modeled by the nonlinear state-space model[4]

$$\dot{x} = \begin{bmatrix} -\frac{1}{k}x_2 + (x_2+k)u \\ \frac{1}{k}x_1 - \alpha x_2 - (x_1 + \alpha k^2)u \end{bmatrix}, \quad y = x_2,$$

where α and k are positive constants.

- Show that the system has an equilibrium point at $x = 0$ when $u = 0$.
- Linearize the system at the equilibrium point.

Solution: Setting $u = 0$ and $\dot{x} = 0$ yields

$$\begin{bmatrix} -\frac{1}{k}x_2 \\ \frac{1}{k}x_1 - \alpha x_2 \end{bmatrix} = \begin{bmatrix} 0 \\ 0 \end{bmatrix} \quad \Rightarrow \quad x_1 = 0 \text{ and } x_2 = 0,$$

$$\left.\frac{\partial f}{\partial x}\right|_{x=0,\,u=0} = \begin{bmatrix} 0 & (-\frac{1}{k}+u) \\ (\frac{1}{k}-u) & -\alpha \end{bmatrix}_{x=0,\,u=0} = \begin{bmatrix} 0 & -\frac{1}{k} \\ \frac{1}{k} & -\alpha \end{bmatrix},$$

$$\left.\frac{\partial f}{\partial u}\right|_{x=0,\,u=0} = \begin{bmatrix} x_2+k \\ -(x_1+\alpha k^2) \end{bmatrix}_{x=0,\,u=0} = \begin{bmatrix} k \\ -\alpha k^2 \end{bmatrix},$$

$$\frac{\partial h}{\partial x} = \begin{bmatrix} 0 & 1 \end{bmatrix}, \quad \frac{\partial h}{\partial u} = 0.$$

The linearized model is given by

$$A = \begin{bmatrix} 0 & -\frac{1}{k} \\ \frac{1}{k} & -\alpha \end{bmatrix}, \quad B = \begin{bmatrix} k \\ -\alpha k^2 \end{bmatrix}, \quad C = \begin{bmatrix} 0 & 1 \end{bmatrix}, \quad D = 0.$$

8-2 Solution of the State Equation

Consider the state equation

$$\dot{x}(t) = Ax(t) + Bu(t) \tag{8.13}$$

with given input $u(t)$ and given initial state $x(0) = x_0$. We use the Laplace transform method to solve for $x(t)$. This requires us to find $X(s)$ and then to take the inverse Laplace transform $x(t) = \mathcal{L}^{-1}\{X(s)\}$. Taking the Laplace transform of a vector is realized by taking the Laplace transform of each element of the vector. Properties of the Laplace transform for scalar variables extend to the Laplace transform of vectors.

Taking the Laplace transform of Eq. (8.13) yields

$$sX(s) - x(0) = AX(s) + BU(s),$$

---
[4]See Appendix A.5 of [22].

$$sIX(s) - AX(s) = x_0 + BU(s),$$
$$(sI - A)X(s) = x_0 + BU(s). \tag{8.14}$$

By multiplying each term, from the left-hand side, by the inverse of $(sI - A)$, we have:

$$X(s) = (sI - A)^{-1}x_0 + (sI - A)^{-1}BU(s),$$

whose time-domain equivalent is

$$x(t) = \mathcal{L}^{-1}\{X(s)\} = \mathcal{L}^{-1}\{(sI - A)^{-1}x_0\} + \mathcal{L}^{-1}\{(sI - A)^{-1}BU(s)\}.$$

Define the *state transition matrix* $\Phi(t)$ by

$$\Phi(t) = \mathcal{L}^{-1}\{(sI - A)^{-1}\}.$$

In this definition, the matrix $(sI - A)^{-1}$ is calculated first, then the inverse Laplace transform is taken for each element of the matrix

$$\mathcal{L}^{-1}\{(sI - A)^{-1}x_0\} = \mathcal{L}^{-1}\{(sI - A)^{-1}\}x_0 = \Phi(t)x_0,$$
$$\mathcal{L}^{-1}\{(sI - A)^{-1}BU(s)\} = \int_0^t \Phi(t - \tau)Bu(\tau)d\tau.$$

Hence,

$$x(t) = \Phi(t)x_0 + \int_0^t \Phi(t - \tau)Bu(\tau)d\tau. \tag{8.15}$$

Substitution of $x(t)$ in the output equation

$$y(t) = Cx(t) + Du(t)$$

yields

$$y(t) = C\Phi(t)x_0 + \int_0^t C\Phi(t - \tau)Bu(\tau)d\tau + Du(t). \tag{8.16}$$

Example 8-6: Computation of the State Transition Matrix

Compute $\Phi(t)$ for

$$A = \begin{bmatrix} -1 & 1 \\ -\frac{1}{2} & -2 \end{bmatrix}.$$

8-2. SOLUTION OF THE STATE EQUATION

Solution:

$$(sI-A)^{-1} = \begin{bmatrix} (s+1) & -1 \\ \frac{1}{2} & (s+2) \end{bmatrix}^{-1} = \frac{1}{(s+1)(s+2)+\frac{1}{2}} \begin{bmatrix} (s+2) & 1 \\ -\frac{1}{2} & (s+1) \end{bmatrix}$$

$$= \frac{1}{\left(s+\frac{3}{2}\right)^2 + \frac{1}{4}} \begin{bmatrix} \left(s+\frac{3}{2}+\frac{1}{2}\right) & 2 \times \frac{1}{2} \\ -\frac{1}{2} & \left(s+\frac{3}{2}-\frac{1}{2}\right) \end{bmatrix},$$

Hence,

$$\Phi(t) = e^{-(3/2)t} \begin{bmatrix} \left(\cos\left(\frac{t}{2}\right) + \sin\left(\frac{t}{2}\right)\right) & 2\sin\left(\frac{t}{2}\right) \\ -\sin\left(\frac{t}{2}\right) & \left(\cos\left(\frac{t}{2}\right) - \sin\left(\frac{t}{2}\right)\right) \end{bmatrix}.$$

Example 8-7: Computation of the State Transition Matrix

Compute $\Phi(t)$ for

$$A = \begin{bmatrix} -3 & 1 \\ -2 & 0 \end{bmatrix}.$$

Solution:

$$(sI-A)^{-1} = \begin{bmatrix} (s+3) & -1 \\ 2 & s \end{bmatrix}^{-1} = \frac{1}{s^2+3s+2} \begin{bmatrix} s & 1 \\ -2 & (s+3) \end{bmatrix}$$

$$= \frac{1}{(s+1)(s+2)} \begin{bmatrix} s & 1 \\ -2 & (s+3) \end{bmatrix} = \frac{1}{s+1} K_1 + \frac{1}{s+2} K_2,$$

where

$$K_1 = (s+1)(sI-A)^{-1}\big|_{s=-1} = \begin{bmatrix} -1 & 1 \\ -2 & 2 \end{bmatrix},$$

$$K_2 = (s+2)(sI-A)^{-1}\big|_{s=-2} = \begin{bmatrix} 2 & -1 \\ 2 & -1 \end{bmatrix}.$$

Therefore,

$$(sI-A)^{-1} = \frac{1}{s+1} \begin{bmatrix} -1 & 1 \\ -2 & 2 \end{bmatrix} + \frac{1}{s+2} \begin{bmatrix} 2 & -1 \\ 2 & -1 \end{bmatrix},$$

$$\Phi(t) = e^{-t} \begin{bmatrix} -1 & 1 \\ -2 & 2 \end{bmatrix} + e^{-2t} \begin{bmatrix} 2 & -1 \\ 2 & -1 \end{bmatrix}$$

$$= \begin{bmatrix} \left(-e^{-t} + 2e^{-2t}\right) & \left(e^{-t} - e^{-2t}\right) \\ \left(-2e^{-t} + 2e^{-2t}\right) & \left(2e^{-t} - e^{-2t}\right) \end{bmatrix}.$$

> **Example 8-8: Solution of the State Equation**

Compute $y(t)$ for

$$\dot{x}(t) = \begin{bmatrix} -3 & 1 \\ -2 & 0 \end{bmatrix} x(t) + \begin{bmatrix} 0 \\ 1 \end{bmatrix} u(t), \quad x(0) = \begin{bmatrix} 1 \\ -1 \end{bmatrix},$$

$$y(t) = \begin{bmatrix} 1 & 0 \end{bmatrix} x(t),$$

when $u(t) = e^t$.

Solution: $\Phi(t)$ was computed in the previous example. Hence,

$$y(t) = Cx(t) = C\Phi(t)\,x(0) + \int_0^t C\Phi(t-\tau)B\,u(\tau)\,d\tau,$$

$$C\Phi(t)\,x(0) = \begin{bmatrix} 1 & 0 \end{bmatrix} \begin{bmatrix} (-e^{-t}+2e^{-2t}) & (e^{-t}-e^{-2t}) \\ (-2e^{-t}+2e^{-2t}) & (2e^{-t}-e^{-2t}) \end{bmatrix} \begin{bmatrix} 1 \\ -1 \end{bmatrix}$$

$$= \begin{bmatrix} (-e^{-t}+2e^{-2t}) & (e^{-t}-e^{-2t}) \end{bmatrix} \begin{bmatrix} 1 \\ -1 \end{bmatrix}$$

$$= -2e^{-t} + 3e^{-2t},$$

$$C\Phi(t)\,B = \begin{bmatrix} 1 & 0 \end{bmatrix} \begin{bmatrix} (-e^{-t}+2e^{-2t}) & (e^{-t}-e^{-2t}) \\ (-2e^{-t}+2e^{-2t}) & (2e^{-t}-e^{-2t}) \end{bmatrix} \begin{bmatrix} 0 \\ 1 \end{bmatrix}$$

$$= \begin{bmatrix} (-e^{-t}+2e^{-2t}) & (e^{-t}-e^{-2t}) \end{bmatrix} \begin{bmatrix} 0 \\ 1 \end{bmatrix} = e^{-t} - e^{-2t},$$

$$\int_0^t C\Phi(t-\tau)\,B\,u(\tau)\,d\tau = \int_0^t [e^{-(t-\tau)} - e^{-2(t-\tau)}]e^\tau\,d\tau$$

$$= \int_0^t [e^{-t+2\tau} - e^{-2t+3\tau}]\,d\tau$$

$$= \left[\tfrac{1}{2} e^{-t+2\tau} - \tfrac{1}{3} e^{-2t+3\tau}\right]_0^t = \tfrac{1}{6} e^t - \tfrac{1}{2} e^{-t} + \tfrac{1}{3} e^{-2t}.$$

Hence,

$$y(t) = (-2e^{-t} + 3e^{-2t}) + (\tfrac{1}{6} e^t - \tfrac{1}{2} e^{-t} + \tfrac{1}{3} e^{-2t}) = \tfrac{1}{6} e^t - \tfrac{5}{2} e^{-t} + \tfrac{10}{3} e^{-2t}.$$

8-3 Stability

For a system represented by the transfer function $G(s)$, stability is defined as bounded-input–bounded-output stability; that is, for every bounded input the output is bounded. For a system represented by the state-space model

$$\dot{x} = Ax + Bu, \tag{8.17a}$$

8-3. STABILITY

$$y = Cx + Du,\qquad(8.17b)$$

stability deals with the response to the initial state as well as the response to the input.

Definition 8.1 The system characterized by Eqs. (8.17a)–(8.17b) is stable if

(1) the solution of $\dot{x} = Ax$, with $x(0) = x_0$, tends to zero as t tends to infinity, and

(2) the solution of $\dot{x} = Ax + Bu$ is bounded for every bounded input.

The equation $y = Cx + Du$ states that a stable system will have a bounded output for every bounded input.

The stability of the system given by Eqs. (8.17a)–(8.17b) is characterized by the *eigenvalues* of the matrix A. The eigenvalues of a real matrix A are the roots of

$$\det(sI - A) = 0.$$

An $n \times n$ matrix has n eigenvalues $\lambda_1, \lambda_2, \ldots \lambda_n$. The eigenvalues of a real matrix can be real or complex, but complex eigenvalues must exist in conjugate pairs ($\sigma \pm j\omega$).

Theorem 8.1 *The system of Eqs. (8.17a)–(8.17b) is stable if and only if all eigenvalues of A have negative real parts.*

Proof: Consider first the special case when the eigenvalues of A are distinct ($\lambda_i \neq \lambda_j$ for all i, j). It follows that

$$\det(sI - A) = (s - \lambda_1)(s - \lambda_2)\ldots(s - \lambda_n)$$

and

$$(sI - A)^{-1} = \frac{1}{\det(sI - A)} \text{Adjoint}(sI - A).$$

By partial fraction expansion,

$$(sI - A)^{-1} = \frac{1}{(s - \lambda_1)} R_1 + \frac{1}{(s - \lambda_2)} R_2 + \cdots + \frac{1}{(s - \lambda_n)} R_n$$

$$= \sum_{i=1}^{n} \frac{1}{(s - \lambda_i)} R_i,$$

where the residue matrices R_1 to R_n are defined by

$$R_i = (s - \lambda_i)(sI - A)^{-1}\Big|_{s=\lambda_i}$$

and

$$\Phi(t) = \mathcal{L}^{-1}\left\{(sI - A)^{-1}\right\} = \sum_{i}^{n} e^{\lambda_i t} R_i.\qquad(8.18)$$

The exponential functions $e^{\lambda_i t}$ are called the *modes* of the system. Then,

$$x(t) = \Phi(t)x(0) + \int_0^t \Phi(t-\tau)Bu(\tau)d\tau$$

$$= \sum_{i=1}^n e^{\lambda_i t}R_i x(0) + \int_0^t \sum_{i=1}^n e^{\lambda_i(t-\tau)}R_iBu(\tau)d\tau.$$

The system is stable if and only if for every eigenvalue λ_i,

$$\lim_{t\to\infty} e^{\lambda_i t} = 0,$$

and

$\int_0^t e^{\lambda_i(t-\tau)}f(\tau)d\tau$ is bounded for every bounded function $f(t)$. For

$$\lambda_i = \sigma_i + j\omega_i \;\longrightarrow\; |e^{\lambda_i t}| = |e^{(\sigma_i + j\omega_i)t}| = |e^{\sigma_i t}||e^{j\omega_i t}| = e^{\sigma_i t},$$

$$\lim_{t\to\infty} e^{\sigma_i t} = 0 \;\longleftrightarrow\; \sigma_i < 0.$$

For $|f(t)| \le k$, $\left|\int_0^t e^{\lambda_i(t-\tau)}f(\tau)d\tau\right| \le \int_0^t e^{\sigma_i(t-\tau)}k\,d\tau = \frac{k}{\sigma_i}\left(e^{\sigma_i t} - 1\right),$

$$\frac{k}{\sigma_i}\left(e^{\sigma_i t} - 1\right) \text{ is bounded } \;\longleftrightarrow\; \sigma_i < 0.$$

If A has multiple eigenvalues, the partial fraction expansion of $(sI - A)^{-1}$ will yield terms where the power of $(s - \lambda_i)$ in the denominator is higher than one. Therefore, $\Phi(t)$ takes the more general form

$$\Phi(t) = \sum_{i=1}^r \sum_{\ell=1}^{m_i} t^{\ell-1} e^{\lambda_i t} R_{ik}, \tag{8.19}$$

where the integer m_i depends on the multiplicity of λ_i, and r is the number of distinct eigenvalues of A. A function of the form $t^k e^{\sigma t}$, with $k \ge 1$, does not converge to zero as $t \to \infty$ if $\sigma \ge 0$. Hence, it is necessary that all the eigenvalues of A have negative real parts.

Now we show that this condition is also sufficient. If $\text{Re}[\lambda_i] = \sigma_i < 0$, there are positive constants α and β such that $|t^{\ell-1}e^{\lambda_i t}| \le \alpha e^{-\beta t}$. Although the polynomial term $t^{\ell-1}$ grows with t, causing the product $|t^{\ell-1}e^{\lambda_i t}|$ to grow for small t, eventually the decaying exponential $e^{\lambda_i t}$ dominates. Therefore, there exist positive constants a and b such that

$$\|\Phi(t)\| \le ae^{-bt}.$$

With this bound, the argument we used for the case of distinct eigenvalues can be repeated and this completes the proof of the theorem.

Example 8-9: Testing Stability of State-Space Models

For each of the following cases of eigenvalues of the matrix A, determine whether or not the system is stable.

8-4. CONTROLLABILITY AND OBSERVABILITY

(a) $-1, -2, 0$

(b) $-1, -2, 1$

(c) $-1, -2, -2$

(d) $-3, -2 \pm j$

Solution:

(a) Not stable because of the 0 eigenvalue.

(b) Not stable because of the 1 eigenvalue.

(c) Stable.

(d) Stable.

Testing stability of a state-space model requires calculation of the eigenvalues of A. This can be done by calculating the roots of $\det(sI - A)$. In MATLAB, eigenvalues are calculated using the command "*eig*".

8-4 Controllability and Observability

Controllability and observability are properties of the state-space model, which play an important role in the design of feedback control. We start by presenting controllability in Section 8-4.1, followed by observability in Section 8-4.2, then we show the duality between these two properties in Section 8-4.3.

8-4.1 Controllability

Definition 8.2 The system $\dot{x} = Ax + Bu$, or the pair (A, B), is *controllable* if for any initial state x_0, there is a continuous control $u(t)$ that steers the state of the system to any desired final state x_f over the time period $[0, t_f]$, for some finite t_f.

We start with a motivating example.

> Example 8-10: Controllability of a Rocket in Vertical Motion

A rocket in vertical motion may be modeled by

$$\dot{h} = v,$$
$$m\dot{v} = -mg + f,$$

where h is the altitude, v the velocity, m the mass, and f the thrust force. Let

$$x_1 = h, \quad x_2 = v, \quad u = f/m - g$$

to obtain the state equation

$$\dot{x} = \underbrace{\begin{bmatrix} 0 & 1 \\ 0 & 0 \end{bmatrix}}_{A} x + \underbrace{\begin{bmatrix} 0 \\ 1 \end{bmatrix}}_{B} u \ .$$

Can we find a control $u(t)$ over the period $[0, t_f]$ to move the state of the system from a given initial state $x(0) = x_0$ to a desired final state $x(t_f) = x_f$?

Solution: The solution of the state equation is

$$x(t_f) = \Phi(t_f)\, x(0) + \int_0^{t_f} \Phi(t_f - \tau)\, B\, u(\tau)\, d\tau , \qquad (8.20)$$

with

$$\Phi(t) = \mathcal{L}^{-1}\{(sI - A)^{-1}\} = \mathcal{L}^{-1}\left\{ \begin{bmatrix} s & -1 \\ 0 & s \end{bmatrix}^{-1} \right\} = \mathcal{L}^{-1}\left\{ \begin{bmatrix} 1/s & 1/s^2 \\ 0 & 1/s \end{bmatrix} \right\} = \begin{bmatrix} 1 & t \\ 0 & 1 \end{bmatrix}$$

and

$$\Phi(t_f - \tau) = \begin{bmatrix} 1 & (t_f - \tau) \\ 0 & 1 \end{bmatrix} = \begin{bmatrix} 1 & t_f \\ 0 & 1 \end{bmatrix} \begin{bmatrix} 1 & -\tau \\ 0 & 1 \end{bmatrix} = \Phi(t_f)\, \Phi(-\tau) \ .$$

We substitute $\Phi(t_f - \tau) = \Phi(t_f)\, \Phi(-\tau)$ in Eq. (8.20), to obtain

$$x_f - \Phi(t_f)\, x_0 = \Phi(t_f) \int_0^{t_f} \Phi(-\tau)\, B\, u(\tau)\, d\tau , \qquad (8.21)$$

where $\Phi(t_f)$ is taken outside the integral because it is not a function of τ. Let

$$u(t) = B^T\, \Phi^T(-t)\, \upsilon , \qquad (8.22)$$

where υ is a constant vector to be chosen. Substitution of Eq. (8.22) in Eq. (8.21) results in

$$x_f - \Phi(t_f)\, x_0 = \Phi(t_f) \int_0^{t_f} \Phi(-\tau)\, BB^T\, \Phi^T(-\tau)\, d\tau\, \upsilon \ . \qquad (8.23)$$

Define the matrix W by

$$W = \int_0^{t_f} \Phi(-\tau)\, BB^T\, \Phi^T(-\tau)\, d\tau$$

$$= \int_0^{t_f} \begin{bmatrix} \tau^2 & -\tau \\ -\tau & 1 \end{bmatrix} d\tau = \begin{bmatrix} t_f^3/3 & -t_f^2/2 \\ -t_f^2/2 & t_f \end{bmatrix} \ .$$

8-4. CONTROLLABILITY AND OBSERVABILITY

With the definition of W, Eq. (8.23) can be rewritten as

$$x_f - \Phi(t_f) x_0 = \Phi(t_f) W v . \tag{8.24}$$

The matrix $\Phi(t_f)$ is nonsingular because $\det \Phi(t_f) = 1$, and the matrix W is nonsingular because $\det W = t_f^4/12 \neq 0$. Hence, Eq. (8.24) has a unique solution:

$$v = W^{-1} \Phi^{-1}(t_f)[x_f - \Phi(t_f) x_0] ,$$

and a control that moves the state from x_0 to x_f is given by

$$u(t) = B^T \Phi^T(-t) v = B^T \Phi^T(-t) W^{-1} \Phi^{-1}(t_f)[x_f - \Phi(t_f) x_0] .$$

To extend the analysis of this example to the general case, we need to establish a few properties of the transition matrix Φ.

Lemma 8.1 The transition matrix $\Phi(t)$ has the following properties:

(i) $\Phi(0) = I$.

(ii) $\Phi(t)$ is nonsingular and its inverse is $\Phi(-t)$.

(iii) $\Phi(t - \tau) = \Phi(t) \Phi(-\tau)$.

Proof:

(i) The solution for $\dot{x} = Ax$ with $x(0) = x_0$ is $x(t) = \Phi(t) x_0$. At $t = 0$,

$$x_0 = \Phi(0) x_0 \quad \longrightarrow \quad [I - \Phi(0)]x_0 = 0.$$

Because this equation holds for any x_0, it must be true that $\Phi(0) = I$.

(ii) Starting from $t = 0$ and solving the equation $\dot{x} = Ax$ in forward time to t, we have $x(t) = \Phi(t) x_0$. Starting at time t from initial state $x(t)$ and solving the equation backward to $t = 0$, we have

$$x(0) = \Phi(-t) x(t) = \Phi(-t) \Phi(t) x_0 \quad \longrightarrow \quad [I - \Phi(-t) \Phi(t)]x_0 = 0.$$

Once again, because this equation holds for any x_0, it must be true that $\Phi(-t) \Phi(t) = I$. Hence, both matrices $\Phi(t)$ and $\Phi(-t)$ are nonsingular and $\Phi(-t)$ is the inverse of $\Phi(t)$.

(iii) Suppose the system starts at time 0 and moves to time τ and from time τ it moves to time t. Then $x(t) = \Phi(t - \tau) x(\tau) = \Phi(t - \tau) \Phi(\tau) x_0$. Alternatively, if the system moves directly from time 0 to time t, $x(t) = \Phi(t) x_0$. Hence,

$$\Phi(t) x_0 = \Phi(t - \tau) \Phi(\tau) x_0 \quad \longrightarrow \quad [\Phi(t) - \Phi(t - \tau) \Phi(\tau)] x_0 = 0$$
$$\longrightarrow \quad \Phi(t) = \Phi(t - \tau) \Phi(\tau) .$$

Multiplying from the right by $\Phi(-\tau)$ and using $\Phi(\tau) \Phi(-\tau) = I$, we arrive at

$$\Phi(t - \tau) = \Phi(t) \Phi(-\tau) .$$

Let us now repeat the analysis of Example 8-10 for the general case. The solution of

$$\dot{x} = Ax + Bu, \quad x(0) = x_0,$$

is

$$x(t_f) = \Phi(t_f) x_0 + \int_0^{t_f} \Phi(t_f - \tau) B u(\tau) d\tau.$$

If we set $x(t_f) = x_f$ and use $\Phi(t - \tau) = \Phi(t) \Phi(-\tau)$, then

$$x_f - \Phi(t_f) x_0 = \Phi(t_f) \int_0^{t_f} \Phi(-\tau) B u(\tau) d\tau.$$

Substitute

$$u(t) = B^T \Phi^T(-\tau) \upsilon$$

with a constant vector υ in the preceding equation to obtain

$$x_f - \Phi(t_f) x_0 = \Phi(t_f) \int_0^{t_f} \Phi(-\tau) BB^T \Phi^T(-\tau) d\tau \, \upsilon.$$

Define the matrix W by[5]

$$W = \int_0^{t_f} \Phi(-\tau) BB^T \Phi^T(-\tau) d\tau.$$

Then,

$$x_f - \Phi(t_f) x_0 = \Phi(t_f) W \upsilon.$$

We already know that $\Phi(t_f)$ is nonsingular. The equation will have a unique solution for υ if W is nonsingular. The next theorem defines the necessary and sufficient condition for this to hold.[6]

Theorem 8.2 *The n-dimensional system $\dot{x} = Ax + Bu$ is controllable if and only if rank $C_M = n$, where C_M is the **controllability matrix**, defined by*

$$C_M = \begin{bmatrix} B & AB & A^2B & \cdots & A^{n-1}B \end{bmatrix}. \tag{8.25}$$

Investigating the controllability of a pair (A, B) is done by forming the controllability matrix C_M and testing its rank. This is performed in MATLAB using the commands "*ctrb*" and "*rank*":

$$CM = ctrb(A, B); \; rank(CM);$$

---

[5]W is called the "controllability Gramian."
[6]The proof of the theorem can be found in any graduate-level textbook on linear systems using state-space models; see, for example, [2] or [32].

8-4. CONTROLLABILITY AND OBSERVABILITY

Example 8-11: Testing Controllability of a Pair (A,B)

Investigate controllability of

$$A = \begin{bmatrix} 0 & 1 & 0 \\ 0 & 0 & 1 \\ -1 & -2 & -3 \end{bmatrix}, \quad B = \begin{bmatrix} 0 \\ 0 \\ 1 \end{bmatrix}.$$

Solution:

$$C_M = \begin{bmatrix} B & AB & A^2B \end{bmatrix} = \begin{bmatrix} 0 & 0 & 1 \\ 0 & 1 & -3 \\ 1 & -3 & 7 \end{bmatrix},$$

rank $C_M = 3 \longrightarrow (A,B)$ is controllable.

Example 8-12: Controllability When A is Diagonal

Investigate controllability of

$$A = \begin{bmatrix} \lambda_1 & 0 & 0 \\ 0 & \lambda_2 & 0 \\ 0 & 0 & \lambda_3 \end{bmatrix}, \quad B = \begin{bmatrix} b_1 \\ b_2 \\ b_3 \end{bmatrix}.$$

Solution:

$$C_M = \begin{bmatrix} B & AB & A^2B \end{bmatrix} = \begin{bmatrix} b_1 & \lambda_1 b_1 & \lambda_1^2 b_1 \\ b_2 & \lambda_2 b_2 & \lambda_2^2 b_2 \\ b_3 & \lambda_3 b_3 & \lambda_3^2 b_3 \end{bmatrix}.$$

(a) If $b_i = 0$ for some i, then rank $C_M < 3$, because for $b_i = 0$, the ith row is zero.

(b) If $\lambda_i = \lambda_j$ for some $i \ne j$, then rank $C_M < 3$, because the ith and jth rows will be the same.

Now we show sufficiency of $b_i \ne 0$ and $\lambda_i \ne \lambda_j$ for C_M to have rank 3. The rank of a matrix does not change after performing elementary row or column operations. We perform the following elementary row operations to simplify the structure of C_M:

(1) Add to the second row the first row multiplied by $-b_2/b_1$ and add to the third row the first row multiplied by $-b_3/b_1$:

$$\longrightarrow \begin{bmatrix} b_1 & \lambda_1 b_1 & \lambda_1^2 b_1 \\ 0 & (\lambda_2 - \lambda_1)b_2 & (\lambda_2^2 - \lambda_1^2)b_2 \\ 0 & (\lambda_3 - \lambda_1)b_3 & (\lambda_3^2 - \lambda_1^2)b_3 \end{bmatrix}.$$

(2) Add to the third row the second row multiplied by $-(\lambda_3 - \lambda_1)b_3/[(\lambda_2 - \lambda_1)b_2]$:

$$\rightarrow \begin{bmatrix} b_1 & \lambda_1 b_1 & \lambda_1^2 b_1 \\ 0 & (\lambda_2 - \lambda_1)b_2 & (\lambda_2^2 - \lambda_1^2)b_2 \\ 0 & 0 & * \end{bmatrix},$$

with $*$ given by

$$* = (\lambda_3^2 - \lambda_1^2)b_3 - (\lambda_2^2 - \lambda_1^2)\frac{(\lambda_3 - \lambda_1)}{(\lambda_2 - \lambda_1)}b_3$$
$$= (\lambda_3 - \lambda_1)(\lambda_3 - \lambda_2)b_3 .$$

Hence,

$$C_M \rightarrow \begin{bmatrix} b_1 & \lambda_1 b_1 & \lambda_1^2 b_1 \\ 0 & (\lambda_2 - \lambda_1)b_2 & (\lambda_2^2 - \lambda_1^2)b_2 \\ 0 & 0 & (\lambda_3 - \lambda_1)(\lambda_3 - \lambda_2)b_3 \end{bmatrix}.$$

The conditions $b_i \neq 0$ and $\lambda_i \neq \lambda_j$ imply that the rank is 3. Hence, the pair (A,B) is controllable if and only if $b_i \neq 0$ for all i and $\lambda_i \neq \lambda_j$ for all $i \neq j$.

8-4.2 Observability

Definition 8.3 The system

$$\dot{x} = Ax, \quad y = Cx$$

or the pair (A,C), is *observable* if any initial state $x(0) = x_0$ can be uniquely determined from $y(t)$ over the period $[0, t_f]$, for some finite t_f.

We start with a motivating example.

Example 8-13: Observability of a Rocket in Vertical Motion

Consider the rocket system from Example 8-10. Suppose we only measure $y = x_1$. When $u = 0$, the system is given by

$$\dot{x} = \underbrace{\begin{bmatrix} 0 & 1 \\ 0 & 0 \end{bmatrix}}_{A} x, \quad y = \underbrace{\begin{bmatrix} 1 & 0 \end{bmatrix}}_{C} x .$$

Can we uniquely determine $x(0) = x_0$ from $y(t)$ over the time period $[0, t_f]$?

Solution: The output is

$$y(t) = Cx(t) = C\Phi(t)x(0) = C\Phi(t)x_0 ,$$

8-4. CONTROLLABILITY AND OBSERVABILITY

and the matrix $\Phi(t)$ was calculated in Example 8-10. Multiply both sides from the left by $\Phi^T(t)\, C^T$ and integrate from 0 to t_f:

$$\int_0^{t_f} \Phi^T(t)\, C^T y(t)\, dt = \int_0^{t_f} \Phi^T(t)\, C^T C\, \Phi(t)\, x_0\, dt = \int_0^{t_f} \Phi^T(t)\, C^T C\, \Phi(t)\, dt\, x_0\, .$$

Define

$$M = \int_0^{t_f} \Phi^T(t)\, C^T C\, \Phi(t)\, dt$$

$$= \int_0^{t_f} \begin{bmatrix} 1 & 0 \\ t & 1 \end{bmatrix} \begin{bmatrix} 1 \\ 0 \end{bmatrix} \begin{bmatrix} 1 & 0 \end{bmatrix} \begin{bmatrix} 1 & t \\ 0 & 1 \end{bmatrix} dt$$

$$= \int_0^{t_f} \begin{bmatrix} 1 & t \\ t & t^2 \end{bmatrix} dt = \begin{bmatrix} t_f & t_f^2/2 \\ t_f^2/2 & t_f^3/3 \end{bmatrix}.$$

The matrix M is nonsingular because $\det M = t_f^4/12$. Hence, the equation

$$\int_0^{t_f} \Phi^T(t)\, C^T\, y(t)\, dt = M x_0$$

has a unique solution

$$x_0 = M^{-1} \int_0^{t_f} \Phi^T(t)\, C^T\, y(t)\, dt\, .$$

Consider now the general case

$$\dot{x} = Ax\, , \qquad y = Cx\, ,$$

with $x(0) = x_0$,

$$y(t) = C\, x(t) = C\, \Phi(t)\, x(0) = C\, \Phi(t)\, x_0\, .$$

Multiply both sides from the left by $\Phi^T(t)\, C^T$ and integrate from 0 to t_f:

$$\int_0^{t_f} \Phi^T(t)\, C^T\, y(t)\, dt = \int_0^{t_f} \Phi^T(t)\, C^T C\, \Phi(t)\, x_0\, dt = \int_0^{t_f} \Phi^T(t)\, C^T C\, \Phi(t)\, dt\, x_0\, .$$

Define the matrix M by[7]

$$M = \int_0^{t_f} \Phi^T(t)\, C^T C\, \Phi(t)\, dt\, .$$

The equation

$$\int_0^{t_f} \Phi^T(t)\, C^T\, y(t)\, dt = M x_0$$

has a unique solution for x_0 if and only if M is nonsingular. The next theorem defines the necessary and sufficient condition for this to hold.[8]

---

[7]M is called the "observability Gramian."

[8]The proof of the theorem can be found in any graduate-level textbook on linear systems using state-space models; see, for example, [2] or [32].

Theorem 8.3 *The n-dimensional system*

$$\dot{x} = Ax, \qquad y = Cx$$

or the pair (A,C), *is observable if and only if rank* $O_M = n$, *where* O_M *is the observability matrix, defined by*

$$O_M = \begin{bmatrix} C \\ CA \\ CA^2 \\ \vdots \\ CA^{n-1} \end{bmatrix}. \qquad (8.26)$$

Investigating the observability of a pair (A,C) is conducted by forming the observability matrix O_M and testing its rank. This is performed in MATLAB using the commands "*obsv*" and "*rank*":

$$OM = obsv(A,C); \; rank(OM);$$

8-4.3 Duality

Duality exists between the controllability and observability properties. The duality follows from the structure of the controllability and observability matrices and the fact that the transpose of a matrix has the same rank as the matrix itself:

$$(A,B) \text{ controllable} \longleftrightarrow \text{rank} \begin{bmatrix} B & AB & A^2B & \cdots & A^{n-1}B \end{bmatrix} = n$$

$$\longleftrightarrow \text{rank} \begin{bmatrix} B^T \\ B^T A^T \\ B^T (A^T)^2 \\ \vdots \\ B^T (A^T)^{n-1} \end{bmatrix} = n$$

$$\longleftrightarrow (A^T, B^T) \text{ observable}.$$

Similarly, observability of (A,C) is equivalent to controllability of (A^T, C^T).

The duality property is useful in designing algorithms that build on the controllability and observability properties. Suppose you want to write numerical algorithms to test controllability and observability. Because of duality, you only need to write an algorithm to test controllability of the pair (A,B). If you want to test observability of the pair (A,C), feed the pair (A^T, C^T) to the controllability algorithm. We shall see other uses of duality in Sections 8-6 and 8-7.

8-5 Transfer Function of a State-Space Model

Our goal here is to find the transfer function of a system represented by the state-space model

$$\dot{x} = Ax + Bu, \qquad y = Cx + Du.$$

8-5. TRANSFER FUNCTION OF A STATE-SPACE MODEL

Take the Laplace transform of the state equation:

$$sX(s) - x(0) = AX(s) + BU(s).$$

Transfer functions are defined for zero initial conditions; so, set $x(0) = 0$, which leads to

$$(sI - A)X(s) = BU(s).$$

Multiply from the left by $(sI - A)^{-1}$ to obtain

$$X(s) = (sI - A)^{-1}BU(s).$$

Take the Laplace transform of the output equation and substitute for $X(s)$ from the preceding equation. The result is

$$Y(s) = CX(s) + DU(s) = C(sI - A)^{-1}BU(s) + DU(s) = [C(sI - A)^{-1}B + D]U(s).$$

Hence,

$$G(s) = C(sI - A)^{-1}B + D. \qquad (8.27)$$

Example 8-14: Computing the Transfer Function of a State-Space Model

Find the transfer function of the system

$$\dot{x} = \begin{bmatrix} -1 & -2 \\ -1 & -2 \end{bmatrix} x + \begin{bmatrix} 1 \\ -1 \end{bmatrix} u, \qquad y = \begin{bmatrix} 0 & 1 \end{bmatrix} x.$$

Solution:

$$sI - A = \begin{bmatrix} (s+1) & 2 \\ 1 & (s+2) \end{bmatrix},$$

$$(sI - A)^{-1} = \frac{1}{(s+1)(s+2) - 2} \begin{bmatrix} (s+2) & -2 \\ -1 & (s+1) \end{bmatrix},$$

$$G(s) = C(sI - A)^{-1}B$$

$$= \begin{bmatrix} 0 & 1 \end{bmatrix} \frac{1}{s(s+3)} \begin{bmatrix} (s+2) & -2 \\ -1 & (s+1) \end{bmatrix} \begin{bmatrix} 1 \\ -1 \end{bmatrix}$$

$$= \frac{-(s+2)}{s(s+3)}.$$

Example 8-15: Series RLC Circuit

Example 8-1 derives a state-space model of the electric circuit of Fig. 8-1. Calculate the transfer function from the state-space model and compare it with the transfer function derived in Example 2-4.

Solution:

$$(sI-A)^{-1} = \begin{bmatrix} (s+\frac{R}{L}) & \frac{1}{L} \\ -\frac{1}{C} & s \end{bmatrix}^{-1} = \frac{1}{s(s+\frac{R}{L})+\frac{1}{CL}} \begin{bmatrix} s & -\frac{1}{L} \\ \frac{1}{C} & (s+\frac{R}{L}) \end{bmatrix},$$

$$G(s) = C(sI-A)^{-1}B = \begin{bmatrix} 1 & 0 \end{bmatrix} \frac{1}{s(s+\frac{R}{L})+\frac{1}{CL}} \begin{bmatrix} s & -\frac{1}{L} \\ \frac{1}{C} & (s+\frac{R}{L}) \end{bmatrix} \begin{bmatrix} \frac{1}{L} \\ 0 \end{bmatrix}$$

$$= \frac{\frac{1}{L}s}{s^2 + \frac{R}{L}s + \frac{1}{CL}}.$$

$G(s)$ is the same as the one derived in Example 2-4.

Example 8-16: Automobile Suspension

Example 8-2 derives a state-space model of the automobile suspension system. Calculate the transfer function from the state-space model and compare it with the transfer function derived in Example 2-11.

Solution:

$$sI - A = \begin{bmatrix} s & -1 & 0 & 0 \\ k_s/M_s & (s+C_s/M_s) & -k_s/M_s & -C_s/M_s \\ 0 & 0 & s & -1 \\ -k_s/M_{us} & -C_s/M_{us} & (k_s+k_t)/M_{us} & (s+C_s/M_{us}) \end{bmatrix},$$

$$G(s) = C(sI-A)^{-1}B = \begin{bmatrix} 1 & 0 & 0 & 0 \end{bmatrix} \begin{bmatrix} * & * & * & \Phi_{14}(s) \\ * & * & * & * \\ * & * & * & * \\ * & * & * & * \end{bmatrix} \begin{bmatrix} 0 \\ 0 \\ 0 \\ k_t/M_{us} \end{bmatrix} = \frac{k_t}{M_{us}} \Phi_{14}(s).$$

The only element of $(sI-A)^{-1}$ that is used in calculating $G(s)$ is the (1,4) element Φ_{14}. All other elements are denoted by $*$. Using the rules of calculating the inverse of a matrix, we have

$$\Phi_{14}(s) = \frac{-1}{\det(sI-A)} \det \begin{bmatrix} -1 & 0 & 0 \\ (s+C_s/M_s) & -k_s/M_s & -C_s/M_s \\ 0 & s & -1 \end{bmatrix} = \frac{(C_s/M_s)s + k_s/M_s}{\det(sI-A)},$$

8-5. TRANSFER FUNCTION OF A STATE-SPACE MODEL

where
$$\det(sI - A) = s \det \begin{bmatrix} (s + C_s/M_s) & -k_s/M_s & -C_s/M_s \\ 0 & s & -1 \\ -C_s/M_{us} & (k_s + k_t)/M_{us} & (s + C_s/M_{us}) \end{bmatrix}$$
$$+ \det \begin{bmatrix} k_s/M_s & -k_s/M_s & -C_s/M_s \\ 0 & s & -1 \\ -k_s/M_{us} & (k_s + k_t)/M_{us} & (s + C_s/M_{us}) \end{bmatrix}$$
$$= s \left\{ \left(s + \frac{C_s}{M_s}\right) \left[s\left(s + \frac{C_s}{M_{us}}\right) + \frac{(k_s + k_t)}{M_{us}} \right] - \frac{C_s}{M_{us}} \left[\frac{k_s}{M_s} + \frac{sC_s}{M_s} \right] \right\}$$
$$+ \left\{ \frac{k_s}{M_s} \left[s\left(s + \frac{C_s}{M_{us}}\right) + \frac{(k_s + k_t)}{M_{us}} \right] - \frac{k_s}{M_{us}} \left[\frac{k_s}{M_s} + \frac{sC_s}{M_s} \right] \right\}$$
$$= s^4 + \left(\frac{C_s}{M_s} + \frac{C_s}{M_{us}} \right) s^3 + \left(\frac{(k_s + k_t)}{M_{us}} + \frac{k_s}{M_s} \right) s^2 + \left(\frac{C_s k_t}{M_s M_{us}} \right) s + \frac{k_s k_t}{M_s M_{us}}.$$

It can be verified that
$$\det(sI - A) = \frac{1}{M_s M_{us}} \Delta(s),$$

where
$$\Delta(s) = [M_s s^2 + C_s s + k_s][M_{us} s^2 + C_s s + (k_s + k_t)] - (C_s s + k_s)^2.$$

Thus,
$$G(s) = \frac{k_t}{M_{us}} \Phi_{14}(s) = \frac{\dfrac{k_t}{M_{us}} \left(\dfrac{C_s}{M_s} s + \dfrac{k_s}{M_s} \right)}{\det(sI - A)} = \frac{k_t(C_s s + k_s)}{M_s M_{us} \det(sI - A)} = \frac{k_t(C_s s + k_s)}{\Delta(s)}.$$

This is the same expression that is derived in Example 2-11.

Example 8-17: dc Motor

Example 8-3 derives a state-space model of an armature-controlled dc motor. Calculate the transfer function from the state-space model and compare with the transfer function derived in Section 2-2.3.

Solution:
$$sI - A = \begin{bmatrix} s & -1 & 0 \\ 0 & (s + b_m/J_m) & -K_t/J_m \\ 0 & K_b/L_a & (s + R_a/L_a) \end{bmatrix}$$

and
$$G(s) = C(sI - A)^{-1} B = \begin{bmatrix} 0 & 1 & 0 \end{bmatrix} \begin{bmatrix} * & * & * \\ * & * & \Phi_{23}(s) \\ * & * & * \end{bmatrix} \begin{bmatrix} 0 \\ 0 \\ 1/L_a \end{bmatrix} = \frac{\Phi_{23}(s)}{L_a}.$$

The only element of $(sI-A)^{-1}$ that is used in the calculation of $G(s)$ is the (2,3) element Φ_{23}:

$$\Phi_{23}(s) = \frac{s(K_t/J_m)}{\det(sI-A)},$$

where

$$\det(sI-A) = s\left[\left(s+\frac{b_m}{J_m}\right)\left(s+\frac{R_a}{L_a}\right)+\frac{K_bK_t}{L_aJ_m}\right].$$

Thus,

$$G(s) = \frac{K_t/(J_mL_a)}{(s+b_m/J_m)(s+R_a/L_a)+K_bK_t/(L_aJ_m)} = \frac{K_t}{(J_ms+b_m)(L_as+R_a)+K_bK_t}.$$

This is the same expression that is derived in Section 2-2.3.

8-6 State-Space Model of a Transfer Function

Given a transfer function $G(s)$, find a state-space model $\{A,B,C,D\}$ such that

$$G(s) = C(sI-A)^{-1}B+D.$$

The state model is not unique. There are many, actually infinitely many, state-space models that have the same transfer function $G(s)$. We shall present three of them: the controllable canonical form, the observable canonical form, and the parallel realization. In all three cases we will deal with strictly proper transfer functions; that is, $G(\infty) = 0$. In this case, $D = 0$ because, from the expression of $G(s)$, $G(\infty) = D$. The case $G(\infty) \neq 0$ is handled as follows: First, we set $D = G(\infty)$, then we subtract $G(\infty)$ from $G(s)$ and proceed to find a state-space model $\{A,B,C\}$ of the strictly proper transfer function

$$G(s)-G(\infty).$$

8-6.1 Controllable Canonical Form

We present the form for the case $n = 3$. The extension to $n > 3$ will be straightforward once we see the structure of the matrices for $n = 3$. Consider the transfer function

$$G(s) = \frac{c_2s^2+c_1s+c_0}{s^3+a_2s^2+a_1s+a_0} = \frac{Y(s)}{U(s)}. \tag{8.28}$$

Represent $G(s)$ as a cascade connection of two transfer functions, as shown in **Fig. 8-5**. The equations of these transfer functions are

$$Z(s) = \frac{1}{s^3+a_2s^2+a_1s+a_0}U(s), \tag{8.29}$$

and

$$Y(s) = c_0\,Z(s)+c_1s\,Z(s)+c_2s^2\,Z(s). \tag{8.30}$$

8-6. STATE-SPACE MODEL OF A TRANSFER FUNCTION

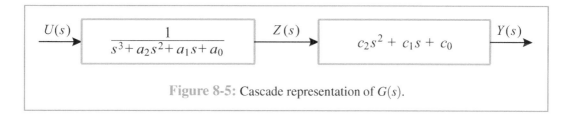

Figure 8-5: Cascade representation of $G(s)$.

Multiplying Eq. (8.29) by $(s^3 + a_2 s^2 + a_1 s + a_0)$ yields

$$(s^3 + a_2 s^2 + a_1 s + a_0) Z(s) = U(s) ,$$

which can be rearranged as

$$s^3 Z(s) = -a_0 Z(s) - a_1 s Z(s) - a_2 s^2 Z(s) + U(s) . \tag{8.31}$$

This equation can be realized by the block diagram shown in Fig. 8-6. Noting that the transfer function of the integrator is $1/s$, we see that with Z as the output of the first integrator from the right, the input to that integrator is sZ, the input to the second integrator is $s^2 Z$, and the input to the third integrator is $s^3 Z$. Hence, Eq. (8.31) is realized by the connections shown in Fig. 8-6. Equation (8.30) shows the output Y as the sum of $c_0 Z$, $c_1 sZ$, and $c_2 s^2 Z$. The signals Z, sZ, and $s^2 Z$ are the outputs of the three integrators. Upon multiplying the output of each integrator by the corresponding coefficient and then adding the three signals together, we obtain Y, as shown in Fig. 8-6.

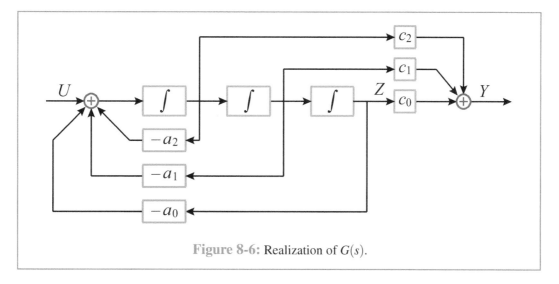

Figure 8-6: Realization of $G(s)$.

To derive a state-space model, set the state variables as the outputs of the three integrators; that is,

$$x_1(t) = z(t), \quad x_2(t) = \dot{z}(t), \quad x_3(t) = \ddot{z}(t) ,$$
$$\dot{x}_1 = x_2 ,$$

$$\dot{x}_2 = x_3,$$
$$\dot{x}_3 = -a_0 x_1 - a_1 x_2 - a_2 x_3 + u,$$

and

$$y = c_0 x_1 + c_1 x_2 + c_2 x_3.$$

From these equations, we write the matrices A, B, and C:

$$A = \begin{bmatrix} 0 & 1 & 0 \\ 0 & 0 & 1 \\ -a_0 & -a_1 & -a_2 \end{bmatrix}, \quad B = \begin{bmatrix} 0 \\ 0 \\ 1 \end{bmatrix}, \quad C = \begin{bmatrix} c_0 & c_1 & c_2 \end{bmatrix}.$$

These matrices have a certain pattern that we should note. The transfer function $G(s)$ is third-order. The dimension of the state-space model is 3. The last row of A is taken from the coefficients of the denominator polynomial of $G(s)$, starting from the absolute term and reversing the signs of the coefficients. The other two rows are zeros everywhere except for ones over the diagonal. The matrix B has 1 in the last row and zeros in the other two rows. The matrix C comes from the coefficients of the numerator polynomial of $G(s)$, starting from the absolute term. This pattern is repeated for higher-order transfer functions. For example, for

$$G(s) = \frac{c_3 s^3 + c_2 s^2 + c_1 s + c_0}{s^4 + a_3 s^3 + a_2 s^2 + a_1 s + a_0}.$$

The A, B, C matrices are

$$A = \begin{bmatrix} 0 & 1 & 0 & 0 \\ 0 & 0 & 1 & 0 \\ 0 & 0 & 0 & 1 \\ -a_0 & -a_1 & -a_2 & -a_3 \end{bmatrix}, \quad B = \begin{bmatrix} 0 \\ 0 \\ 0 \\ 1 \end{bmatrix}, \quad C = \begin{bmatrix} c_0 & c_1 & c_2 & c_3 \end{bmatrix}.$$

Why is the name "controllable canonical form" associated with this model? To answer this question, let us calculate the controllability matrix. We do it for the case $n = 3$:

$$C_M = \begin{bmatrix} B & AB & A^2 B \end{bmatrix} = \begin{bmatrix} 0 & 0 & 1 \\ 0 & 1 & -a_2 \\ 1 & -a_2 & (-a_1 + a_2^2) \end{bmatrix}.$$

The rank of this matrix is 3 irrespective of the coefficients of the transfer function. Hence, the state-space model is always controllable.

Example 8-18: Controllable-Canonical-Form Realization

Find a controllable-canonical-form realization of the transfer function

$$G(s) = \frac{2s^2 + 1}{s^3 + 3s^2 - s + 1}.$$

8-6. STATE-SPACE MODEL OF A TRANSFER FUNCTION

Solution:

$$A = \begin{bmatrix} 0 & 1 & 0 \\ 0 & 0 & 1 \\ -1 & 1 & -3 \end{bmatrix}, \quad B = \begin{bmatrix} 0 \\ 0 \\ 1 \end{bmatrix}, \quad C = \begin{bmatrix} 1 & 0 & 2 \end{bmatrix}, \quad D = 0.$$

Example 8-19: Controllable-Canonical-Form Realization When $G(\infty) \neq 0$

Find a controllable-canonical-form realization of the transfer function

$$G(s) = \frac{2s^3 + 2s^2 + 1}{s^3 + 3s^2 - s + 1}.$$

Solution: The transfer function is not strictly proper because $G(\infty) = 2$. So

$$D = G(\infty) = 2,$$

and our modified function is

$$G(s) - G(\infty) = \frac{2s^3 + 2s^2 + 1 - 2[s^3 + 3s^2 - s + 1]}{s^3 + 3s^2 - s + 1} = \frac{-4s^2 + 2s - 1}{s^3 + 3s^2 - s + 1}.$$

Hence,

$$A = \begin{bmatrix} 0 & 1 & 0 \\ 0 & 0 & 1 \\ -1 & 1 & -3 \end{bmatrix}, \quad B = \begin{bmatrix} 0 \\ 0 \\ 1 \end{bmatrix}, \quad C = \begin{bmatrix} -1 & 2 & -4 \end{bmatrix}, \quad D = 2.$$

8-6.2 Observable Canonical Form

The state-space model in this case should be observable irrespective of the coefficients of the transfer function. We use duality to derive such a state-space model. Consider the transfer function

$$G(s) = \frac{c_2 s^2 + c_1 s + c_0}{s^3 + a_2 s^2 + a_1 s + a_0},$$

and let $\{A_c, B_c, C_c\}$ be a controllable-canonical-form realization:

$$A_c = \begin{bmatrix} 0 & 1 & 0 \\ 0 & 0 & 1 \\ -a_0 & -a_1 & -a_2 \end{bmatrix}, \quad B_c = \begin{bmatrix} 0 \\ 0 \\ 1 \end{bmatrix}, \quad C_c = \begin{bmatrix} c_0 & c_1 & c_2 \end{bmatrix},$$

which leads to

$$G(s) = C_c (sI - A_c)^{-1} B_c.$$

Because $G(s)$ is scalar, $G^T(s) = G(s)$. That is,

$$\begin{aligned} G(s) = G^T(s) &= [C_c(sI - A_c)^{-1}B_c]^T \\ &= B_c^T[(sI - A_c)^{-1}]^T C_c^T \\ &= B_c^T[(sI - A_c)^T]^{-1} C_c^T \\ &= B_c^T(sI - A_c^T)^{-1} C_c^T \\ &= C(sI - A)^{-1}B, \end{aligned}$$

where

$$A = A_c^T = \begin{bmatrix} 0 & 0 & -a_0 \\ 1 & 0 & -a_1 \\ 0 & 1 & -a_2 \end{bmatrix}, \quad B = C_c^T = \begin{bmatrix} c_0 \\ c_1 \\ c_2 \end{bmatrix}, \quad C = B_c^T = \begin{bmatrix} 0 & 0 & 1 \end{bmatrix}.$$

By duality, the pair (A,C) is observable irrespective of the coefficients of the transfer function. There is a pattern for the matrices A, B, C. The last column of A comes from the coefficients of the denominator polynomial of $G(s)$, starting from the absolute term and reversing the signs of the coefficients. The other two columns are zeros everywhere except for ones under the diagonal. The matrix B comes from the coefficients of the numerator polynomial of $G(s)$, starting from the absolute term. The matrix C has 1 in the last column and zeros in the other two columns.

Example 8-20: Observable-Canonical-Form Realization

Find an observable-canonical-form realization of the transfer function

$$G(s) = \frac{2s^2 + 1}{s^3 + 3s^2 - s + 1}.$$

Solution:

$$A = \begin{bmatrix} 0 & 0 & -1 \\ 1 & 0 & 1 \\ 0 & 1 & -3 \end{bmatrix}, \quad B = \begin{bmatrix} 1 \\ 0 \\ 2 \end{bmatrix}, \quad C = \begin{bmatrix} 0 & 0 & 1 \end{bmatrix}, \quad D = 0.$$

Example 8-21: Observable-Canonical-Form Realization

Find an observable-canonical-form realization of the transfer function

$$G(s) = \frac{2s^2 + 1}{s^4 + s^3 + 3s^2 - s + 1}.$$

8-6. STATE-SPACE MODEL OF A TRANSFER FUNCTION

Solution:

$$A = \begin{bmatrix} 0 & 0 & 0 & -1 \\ 1 & 0 & 0 & 1 \\ 0 & 1 & 0 & -3 \\ 0 & 0 & 1 & -1 \end{bmatrix}, \qquad B = \begin{bmatrix} 1 \\ 0 \\ 2 \\ 0 \end{bmatrix}, \qquad C = \begin{bmatrix} 0 & 0 & 0 & 1 \end{bmatrix}, \qquad D = 0.$$

8-6.3 Parallel Realization

Consider the parallel connection of Fig. 8-7. Let

$$\dot{x}_a = A_1 x_a + B_1 u, \tag{8.32a}$$

$$y_1 = C_1 x_a + D_1 u, \tag{8.32b}$$

be a state-space model of $G_1(s)$ and

$$\dot{x}_b = A_2 x_b + B_2 u, \tag{8.33a}$$

$$y_2 = C_2 x_b + D_2 u, \tag{8.33b}$$

be a state-space model of $G_2(s)$. Set $x = \begin{bmatrix} x_a \\ x_b \end{bmatrix}$ as a state vector for the parallel connection:

$$\begin{bmatrix} \dot{x}_a \\ \dot{x}_b \end{bmatrix} = \begin{bmatrix} A_1 & 0 \\ 0 & A_2 \end{bmatrix} \begin{bmatrix} x_a \\ x_b \end{bmatrix} + \begin{bmatrix} B_1 \\ B_2 \end{bmatrix} u$$

and

$$y = y_1 + y_2 = \begin{bmatrix} C_1 & C_2 \end{bmatrix} \begin{bmatrix} x_a \\ x_b \end{bmatrix} + (D_1 + D_2) u.$$

Hence, a state-space model of the parallel connection is given by

$$A = \begin{bmatrix} A_1 & 0 \\ 0 & A_2 \end{bmatrix}, \qquad B = \begin{bmatrix} B_1 \\ B_2 \end{bmatrix}, \qquad C = \begin{bmatrix} C_1 & C_2 \end{bmatrix}, \qquad D = D_1 + D_2.$$

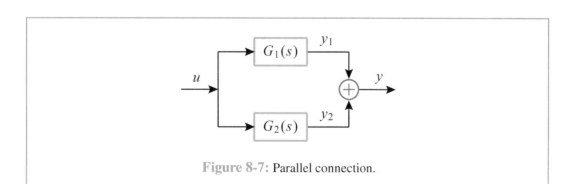

Figure 8-7: Parallel connection.

Parallel Realization of $G(s)$

The procedure for obtaining a parallel realization of a transfer function consists of the following steps:

Step 1: Factor the denominator polynomial of $G(s)$ as

$$d(s) = d_1(s)\, d_2(s) \ldots d_r(s)$$

such that

(a) $d_i(s)$ has real coefficients, and

(b) there are no common factors between $d_i(s)$ and $d_j(s)$ for $i \neq j$.

Step 2: Use partial fraction expansion to express $G(s) - G(\infty)$ as

$$G(s) - G(\infty) = G_1(s) + G_2(s) + \cdots + G_r(s),$$

where $G_i(s)$ is strictly proper and its denominator is $d_i(s)$.

Step 3: Find a state-space model $\{A_i, B_i, C_i\}$ of each $G_i(s)$.

Step 4: Connect the r realizations in parallel. A state-space model of $G(s)$ is then given by

$$A = \begin{bmatrix} A_1 & & & \\ & A_2 & & \\ & & \ddots & \\ & & & A_r \end{bmatrix}, \quad B = \begin{bmatrix} B_1 \\ B_2 \\ \vdots \\ B_r \end{bmatrix}$$

$$C = \begin{bmatrix} C_1 & C_2 & \cdots & C_r \end{bmatrix}, \qquad D = G(\infty).$$

If $G(s)$ is strictly proper and has real distinct poles, partial fraction expansion yields

$$G(s) = \sum_{i=1}^{n} \frac{K_i}{s - p_i}, \quad \text{where } K_i = (s - p_i)\, G(s)|_{s=p_i}.$$

The first-order transfer function $K_i/(s - p_i)$ can be realized by the state-space model

$$\dot{x}_i = p_i x_i + u, \qquad y_i = K_i x_i,$$

which is a special case of the controllable canonical form, or by

$$\dot{x}_i = p_i x_i + K_i u, \qquad y_i = x_i,$$

which is a special case of the observable canonical form.

There are two cases where partial fraction expansion will produce terms of order higher than 1. If $G(s)$ has multiple poles, partial fraction expansion will yield terms of the form

$$\frac{N_i(s)}{(s - p_i)^k}, \text{ with } k > 1.$$

8-6. STATE-SPACE MODEL OF A TRANSFER FUNCTION

The second case is when $G(s)$ has complex poles. The polynomial $s^2 + 2\zeta\omega_n s + \omega_n^2$, with $\zeta < 1$, can be factored as $(s + \sigma + j\omega)(s + \sigma - j\omega)$. This factorization is not allowed because all the coefficients of the partial fraction expansion terms must be real because the matrices A, B, C, are real. Therefore, the second-order polynomial must be kept without factorization. Such higher-order terms of the partial fraction expansion can be realized by the controllable canonical form or the observable canonical form.

Example 8-22: Parallel Realization When $G(s)$ Has Distinct Real Poles

Find a parallel realization of the transfer function

$$G(s) = \frac{1}{(s+1)(s+2)(s+3)}.$$

Solution:

$$G(s) = \frac{K_1}{s+1} + \frac{K_2}{s+2} + \frac{K_3}{s+3},$$

with

$$K_1 = \frac{1}{(s+2)(s+3)}\bigg|_{s=-1} = \frac{1}{2},$$

$$K_2 = \frac{1}{(s+1)(s+3)}\bigg|_{s=-2} = -1,$$

and

$$K_3 = \frac{1}{(s+1)(s+2)}\bigg|_{s=-3} = \frac{1}{2},$$

which leads to

$$G(s) = \frac{\frac{1}{2}}{s+1} + \frac{-1}{s+2} + \frac{\frac{1}{2}}{s+3}.$$

The parallel realization is given by

$$A = \begin{bmatrix} -1 & 0 & 0 \\ 0 & -2 & 0 \\ 0 & 0 & -3 \end{bmatrix}, \quad B = \begin{bmatrix} 1 \\ 1 \\ 1 \end{bmatrix}, \quad C = \begin{bmatrix} \frac{1}{2} & -1 & \frac{1}{2} \end{bmatrix}, \quad D = 0.$$

Example 8-23: Parallel Realization When $G(s)$ Has Multiple and Complex Poles

Find a parallel realization of the transfer function

$$G(s) = \frac{5s^4 + 19s^3 + 30s^2 + 24s + 9}{(s+1)^2(s+2)(s^2+s+1)}.$$

Solution: Here, $G(\infty) = 0$. Hence,

$$G(s) = \underbrace{\frac{s+2}{(s+1)^2}}_{G_1(s)} + \underbrace{\frac{3}{s+2}}_{G_2(s)} + \underbrace{\frac{s+1}{s^2+s+1}}_{G_3(s)}.$$

For

$$G_1(s) = \frac{s+2}{(s+1)^2} = \frac{s+2}{s^2+2s+1},$$

a controllable-canonical-form realization of $G_1(s)$ is given by

$$A_1 = \begin{bmatrix} 0 & 1 \\ -1 & -2 \end{bmatrix}, \quad B_1 = \begin{bmatrix} 0 \\ 1 \end{bmatrix}, \quad C_1 = \begin{bmatrix} 2 & 1 \end{bmatrix}.$$

Similarly,

$$G_2(s) = \frac{3}{s+2}$$

is realized by

$$A_2 = -2, \quad B_2 = 1, \quad C_2 = 3,$$

and for

$$G_3(s) = \frac{s+1}{s^2+s+1},$$

a controllable-canonical-form realization of $G_3(s)$ is given by

$$A_3 = \begin{bmatrix} 0 & 1 \\ -1 & -1 \end{bmatrix}, \quad B_3 = \begin{bmatrix} 0 \\ 1 \end{bmatrix}, \quad C_3 = \begin{bmatrix} 1 & 1 \end{bmatrix}.$$

Thus, a parallel realization of $G(s)$ is given by

$$A = \left[\begin{array}{cc|c|cc} 0 & 1 & 0 & 0 & 0 \\ -1 & -2 & 0 & 0 & 0 \\ \hline 0 & 0 & -2 & 0 & 0 \\ \hline 0 & 0 & 0 & 0 & 1 \\ 0 & 0 & 0 & -1 & -1 \end{array}\right], \quad B = \begin{bmatrix} 0 \\ 1 \\ 1 \\ 0 \\ 1 \end{bmatrix},$$

$$C = \begin{bmatrix} 2 & 1 & | & 3 & | & 1 & 1 \end{bmatrix}, \quad D = 0.$$

8-7 State-Space Design

The stability of the *n*-dimensional system

$$\dot{x} = Ax + Bu, \quad y = Cx + Du$$

and the shape of its transient response are determined by the location of the eigenvalues of A. If the system is not stable or if it is stable but the transient response does not meet the design specifications, we can re-assign the eigenvalues by feedback control.

Feedback control is classified into

- *State Feedback*: All state variables are measured and can be used in feedback.

- *Output Feedback*: Only the output y is measured and can be used in feedback.

8-7.1 State Feedback Control

Consider a control defined by

$$u = -Kx + v, \tag{8.34}$$

where K is a $1 \times n$ matrix and $v(t)$ is an additional input. The closed-loop system is given by

$$\dot{x} = (A - BK)x + Bv, \tag{8.35a}$$
$$y = (C - DK)x + Dv. \tag{8.35b}$$

The stability and transient response of the closed-loop system are determined by the eigenvalues of $(A - BK)$, which are called the *closed-loop eigenvalues*, in contrast to the *open-loop eigenvalues*, which are the eigenvalues of A.

There are two issues that we need to address regarding the design of K:

(1) How to choose the closed-loop eigenvalues.

(2) How to compute K.

We start with the second issue.

Computation of K:

Consider a system in the controllable canonical form:

$$A = \begin{bmatrix} 0 & 1 & \cdots & 0 \\ \vdots & \vdots & \ddots & \vdots \\ 0 & 0 & \cdots & 1 \\ -a_0 & -a_1 & \cdots & -a_{n-1} \end{bmatrix}, \quad B = \begin{bmatrix} 0 \\ \vdots \\ 0 \\ 1 \end{bmatrix},$$

and

$$\det(sI - A) = s^n + a_{n-1}s^{n-1} + \cdots + a_1 s + a_0. \tag{8.36}$$

Let

$$K = \begin{bmatrix} k_0 & k_1 & \cdots & k_{n-1} \end{bmatrix},$$

$$A - BK = \begin{bmatrix} 0 & 1 & \cdots & 0 \\ \vdots & \vdots & \ddots & \vdots \\ 0 & 0 & \cdots & 1 \\ -(a_0 + k_0) & -(a_1 + k_1) & \cdots & -(a_{n-1} + k_{n-1}) \end{bmatrix},$$

and

$$\det[sI - (A - BK)] = s^n + (a_{n-1} + k_{n-1})s^{n-1} + \cdots + (a_1 + k_1)s + (a_0 + k_0). \tag{8.37}$$

If $\{\lambda_1, \ldots, \lambda_n\}$ are the desired eigenvalues, and are such that those with complex eigenvalues are in conjugate pairs, then the desired characteristic polynomial is

$$(s - \lambda_1) \times \cdots \times (s - \lambda_n) = s^n + f_{n-1}s^{n-1} + \cdots + f_0.$$

By specifying $k_i = f_i - a_i,$ for $0 \leq i \leq n-1,$

we ensure that the matrix $(A - BK)$ has the desired eigenvalues $\{\lambda_1, \ldots, \lambda_n\}$.

Example 8-24: Computation of K to Assign the Closed-Loop Eigenvalues

Let
$$A = \begin{bmatrix} 0 & 1 & 0 \\ 0 & 0 & 1 \\ 1 & 0 & 1 \end{bmatrix}, \quad B = \begin{bmatrix} 0 \\ 0 \\ 1 \end{bmatrix},$$

and
$$\det(sI - A) = s^3 - s^2 - 1.$$

By application of the Routh-Hurwitz criterion or calculating the roots of $s^3 - s^2 - 1$, it can be shown that the system is not stable. Design K to assign the eigenvalues of $(A - BK)$ at $-1, -1 \pm j$.

Solution: The desired characteristic polynomial is

$$(s+1)(s+1+j)(s+1-j) = s^3 + 3s^2 + 4s + 2. \tag{8.38}$$

Comparing the individual terms in Eq. (8.38) with the corresponding terms in Eq. (8.37) leads to
$$K = \begin{bmatrix} 3 & 4 & 4 \end{bmatrix}.$$

What if the pair (A,B) is not in the controllable canonical form? We can use the same procedure but solving for K requires solving n simultaneous equations in n unknowns.

Example 8-25: Computation of K When (A,B) Is Not in the Controllable-Canonical Form

Let
$$A = \begin{bmatrix} -1 & 0 & 0 \\ 0 & 1 & 0 \\ 0 & 0 & -2 \end{bmatrix}, \quad B = \begin{bmatrix} 1 \\ 2 \\ 1 \end{bmatrix}.$$

Design K such that the eigenvalues of $(A - BK)$ are $-2, -2 \pm j$.

Solution:

$$K = \begin{bmatrix} k_0 & k_1 & k_2 \end{bmatrix}, \quad A - BK = \begin{bmatrix} (-1-k_0) & -k_1 & -k_2 \\ -2k_0 & (1-2k_1) & -2k_2 \\ -k_0 & -k_1 & (-2-k_2) \end{bmatrix},$$

8-7. STATE-SPACE DESIGN

$$sI - (A - BK) = \begin{bmatrix} (s+1+k_0) & k_1 & k_2 \\ 2k_0 & (s-1+2k_1) & 2k_2 \\ k_0 & k_1 & (s+2+k_2) \end{bmatrix},$$

and therefore

$$\det[sI - (A - BK)] = s^3 + (k_0 + 2k_1 + k_2 + 2)s^2 + (k_0 + 6k_1 - 1)s + (-2k_0 + 4k_1 - k_2 - 2).$$

The desired characteristic polynomial is

$$(s+2)(s+2+j)(s+2-j) = s^3 + 6s^2 + 13s + 10.$$

Matching coefficients of like powers of s leads to

$$\left. \begin{array}{r} k_0 + 2k_1 + k_2 + 2 = 6 \\ k_0 + 6k_1 - 1 = 13 \\ -2k_0 + 4k_1 - k_2 - 2 = 10 \end{array} \right\} \rightarrow K = \begin{bmatrix} -1 & 2.5 & 0 \end{bmatrix}.$$

Since for general A and B we need to solve n equations in n unknowns, is there a guarantee these equations will have a solution? The answer is yes if the pair (A,B) is controllable, as stated in the following theorem.[9]

Theorem 8.4 *Given the n-dimensional controllable pair (A,B), and the set of desired eigenvalues $\{\lambda_1, \ldots, \lambda_n\}$, where complex eigenvalues are in conjugate pairs, there is a unique matrix K that assigns the eigenvalues of $(A - BK)$ at $\{\lambda_1, \ldots, \lambda_n\}$.*

The computation of K in MATLAB is realized using the commands "*acker*" or "*place*":

(1) Define A, B, and p (a vector of desired eigenvalues);

(2) K = acker(A,B,p);

(3) K = place(A,B,p);   (does not accept multiple eigenvalues).

Selection of the eigenvalues

The transient response of the closed-loop system

$$\dot{x} = (A - BK)x + Bv, \qquad (8.39a)$$
$$y = (C - DK)x + Dv, \qquad (8.39b)$$

is examined by simulating the step response or the zero input response. The locations of the eigenvalues are adjusted to meet the design specifications.

---

[9]The proof of the theorem can be found in any graduate-level textbook on linear systems using state-space models; see, for example, [2] or [32]. It is based on the fact that for a controllable pair (A,B) there is a nonsingular matrix P such that $P^{-1}AP = A_c$ and $P^{-1}B = B_c$, where the pair (A_c, B_c) is in the controllable canonical form. A matrix K_c is designed to assign the eigenvalues of $(A_c - B_c K_c)$ at the desired eigenvalues. Taking $K = K_c P^{-1}$ yields $(A - BK) = P(A_c - B_c K_c)P^{-1}$. Hence, the eigenvalues of $(A - BK)$ are assigned at the desired eigenvalues because the similarity transformation $P(A_c - B_c K_c)P^{-1}$ does not change the eigenvalues.

Guidelines:

(a) The settling time of each mode is $T_s = 4/|\text{Re}[\lambda_i]|$. To achieve a faster response, the eigenvalues should be moved to the left in the complex plane.

(b) For complex eigenvalues, the response is more oscillatory for smaller values of the damping ratio ζ.

(c) If the eigenvalues are clustered into slow and fast ones, the slow eigenvalues are dominant.

Can we make the response arbitrarily fast by moving the eigenvalues far enough to the left? In theory, yes, but this will require large feedback gains and hence large control magnitude since $u(t) = -Kx(t)$.

The choice of eigenvalues is a trade-off between the transient response and the control effort.

Example 8-26: Assignment of the Closed-Loop Eigenvalues for a Desired Settling Time

---

Let
$$A = \begin{bmatrix} 0 & 1 & 0 \\ 0 & 0 & 1 \\ -0.4 & -4.2 & -2.1 \end{bmatrix}, \quad B = \begin{bmatrix} 0 \\ 0 \\ 1 \end{bmatrix}.$$

The open-loop eigenvalues are $-0.1, -1 \pm \sqrt{3}j$. Because of the eigenvalue at -0.1, the settling time is estimated to be 40 s. Design feedback gain K such that the settling time is ≤ 4 s.

Solution: Choose the closed-loop eigenvalues as $-2, -2 \pm 2\sqrt{3}j$ to achieve a settling time of about 2 s. Hence,
$$K = \begin{bmatrix} 31.6 & 19.8 & 3.9 \end{bmatrix}.$$

The zero-input response to $x(0) = \begin{bmatrix} 1 & 1 & 1 \end{bmatrix}^T$ is shown in Fig. 8-8(a). The settling time is indeed about 2 s, but the magnitude of the control signal is very large during the initial transient.

To address this shortcoming, move the real parts of the eigenvalues to -1 to achieve a settling time of about 4 s, while keeping the damping ratio of the complex eigenvalues the same. The new choice is $-1, -1 \pm \sqrt{3}j$ and
$$K = \begin{bmatrix} 3.6 & 1.8 & 0.9 \end{bmatrix}.$$

In comparison with the previous K, there is now significant reduction in the feedback gain.

Figure 8-8(b) shows the zero-input response to $x(0) = \begin{bmatrix} 1 & 1 & 1 \end{bmatrix}^T$. The control magnitude at the initial transient has been reduced by a factor close to 10 and the settling time is about 4 s. This is an example of how the transient response and control effort are traded off against each other.

8-7. STATE-SPACE DESIGN

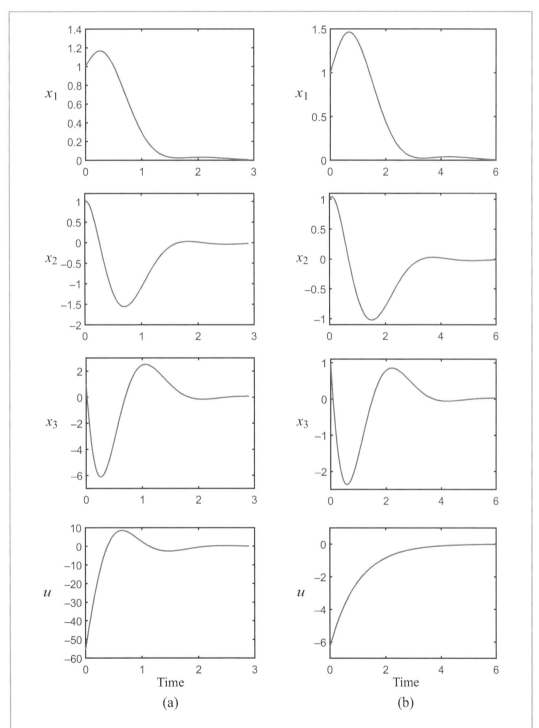

Figure 8-8: (a) Simulation of Example 8-26 with $K = [31.6 \ \ 19.8 \ \ 3.9]$. (b) Simulation of Example 8-26 with $K = [3.6 \ \ 1.8 \ \ 0.9]$.

8-7.2 Observers and Output Feedback Control

Observers

Consider the system

$$\dot{x} = Ax + Bu, \qquad (8.40a)$$
$$y = Cx + Du, \qquad (8.40b)$$

where the initial state $x(0)$ is unknown, but we can measure the output y. We want to estimate the state x from the available signals u and y; u is available because it is the control that we implement.

Had $x(0)$ been known, we could have estimated x by solving the equation

$$\dot{\hat{x}} = A\hat{x} + Bu, \qquad \hat{x}(0) = x(0)$$

and we would have obtained $\hat{x}(t) \equiv x(t)$. This can be seen by defining the estimation error

$$e = x - \hat{x},$$

which satisfies the equation

$$\dot{e} = Ax + Bu - A\hat{x} - Bu = Ae, \qquad e(0) = 0,$$

whose solution is $e(t) \equiv 0$.

If $x(0)$ is unknown, we cannot choose $\hat{x}(0) = x(0)$. Hence $e(0) \neq 0$ and the estimation error will be different from zero. Note, however, that if $\text{Re}[\lambda(A)] < 0$, that is, if all eigenvalues of A have negative real parts, then

$$\lim_{t \to \infty} e(t) = 0.$$

Hence, we can asymptotically estimate the state x.

What if A has some eigenvalues with nonnegative real parts? Or what if all the eigenvalues have negative real parts, but convergence is slow? Can we use feedback to stabilize the observer? What signal can we feedback?

We measure $y(t) = Cx(t) + Du(t)$. If an estimate of the state $\hat{x}(t)$ is available, we can use it to estimate the output by $\hat{y}(t) = C\hat{x}(t) + Du(t)$. Then,

$$y(t) - \hat{y}(t) = Cx(t) + Du(t) - C\hat{x}(t) - Du(t) = Ce(t). \qquad (8.41)$$

The signal $y - \hat{y} = Ce$ is the feedback signal to the observer.

Consider the *observer* (*estimator*):

$$\dot{\hat{x}} = A\hat{x} + Bu + L[y - \hat{y}] = A\hat{x} + Bu + L[y - C\hat{x} - Du], \qquad (8.42)$$

where L is the *observer gain* (an $n \times 1$ matrix). The estimation error $e = x - \hat{x}$ satisfies the equation

$$\dot{e} = \dot{x} - \dot{\hat{x}} = Ax + Bu - A\hat{x} - Bu - L[Cx - C\hat{x}] = (A - LC)e.$$

8-7. STATE-SPACE DESIGN

We would like to design L such that all the eigenvalues of $(A-LC)$ have negative real parts, because then $\lim_{t\to\infty} e(t) = 0$. When is that possible? The design of L to assign the eigenvalues of $(A-LC)$ is dual to the design of K to assign the eigenvalues of $(A-BK)$. In particular, $(A-LC)$ and $(A-LC)^T$ have the same eigenvalues. Then,

$$(A-LC)^T = A^T - C^T L^T.$$

Set $\quad \hat{A} = A^T, \quad \hat{B} = C^T, \quad \hat{K} = L^T \quad \longrightarrow \quad (A-LC)^T = \hat{A} - \hat{B}\hat{K}.$

Thus, the design of L to assign the eigenvalues of $(A-LC)$ can be realized by the design of $\hat{K}$ to assign the eigenvalues of $(\hat{A} - \hat{B}\hat{K})$.

Observer design:

Given the pair (A,C) and the set of eigenvalues $\{\lambda_1, \ldots, \lambda_n\}$, where complex eigenvalues are in conjugate pairs, design L such that the eigenvalues of $(A-LC)$ are assigned at $\{\lambda_1, \ldots, \lambda_n\}$.

Design procedure:

Step 1: Specify $\hat{A} = A^T$ and $\hat{B} = C^T$.

Step 2: Design $\hat{K}$ to assign the eigenvalues of $(\hat{A} - \hat{B}\hat{K})$ at $\{\lambda_1, \ldots, \lambda_n\}$.

Step 3: Set $L = \hat{K}^T$.

We know from Theorem 8.4 that $\hat{K}$ exists if $(\hat{A}, \hat{B})$ is controllable. By duality, L exists if (A,C) is observable.

Output feedback control:

Design an *output feedback controller* to stabilize the system

$$\dot{x} = Ax + Bu, \quad y = Cx + Du.$$

Design procedure:

Step 1: Design a state feedback controller $u = -Kx + v$ such that $\text{Re}[\lambda(A-BK)] < 0$.

Step 2: Design an observer

$$\dot{\hat{x}} = A\hat{x} + Bu + L[y - C\hat{x} - Du],$$

where L is chosen such that $\text{Re}[\lambda(A-LC)] < 0$.

Step 3: Set the output feedback controller as $u = -K\hat{x} + v$.

To derive the state equation of the closed-loop system under output feedback, select the state vector as

$$\begin{bmatrix} x \\ e \end{bmatrix} = \begin{bmatrix} x \\ x - \hat{x} \end{bmatrix}.$$

Then

$$\dot{x} = Ax - BK\hat{x} + Bv = Ax - BK(x-e) + Bv = (A-BK)x + BKe + Bv.$$

We have already seen that $\dot{e} = (A - LC)e$ for any control u. Hence,

$$\begin{bmatrix} \dot{x} \\ \dot{e} \end{bmatrix} = \begin{bmatrix} (A-BK) & BK \\ 0 & (A-LC) \end{bmatrix} \begin{bmatrix} x \\ e \end{bmatrix} + \begin{bmatrix} B \\ 0 \end{bmatrix} v. \tag{8.43}$$

The closed-loop matrix is block triangular. Block triangular matrices have the property that their eigenvalues are the eigenvalues of their diagonal blocks. Hence, the closed-loop eigenvalues are the eigenvalues of $(A - BK)$ and $(A - LC)$. Since both matrices are designed to have eigenvalues with negative real parts, the closed-loop system is stable.

It is useful to note that the closed-loop transfer function from v to y is

$$(C - DK)[sI - (A - BK)]^{-1}B + D,$$

which is the same closed-loop transfer function under state feedback. This is so because when the Laplace transform is taken of the equation $\dot{e} = (A - LC)e$ with zero initial condition, it results in

$$sE(s) = (A - LC)E(s) \quad \rightarrow \quad [sI - (A - LC)]E(s) = 0 \quad \rightarrow \quad E(s) = 0.$$

Example 8-27: Output Feedback Stabilization

Design an output feedback controller to stabilize the system

$$\dot{x} = Ax + Bu, \qquad y = Cx,$$

where

$$A = \begin{bmatrix} 0 & 1 \\ 0 & 0 \end{bmatrix}, \qquad B = \begin{bmatrix} 0 \\ 1 \end{bmatrix}, \qquad C = \begin{bmatrix} 1 & 0 \end{bmatrix}.$$

Solution:

State feedback control: $u = -Kx + v$

$$K = \begin{bmatrix} k_0 & k_1 \end{bmatrix} \quad \rightarrow \quad A - BK = \begin{bmatrix} 0 & 1 \\ -k_0 & -k_1 \end{bmatrix}$$

$$\det[sI - (A - BK)] = s^2 + k_1 s + k_0.$$

Specify the desired eigenvalues to be a pair of complex eigenvalues with $\omega_n = 2$ and $\zeta = 0.5$, which leads to

$$s^2 + 2\zeta \omega_n s + \omega_n^2 = s^2 + 2s + 4.$$

Matching coefficients of like powers of s, we obtain:

$$K = \begin{bmatrix} 4 & 2 \end{bmatrix}.$$

8-7. STATE-SPACE DESIGN

Observer:

$$L = \begin{bmatrix} \ell_0 \\ \ell_1 \end{bmatrix} \quad \rightarrow \quad A - LC = \begin{bmatrix} -\ell_0 & 1 \\ -\ell_1 & 0 \end{bmatrix}$$

and

$$\det[sI - (A - LC)] = s^2 + \ell_0 s + \ell_1 .$$

Specify the desired eigenvalues to be a pair of complex eigenvalues with $\omega_n = 10$ and $\zeta = 0.5$, which yields

$$s^2 + 2\zeta\omega_n s + \omega_n^2 = s^2 + 10s + 100.$$

Matching coefficients of like powers of s, we obtain:

$$L = \begin{bmatrix} 10 \\ 100 \end{bmatrix}.$$

Output feedback controller:

$$u = -4\hat{x}_1 - 2\hat{x}_2 + v ,$$
$$\dot{\hat{x}}_1 = \hat{x}_2 + 10(y - \hat{x}_1) ,$$
$$\dot{\hat{x}}_2 = u + 100(y - \hat{x}_1) .$$

8-7.3 Set-Point Regulation

Consider the system

$$\dot{x} = Ax + Bu , \qquad y = Cx .$$

Design feedback control to:

(a) stabilize the system, and

(b) regulate the output y to a set point r; i.e.,

$$\lim_{t \to \infty} y(t) = r .$$

There are two approaches to design the feedback control: *feedforward control* and *integral control*.

Feedforward Control

We start with the state feedback control:

$$u = -Kx + v ,$$

where K is designed such that $\text{Re}[\lambda(A - BK)] < 0$. The closed-loop system is

$$\dot{x} = (A - BK)x + Bv , \qquad y = Cx .$$

The transfer function from v to y is

$$C[sI - (A - BK)]^{-1}B .$$

Set $v = Nr$. The transfer function from r to y is

$$G(s) = C[sI - (A - BK)]^{-1}BN . \tag{8.44}$$

How would we choose N to achieve $\lim_{t \to \infty} y(t) = r$? From the final value theorem of the Laplace transform

$$\lim_{t \to \infty} y(t) = \lim_{s \to 0} sY(s) = \lim_{s \to 0} s\, G(s) \frac{r}{s} = G(0)\, r ,$$

which gives

$$\lim_{t \to \infty} y(t) = -C(A - BK)^{-1}BNr .$$

Here, $(A - BK)$ is nonsingular because all its eigenvalues have negative real parts. Assuming that $C(A - BK)^{-1}B \neq 0$,[10] set

$$N = -\frac{1}{C(A - BK)^{-1}B} . \tag{8.45}$$

With this choice of N, the state feedback control is given by

$$u = -Kx + Nr . \tag{8.46}$$

For output feedback control, we design an observer to estimate x by $\hat{x}$ and implement the state feedback control replacing x by $\hat{x}$. For the observer, we design L such that $\operatorname{Re}[\lambda(A - LC)] < 0$. The output feedback control is given by:

$$u = -K\hat{x} + Nr ,$$

$$\dot{\hat{x}} = A\hat{x} + Bu + L(y - C\hat{x}) .$$

The transfer function from r to y under output feedback is the same as the one under state feedback. Therefore $\lim_{t \to \infty} y(t) = r$.

Example 8-28: Set-Point Regulation by Feedforward Control

Consider the system

$$\dot{x}_1 = x_2 , \qquad \dot{x}_2 = x_1 - x_2 + u, \qquad y = x_1 .$$

---

[10] $C(A - BK)^{-1}B \neq 0$ if and only if

$$\operatorname{rank} \begin{bmatrix} A & B \\ C & 0 \end{bmatrix} = n + 1.$$

This can be shown by performing the following elementary operations:

$$\begin{bmatrix} A & B \\ C & 0 \end{bmatrix} \to \begin{bmatrix} A - BK & B \\ C & 0 \end{bmatrix} \to \begin{bmatrix} I & (A-BK)^{-1}B \\ C & 0 \end{bmatrix} \to \begin{bmatrix} I & (A-BK)^{-1}B \\ 0 & -C(A-BK)^{-1}B \end{bmatrix} .$$

8-7. STATE-SPACE DESIGN

Design an output feedback controller to stabilize the closed-loop system and regulate y to a set point r.

Solution:

$$A = \begin{bmatrix} 0 & 1 \\ 1 & -1 \end{bmatrix}, \quad B = \begin{bmatrix} 0 \\ 1 \end{bmatrix}, \quad C = \begin{bmatrix} 1 & 0 \end{bmatrix}.$$

Here, (A, B) is controllable and (A, C) is observable. We choose $K = \begin{bmatrix} 2 & 1 \end{bmatrix}$ so as to assign the eigenvalues of $(A - BK)$ at $-1, -1$. Hence,

$$C(A - BK)^{-1} B = \begin{bmatrix} 1 & 0 \end{bmatrix} \begin{bmatrix} -2 & -1 \\ 1 & 0 \end{bmatrix} \begin{bmatrix} 0 \\ 1 \end{bmatrix} = -1 \quad \rightarrow \quad N = 1.$$

We choose $L = \begin{bmatrix} 9 \\ 17 \end{bmatrix}$ so as to assign the eigenvalues of $(A - LC)$ at $-5, -5$. The output feedback controller is then given by

$$u = -2\hat{x}_1 - \hat{x}_2 + r,$$

$$\dot{\hat{x}} = \begin{bmatrix} 0 & 1 \\ 1 & -1 \end{bmatrix} \hat{x} + \begin{bmatrix} 0 \\ 1 \end{bmatrix} u + \begin{bmatrix} 9 \\ 17 \end{bmatrix} (y - \hat{x}_1).$$

Integral Control

We augment the system with an integrator driven by the regulation error:

$$\dot{z} = r - y.$$

The augmented system is described by

$$\begin{bmatrix} \dot{x} \\ \dot{z} \end{bmatrix} = \begin{bmatrix} A & 0 \\ -C & 0 \end{bmatrix} \begin{bmatrix} x \\ z \end{bmatrix} + \begin{bmatrix} B \\ 0 \end{bmatrix} u + \begin{bmatrix} 0 \\ 1 \end{bmatrix} r.$$

Set $\mathcal{X} = \begin{bmatrix} x \\ z \end{bmatrix}, \quad \mathcal{A} = \begin{bmatrix} A & 0 \\ -C & 0 \end{bmatrix}, \quad \mathcal{B} = \begin{bmatrix} B \\ 0 \end{bmatrix}, \quad \Gamma = \begin{bmatrix} 0 \\ 1 \end{bmatrix},$

to obtain

$$\dot{\mathcal{X}} = \mathcal{A}\mathcal{X} + \mathcal{B}u + \Gamma r.$$

We design the state feedback control:[11]

$$u = -K\mathcal{X} = \begin{bmatrix} -K_1 & -K_2 \end{bmatrix} \begin{bmatrix} x \\ z \end{bmatrix} = -K_1 x - K_2 z$$

such that

$$\operatorname{Re}[\lambda(\mathcal{A} - \mathcal{B}K)] < 0,$$

---

[11] The existence of K is guaranteed if $(\mathcal{A}, \mathcal{B})$ is controllable. It can be shown that $(\mathcal{A}, \mathcal{B})$ is controllable if and only if (A, B) is controllable and rank $\begin{bmatrix} A & B \\ C & 0 \end{bmatrix} = n + 1.$

which ensures that the closed-loop system

$$\dot{x} = Ax - BK_1 x - BK_2 z$$
$$\dot{z} = r - Cx$$

is stable. The closed-loop system has an equilibrium point at $(\bar{x}, \bar{z})$, which satisfies the equlibrium equations[12]

$$0 = A\bar{x} - BK_1\bar{x} - BK_2\bar{z},$$
$$0 = r - C\bar{x}.$$

Because the system is stable, its solution converges to the equilibrium point as $t \to \infty$. Consequently,

$$\lim_{t \to \infty} y(t) = \bar{y} = C\bar{x} = r.$$

For output feedback control, we design an observer to estimate x by $\hat{x}$ and implement the state feedback control replacing x with $\hat{x}$. For the observer, we design L such that $\mathrm{Re}[\lambda(A - LC)] < 0$. We do not need to estimate z because it is available by integrating $(r - y)$. The output feedback control is given by

$$u = -K_1\hat{x} - K_2 z,$$
$$\dot{z} = r - y,$$
$$\dot{\hat{x}} = A\hat{x} + Bu + L(y - C\hat{x}).$$

The closed-loop system given by

$$\dot{x} = (A - BK_1)x + BK_1 e - BK_2 z,$$
$$\dot{z} = r - Cx,$$

and

$$\dot{e} = (A - LC)e,$$

is stable and its solution converges to the equilibrium point $(x = \bar{x},\ z = \bar{z},\ e = 0)$. Therefore,

$$\lim_{t \to \infty} y(t) = \bar{y} = C\bar{x} = r.$$

The main advantage of integral control is its robustness to parameters perturbation and constant disturbances. Such perturbation and disturbance may change the equilibrium point but at equilibrium, it is always true that $\bar{y} = r$. Regulation is achieved for all parameter perturbations for which the closed-loop system remains stable.

---

[12] The condition rank $\begin{bmatrix} A & B \\ C & 0 \end{bmatrix} = n + 1$ ensures there there is a unique equilibrium point.

8-7. STATE-SPACE DESIGN

Example 8-29: Set-Point Regulation by Integral Control

Reconsider the regulation problem of Example 8-28 and design the output feedback controller using integral control.

Solution:

$$\mathcal{A} = \begin{bmatrix} A & 0 \\ -C & 0 \end{bmatrix} = \begin{bmatrix} 0 & 1 & 0 \\ 1 & -1 & 0 \\ -1 & 0 & 0 \end{bmatrix}, \qquad \mathcal{B} = \begin{bmatrix} B \\ 0 \end{bmatrix} = \begin{bmatrix} 0 \\ 1 \\ 0 \end{bmatrix}.$$

Vector $K = \begin{bmatrix} 6 & 3 & -2 \end{bmatrix}$ assigns the eigenvalues at $-1, -1, -2$. We use the same observer as in Example 8-28. The output feedback controller is given by

$$u = -6\hat{x}_1 - 3\hat{x}_2 + 2z, \qquad \dot{z} = r - y.$$

The unit step responses of the closed-loop system under integral control and feedforward control are shown in Fig. 8-9. Parts (a) and (b) of the figure show the output and control signals for nominal parameters, and parts (c) and (d) show them when the control coefficient is perturbed from 1 to 1.5. For nominal parameters both controllers regulate the output to the set point, with the response of the integral control slightly slower than the response of the feedforward control. When the control coefficient is perturbed, the integral control still regulates the output to the set point but the feedforward control fails to do so.

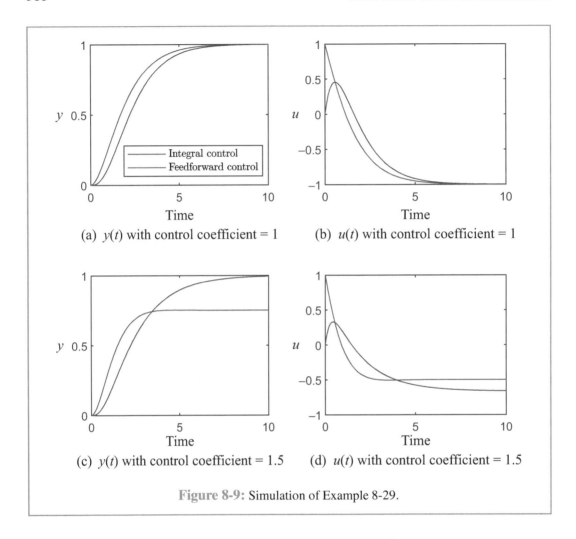

Figure 8-9: Simulation of Example 8-29.

Summary

Concepts

- A state-space model captures both the internal behavior due to initial conditions and the external behavior due to inputs.
- State-space models are well suited to represent multi-input–multi-output and nonlinear systems.
- In deriving state-space models of electric circuits, a typical choice of state variables is the voltages across capacitors and the currents through inductors.
- In deriving state-space models of mechanical systems, a typical choice of state variables is positions and velocities.
- The solution of the state equation is expressed in terms of the state transition matrix.
- The system is stable if its natural response converges to zeros as time tends to infinity and its forced response is bounded for every bounded input. The system is stable if and only if all the eigenvalues of A have negative real parts.
- The system is controllable if there is a continuous control that moves the state of the system from any initial state to any desired final state over a finite time period. The system is controllable if and only if the controllability matrix has full rank.
- The system is observable if the initial state can be uniquely determined by observing the output over a finite time period. The system is observable if and only if the observability matrix has full rank.
- The transfer function of a state-space model is given by an algebraic expression.
- There are many state-space models for a given transfer function. They include the controllable canonical form, the observable canonical form, and the parallel realizations.
- The closed-loop eigenvalues can be arbitrarily assigned by state feedback if the system is controllable.
- The observable eigenvalues can be arbitrarily assigned if the system is observable.
- The closed-loop eigenvalues under observer-based output feedback control are the eigenvalues under state feedback and the observer eigenvalues.
- Set-point regulation can be achieved by design of a feedforward gain or by augmenting an integrator driven by the tracking error. The integral control design is robust to parameter perturbations that do not destroy the stability of the system.

Mathematical Models

State transition matrix: $\Phi(t) = \mathcal{L}^{-1}\{(sI-A)^{-1}\}$

Solution of the state equation: $x(t) = \Phi(t)\,x_0 + \int_0^t \Phi(t-\tau)\,B\,u(\tau)\,d\tau$

Characteristic equation of A: $\det(sI-A) = 0$

Controllability matrix of (A,B): $C_M = \begin{bmatrix} B & AB & A^2B & \cdots & A^{n-1}B \end{bmatrix}$

Observability matrix of (A,C): $O_M = \begin{bmatrix} C \\ CA \\ CA^2 \\ \vdots \\ CA^{n-1} \end{bmatrix}$

Transfer function of $\{A,B,C,D\}$: $G(s) = C(sI-A)^{-1}B + D$

Important Terms Provide definitions or explain the meaning of the following terms:

closed-loop eigenvalues	linearization	parallel realization
controllability	observability	set-point regulation
controllable canonical form	observable canonical form	stability
duality	observer	state equation
eigenvalues	open-loop eigenvalues	state feedback control
feedforward control	output equation	state-space model
integral control	output feedback control	state transition matrix

PROBLEMS

8.1 Find a state-space model of each of the electric circuits shown in the following figures from Chapter 2. In each case, take the input and output as described in Chapter 2. For example if the transfer function of the circuit is described as V/W, take w to be the input and v to be the output. For Figs. P2.4 and P2.7, take the inputs and outputs as in Problems 2.5 and 2.8, respectively.

(a) Fig. 2-3 (b) Fig. 2-10 (c) Fig. P2.4(a)
(d) Fig. P2.4(b) (e) Fig. P2.7 (f) Fig. P2.9(a)
(g) Fig. P2.9(b) (h) Fig. P2.10(a) (i) Fig. P2.10(b)

8.2 Find a state-space model of each of the mechanical shown in the following figures from Chapter 2. In each case, take the input and output as described in Chapter 2. For example if

PROBLEMS 369

the transfer function of the system is described as V/W, take w to be the input and v to be the output.

(a) Fig. 2-15 (b) Fig. 2-16 (c) Fig. 2-21
(d) Fig. 2-22 (e) Fig. 2-24 (f) Fig. 2-27
(g) Fig. P2.11 (h) Fig. P2.12 (i) Fig. P2.13
(j) Fig. P2.14 (k) Fig. P2.15 (l) Fig. P2.17
(m) Fig. P2.18 (n) Fig. P2.19 (o) Fig. P2.20
(p) Fig. P2.21 (q) Fig. P2.22

8.3 For each of the following systems,

(i) determine if the system is stable;

(ii) find the transition matrix;

(iii) find the transfer function.

(a)
$$\dot{x} = \begin{bmatrix} 0 & 1 \\ 1 & 0 \end{bmatrix} x + \begin{bmatrix} 0 \\ 1 \end{bmatrix} u, \qquad y = \begin{bmatrix} 0 & 1 \end{bmatrix} x + u$$

(b)
$$\dot{x} = \begin{bmatrix} -1 & 0 \\ 1 & -1 \end{bmatrix} x + \begin{bmatrix} 1 \\ 1 \end{bmatrix} u, \qquad y = \begin{bmatrix} 1 & 0 \end{bmatrix} x$$

(c)
$$\dot{x} = \begin{bmatrix} 0 & 1 \\ -1 & 0 \end{bmatrix} x + \begin{bmatrix} 0 \\ 1 \end{bmatrix} u, \qquad y = \begin{bmatrix} 1 & 1 \end{bmatrix} x$$

(d)
$$\dot{x} = \begin{bmatrix} 0 & 1 \\ -1 & -2 \end{bmatrix} x + \begin{bmatrix} 0 \\ 1 \end{bmatrix} u, \qquad y = \begin{bmatrix} 1 & 0 \end{bmatrix} x + u$$

(e)
$$\dot{x} = \begin{bmatrix} -1 & 1 \\ -4 & -1 \end{bmatrix} x + \begin{bmatrix} 1 \\ 1 \end{bmatrix} u, \qquad y = \begin{bmatrix} 1 & 2 \end{bmatrix} x + u$$

(f)
$$\dot{x} = \begin{bmatrix} 0 & 1 & 0 \\ 0 & -1 & 1 \\ 0 & -4 & -1 \end{bmatrix} x + B = \begin{bmatrix} 0 \\ 0 \\ 1 \end{bmatrix} u, \qquad y = \begin{bmatrix} 0 & 1 & 0 \end{bmatrix} x$$

8.4 For each of the following transfer functions, find

(i) a controllable-canonical-form realization;

(ii) an observable-canonical-form realization;

(iii) a parallel realization.

(a)
$$G(s) = \frac{s^3 + s + 2}{s^3 + 3s^2 + 3s + 2}$$

(b)
$$G(s) = \frac{1}{s(s^2 + 1)}$$

(c)
$$G(s) = \frac{2s^3 + 2s + 1}{s^3 + 6s^2 + 11s + 6}$$

(d)
$$G(s) = \frac{s^2 + 1}{s^3 + 2s^2 + 3s + 4}$$

(e)
$$G(s) = \frac{3s^3 + s + 1}{s^3 + 6s^2 + 12s + 8}$$

(f)
$$G(s) = \frac{2s^3 + 2s + 1}{s^3 + 2s^2 + 3s + 4}$$

8.5 For each of the systems in Problem 8.3, design state feedback control $u = -Kx$ to assign the closed-loop eigenvalues at the desired locations shown below. Do your calculations without using MATLAB.

(a) $\lambda = -1, -1$

(b) $\lambda = -4, -5$

(c) $\lambda = -1, -2$

(d) $\lambda = -3, -3$

(e) $\lambda = -2 \pm 2j$

(f) $\lambda = -1, -1 \pm 2j$

PROBLEMS

8.6 Consider the system

$$\dot{x} = \begin{bmatrix} 0 & 1 \\ -2 & -3 \end{bmatrix} x + \begin{bmatrix} 1 \\ 1 \end{bmatrix} u, \qquad y = \begin{bmatrix} 1 & 1 \end{bmatrix} x + u.$$

(a) Find the transition matrix.
(b) Is the system stable? Is it controllable? Is it observable?
(c) Find the transfer function.
(d) Find $y(t)$ when $x(0) = \begin{bmatrix} 1 \\ 0 \end{bmatrix}$ and $u(t) = 0$.
(e) Find $y(t)$ when $x(0) = 0$ and $u(t)$ is the unit step input.
(f) Find $y(t)$ when $x(0) = \begin{bmatrix} 1 \\ 0 \end{bmatrix}$ and $u(t)$ is the unit step input.

8.7 Repeat Problem 8.6 for the system

$$\dot{x} = \begin{bmatrix} -1 & 0 \\ 0 & -2 \end{bmatrix} x + \begin{bmatrix} 1 \\ 1 \end{bmatrix} u, \qquad y = \begin{bmatrix} 2 & 1 \end{bmatrix} x. \qquad \text{Take } x(0) = \begin{bmatrix} 1 \\ -1 \end{bmatrix}.$$

8.8 Repeat Problem 8.6 for the system

$$\dot{x} = \begin{bmatrix} -1 & 0 & 0 \\ 0 & -2 & 0 \\ 0 & 0 & -3 \end{bmatrix} x + \begin{bmatrix} 1 \\ 1 \\ 1 \end{bmatrix} u, \qquad y = \begin{bmatrix} 3 & 2 & 1 \end{bmatrix} x. \qquad \text{Take } x(0) = \begin{bmatrix} 1 \\ 1 \\ 0 \end{bmatrix}.$$

8.9 Repeat Problem 8.6 for the system

$$\dot{x} = \begin{bmatrix} 10 & 3 \\ -37 & -11 \end{bmatrix} x + \begin{bmatrix} -1 \\ 4 \end{bmatrix} u, \qquad y = \begin{bmatrix} 4 & 1 \end{bmatrix} x + u. \qquad \text{Take } x(0) = \begin{bmatrix} 1 \\ 1 \end{bmatrix}.$$

8.10 Repeat Problem 8.6 for the system

$$\dot{x} = \begin{bmatrix} -3 & 0 & 0 \\ 0 & 0 & 1 \\ 0 & -2 & -3 \end{bmatrix} x + \begin{bmatrix} 2 \\ 0 \\ 1 \end{bmatrix} u, \qquad y = \begin{bmatrix} 0 & 0 & 1 \end{bmatrix} x. \qquad \text{Take } x(0) = \begin{bmatrix} 1 \\ 1 \\ 0 \end{bmatrix}.$$

8.11 Consider the position control of a dc motor whose state model is

$$\dot{x} = \begin{bmatrix} 0 & 1 \\ 0 & -\frac{1}{\tau} \end{bmatrix} x + \begin{bmatrix} 0 \\ \frac{k}{\tau} \end{bmatrix} u, \quad y = \begin{bmatrix} 1 & 0 \end{bmatrix} x,$$

where $x_1 = \theta$, $x_2 = \omega$, and u is the voltage input.

(a) Is the system stable? Find the transition matrix and the transfer function.

(b) Find $K = \begin{bmatrix} k_1 & k_2 \end{bmatrix}$ to assign the eigenvalues of $(A - BK)$ at the roots of $s^2 + 2\zeta\omega_n s + \omega_n^2$ with $\zeta = 0.5$. Choose ω_n such that the settling time ≤ 0.4 s.

(c) Let $u = -k_1 x_1 - k_2 x_2 + k_1 \theta_r$. Find the transition matrix and transfer function of the closed-loop system with input θ_r and output θ. What is the steady-state output to a unit step input?

(d) Design an observer whose settling time is less than 0.1 s. Implement the controller of part (c) using output feedback.

(e) Let $k = 2$ rad/s/V and $\tau = 0.2$ s. Simulate the step response of the closed-loop system under output feedback control.

8.12 It is shown in Example 8-5 that a linearized model of a dc-to-dc power converter is given by

$$A = \begin{bmatrix} 0 & -\frac{1}{k} \\ \frac{1}{k} & -\alpha \end{bmatrix}, \quad B = \begin{bmatrix} k \\ -\alpha k^2 \end{bmatrix}, \quad C = \begin{bmatrix} 0 & 1 \end{bmatrix}, \quad D = 0,$$

where α and k are positive constants. Assume $\alpha < 2/k$.

(a) Is the system stable? Is it controllable? Is it observable?

(b) Find the transition matrix.

(c) Find the transfer function.

For the rest of this problem, set $\alpha = 0.1$ and $k = 2$.

(d) Find the open-loop eigenvalues. Estimate the settling time of the open-loop response.

(e) Design state feedback control to assign the closed-loop eigenvalues at the roots of $s^2 + 2\zeta\omega_n s + \omega_n^2$ with $\zeta = 0.5$. Choose ω_n such that the settling time of the closed-loop response is 10 times smaller than the settling time of the open-loop response.

(f) Simulate the open-loop and closed-loop responses for $x(0) = \begin{bmatrix} 1 & 1 \end{bmatrix}^T$.

8.13 Figure P8.13 shows a schematic diagram of a magnetic levitation system, where a ball of magnetic material is suspended by means of electromagnet whose current is controlled by

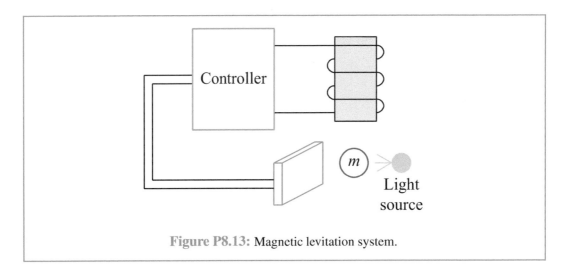

Figure P8.13: Magnetic levitation system.

feedback from the optically measured ball position. This system has the basic ingredients of systems constructed to levitate mass, used in gyroscopes, accelerometers, and fast trains. The system can be modeled by the nonlinear state equation[13]

$$\dot{x}_1 = x_2,$$
$$\dot{x}_2 = 1 - \frac{4x_3^2}{(1+x_1)^2},$$
$$\dot{x}_3 = \frac{0.2(1+x_1)}{1+0.6x_1}\left[-x_3 + u + \frac{2x_2 x_3}{(1+x_1)^2}\right],$$

where x_1 is a normalized position of the ball and x_3 is a normalized electric current of the electromagnet.

(a) Find the steady-state control u_{ss} that maintains the system at an equilibrium point $\bar{x}$ with $\bar{x}_1 = 1$.

(b) Linearize the state equation at $(x = \bar{x}, u = u_{ss})$. Let $x_\delta = x - \bar{x}$ and $u_\delta = u - u_{ss}$.

(c) Is the linear state equation stable? Is it controllable?

(d) Design state feedback control $u_\delta = -Kx_\delta$ that assigns the closed-loop eigenvalues at $-\frac{1}{4}, -\frac{1}{4} \pm \frac{1}{4}j$.

(e) If the only measured signal is the output $y = x_1 - \bar{x}_1$. Is the system observable?

(f) Design an observer to estimate x_δ, with the observer eigenvalues located at $-1, -1 \pm j$.

(g) Combine the state feedback and observer to obtain an output feedback controller that stabilizes the system. Simulate the response of the system with the initial conditions $x_\delta(0) = \begin{bmatrix} 1 & 0 & 0 \end{bmatrix}^T$.

8.14 Consider the automobile suspension system of Examples 2-11 and 8-2. In active suspension systems a force actuator is mounted between the sprung and unsprung masses, applying equal and opposite forces to theses two masses. The equations of motion are given by

$$M_s \ddot{y}_s = -k_s(y_s - y_{us}) - C_s(\dot{y}_s - \dot{y}_{us}) + u$$
$$M_{us} \ddot{y}_{us} = k_s(y_s - y_{us}) + C_s(\dot{y}_s - \dot{y}_{us}) - k_t(y_{us} - S) - u,$$

where u is the force applied by the actuator.

(a) Find the state-space model with

$$x_1 = y_s - y_{us}, \quad x_2 = y_{us} - S, \quad x_3 = \dot{y}_s, \quad x_4 = \dot{y}_{us}$$

as the state variables, u as the control input, and $\dot{S}$ as a disturbance input.

For the rest of this problem, the parameters of the system are taken as [13] $M_s = 973$ kg, $M_{us} = 114$ kg, $k_s = 42720$ N/m, $k_t = 101115$ N/m, and $C_s = 1095$ N s/m.

---

[13] See Appendix A.8 of [22].

(b) Is the system stable? Is it controllable?

(c) Is the system observable if the output is $\ddot{y}_s$?

(d) Is the system observable if the output is $y_s - y_{us}$?

Consider the state feedback control $u = -Kx$, where K is given by [13]

$$K = -10^3 \times \begin{bmatrix} 1.3308 & 3.9564 & -5.5819 & 0.5438 \end{bmatrix}$$

(e) Find the closed-loop eigenvalues and compare them with the open-loop eigenvalues.

(f) Compare the responses of the closed-loop and open-loop systems to an isolated bump in an otherwise smooth surface. The road displacement is defined by

$$S(t) = \begin{cases} \frac{A}{2}[1 - \cos(2\pi v t/L)] & \text{if } 0 \leq t \leq L/v, \\ 0 & \text{if } t > L/v, \end{cases}$$

where A and L are the height and length of the bump and v is the forward speed of the automobile. Assume $A = 0.06$ m, $L = 5$ m, and $v = 30$ m/s. For the comparison, plot the responses of the body acceleration $\ddot{y}_s$ and the suspension stroke $y_s - y_{us}$. Take $x(0) = 0$ in your simulation.

(g) If the measured output is $\ddot{y}_s$, build an observer to implement the state feedback control.

(h) Repeat (g) if the measured output is $y_s - y_{us}$.

8.15 Consider the inverted pendulum on a cart of Fig. P8.15. The pivot of the pendulum is mounted on a cart that can move in a horizontal direction. The cart is driven by a motor that exerts a horizontal force F. The system can be modeled by the equations [22]

$$J\ddot{\theta} = -mL^2\ddot{\theta} + mgL\sin\theta - mL\ddot{y}\cos\theta,$$

$$M\ddot{y} = F - m(\ddot{y} + L\ddot{\theta}\cos\theta - L\dot{\theta}^2\sin\theta) - k\dot{y},$$

where θ is the angular rotation of the pendulum (measured clockwise from the vertical position), y the displacement of the cart, m the mass of the pendulum, M the mass of the

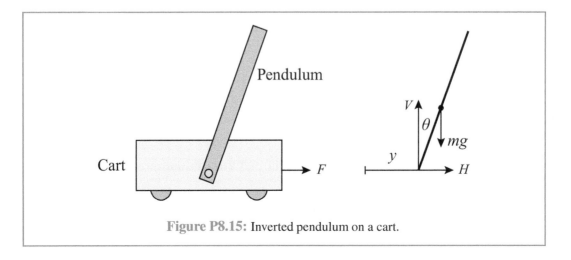

Figure P8.15: Inverted pendulum on a cart.

cart, L the distance from the center of gravity to the pivot, J the moment of inertia of the pendulum with respect to the center of gravity, k a friction coefficient, and g the acceleration due to gravity.

When $F = 0$, the system has equilibrium with the pendulum at the vertical position. Linearization of the equations of motion at the equilibrium point yields

$$J\ddot{\theta} = -mL^2\ddot{\theta} + mgL\theta - mL\ddot{y},$$

$$M\ddot{y} = F - m(\ddot{y} + L\ddot{\theta}) - k\dot{y}.$$

(a) Find the state equation of the system with $x_1 = \theta$, $x_2 = \dot{\theta}$, $x_3 = y$, and $x_4 = \dot{y}$ as the state variables and $u = F$ as the control input.

(b) Show that the system is unstable.

For the rest of this problem, let $m = 0.1$ kg, $M = 1$ kg, $k = 0.1$ kg·m/s², $J = 0.008$ kg·m², $g = 9.81$ m/s², and $L = 0.5$ m.

(c) Verify that the system is controllable.

(d) Design state feedback control $u = -Kx$ to stabilize the system.

(e) Plot the state variables of the closed-loop system when $x(0) = \begin{bmatrix} 1 & 0 & 0 & 0 \end{bmatrix}^T$.

8.16 Under certain assumptions, a satellite in circular orbit is modeled by

$$\frac{d^2 r}{dt^2} = r\left(\frac{d\varphi}{dt}\right)^2 - \frac{\omega_0^2 r_0^3}{r^2} + \frac{F_r}{m},$$

$$\frac{d^2 \varphi}{dt^2} = -2\left(\frac{d\varphi}{dt}\right)\left(\frac{dr}{dt}\right)\frac{1}{r} + \frac{F_\varphi}{mr},$$

where r is the distance from the earth center to the satellite, φ the orbit angle of the satellite, m the mass of the satellite, F_r the radial force, F_φ the tangential force, t the time in seconds, and r_0 and ω_0 are constants.

(a) Verify that when $F_r = F_\varphi = 0$, the circular orbit $\{r(t) = r_0, \varphi(t) = \omega_0 t\}$ satisfies the equations of motion.

(b) Show that, with

$$x_1 = \frac{r - r_0}{r_0}, \quad x_2 = \frac{1}{\omega_0 r_0}\frac{dr}{dt}, \quad x_3 = \varphi - \omega_0 t, \quad x_4 = \frac{1}{\omega_0}\frac{d\varphi}{dt} - 1$$

as the state variables, $u_1 = F_r/(mr_0\omega_0^2)$ and $u_2 = F_\varphi/(mr_0\omega_0^2)$ as the control inputs, and the change of the time variable from t to $\tau = \omega_0 t$, the state equation is given by

$$\dot{x}_1 = x_2, \tag{8.47}$$

$$\dot{x}_2 = (x_1 + 1)(x_4 + 1)^2 - \frac{1}{(x_1 + 1)^2} + u_1, \tag{8.48}$$

$$\dot{x}_3 = x_4, \tag{8.49}$$

$$\dot{x}_4 = -\frac{2(x_4+1)x_2}{x_1+1} + \frac{u_2}{x_1+1}, \tag{8.50}$$

where $\dot{x}_i = dx_i/d\tau$.

(c) Verify that when $u_1 = u_2 = 0$, the system has an equilibrium point at $x = 0$ and linearization at the equilibrium point yields the linear state equation

$$\dot{x} = \begin{bmatrix} 0 & 1 & 0 & 0 \\ 3 & 0 & 0 & 2 \\ 0 & 0 & 0 & 1 \\ 0 & -2 & 0 & 0 \end{bmatrix} x + \begin{bmatrix} 0 \\ 1 \\ 0 \\ 0 \end{bmatrix} u_1 + \begin{bmatrix} 0 \\ 0 \\ 0 \\ 1 \end{bmatrix} u_2.$$

(d) Show that the system is unstable.
(e) Investigate controllability from u_1.
(f) Investigate controllability from u_2.
(g) Investigate observability if the measured output is x_1.
(h) Investigate observability if the measured output is x_3.
(i) Using one of the two control inputs, design a state feedback controller to stabilize the system.
(j) Using one of the two control inputs and one of the two measured outputs x_1 and x_3, design an output feedback controller to stabilize the system.[14]

---

[14] Using both control inputs or both measured outputs will turn the problem into multi-input–multi-output, which is beyond the discussion in this chapter.

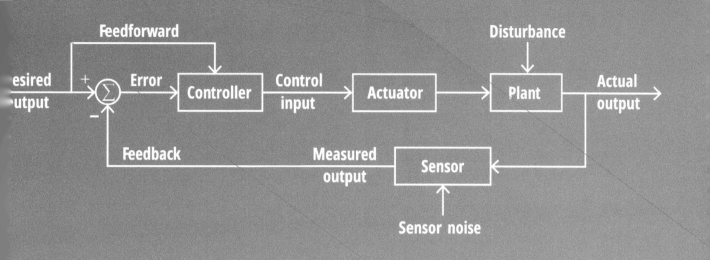

Chapter 9
Digital Control

Chapter Contents

    Overview, 378
9-1 Digital Control Systems, 378
9-2 Discrete-Time Systems, 381
9-3 Discrete-Time Equivalent of Digital Control System, 387
9-4 Stability of Digital Control Systems, 393
9-5 Steady-State Error, 403
9-6 Sample-Frequency Selection, 405
9-7 Design by Emulation, 411
9-8 Discrete-Time Design, 416
9-9 Quantization, 422
    Chapter Summary, 430
    Problems, 432

Objectives

Upon learning the material presented in this chapter, you should be able to:

1. Understand why the controller is implemented digitally.
2. Recognize the basic components of a digital control system.
3. Review discrete-time systems and the z-transform.
4. Derive the discrete-time equivalent of a continuous-time transfer function preceded by zero-order hold.
5. Represent the closed-loop control system in discrete time.
6. Study the stability of digital control systems.
7. Select the sampling frequency.
8. Design a digital controller by converting a continuous-time controller.
9. Design a digital controller in the discrete-time domain.
10. Study the effect of quantization on the performance of the digital control system.

Overview

Controllers can be implemented using analog circuits, similar to the ones shown earlier in Fig. 5-7, or they can be implemented using digital computers. There are several advantages to implementing the controllers digitally:

(a) Flexibility in control implementation. Modification of analog controllers requires rewiring and changing circuit components, while modification of a digital controller is realized by changing a computer program.

(b) Higher control accuracy. The control accuracy in analog control is limited by the accuracy of the devices, temperature drift and other factors.

(c) Cost-effectiveness due to the rapid development of VLSI technology.

(d) Improved user interface because information can be displayed graphically on a monitor.

(e) The digital control mode can avoid the distortion and interference of spurious signals associated with the transmission of analog signals.

Section 9-1 describes digital control systems. Digital signals are discrete-time signals whose values are quantized because of the finite wordlength of digital computers. It is standard practice in the analysis and design of digital systems to ignore quantization and to treat the digital signals as if they were discrete-time signals whose values are defined continuously. Therefore, the digital system is modeled as a discrete-time system. Section 9-2 reviews discrete-time systems. Section 9-3 derives the discrete-time equivalent of digital control systems. Sections 9-4 and 9-5 discuss stability and the steady-state error of discrete-time feedback systems. Section 9-6 demonstrates how the sampling frequency is chosen and Sections 9-7 and 9-8 discuss the design of digital controllers. In Section 9-7 the controller is designed in continuous time, then discretized, a process known as emulation, while in Section 9-8 the controller is designed in discrete time entirely. Section 9-9 shows how to analyze the effect of quantization on the performance of the control system.

9-1 Digital Control Systems

The inputs and outputs of an analog system are analog signals, while the inputs and outputs of a digital system are digital signals. An analog signal $x(t)$ is defined for all time instants t over a given time interval and x can assume any value in a certain interval. In other words, the signal is continuous in time and continuous in value. Physical signals, such as temperature, humidity, velocity, etc., are analog signals. Digital signals, on the other hand, are discrete in time and discrete in value. They are discrete in time because they are defined only at discrete instants of time, usually synchronized with a clock. They are discrete in value because numbers in digital computers are represented by words of finite wordlength. For example, in a digital system in which the magnitude of a number is represented by 7 bits, there are only $2^7 = 128$ values the number can take. Other values will have to be rounded off to the nearest one of those 128 values. This process is known as *quantization*.

9-1. DIGITAL CONTROL SYSTEMS

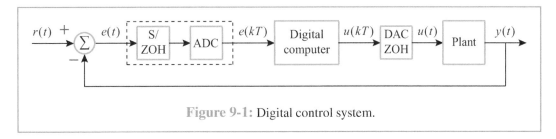

Figure 9-1: Digital control system.

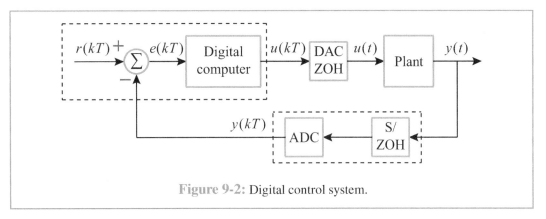

Figure 9-2: Digital control system.

When a digital computer is used to control an analog system, it is necessary to use an analog-to-digital interface between the analog and digital systems. Figure 9-1 shows the basic components of a typical digital control system, and Fig. 9-2 shows a slightly different configuration of Fig. 9-1. The basic components of the system are :

(a) Sample/Zero-Order Hold (S/ZOH)

(b) Analog-to-Digital Converter (ADC)

(c) Digital computer

(d) Digital-to-Analog Converter with Zero-Order Hold (DAC/ZOH)

(e) Plant

- The S/ZOH samples the continuous-time signal to produce a discrete-time signal. The samples are taken at points that are uniformly spaced with sampling period T, as shown in Fig. 9-3. The samples are held constant over the sampling period, producing a *staircase continuous-time function*. Holding the sample constant over the sampling period is called *zero-order hold* (*ZOH*).[1]

---

[1] Digital control books, e.g., [4, 9, 12, 31], discuss other types of hold, especially first-order hold where two consecutive samples are connected by a straight line. However, zero-order hold is typical in practical digital control systems.

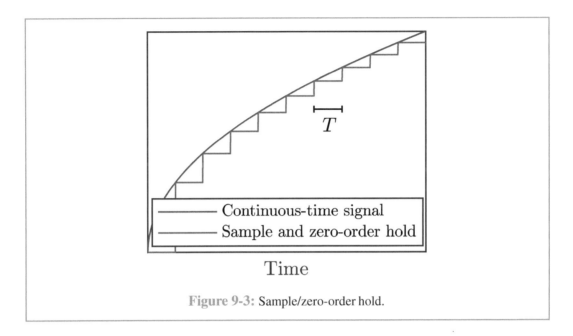

Figure 9-3: Sample/zero-order hold.

- The ADC receives the samples from the S/ZOH and quantizes their values to produce the digital signal that is fed to the computer. The quantization is usually done by rounding off numbers to the nearest quantization level. In a binary system with ℓ magnitude bits, the difference between two quantization levels is $q = 2^{-\ell}$. The maximum quantization error is $q/2 = 2^{-(\ell+1)}$. In Fig. 9-1, the S/ZOH and ADC are boxed together in a dashed box to indicate that these two blocks are usually combined into one device (chip).

- The digital computer implements the control algorithm, which is usually represented by difference equations.

- The DAC converts the digital signal coming out of the computer into an analog signal in the form of a staircase continuous-time function, and that is why it is said that the ADC is combined with zero-order hold.

- The plant is the system to be controlled. It is modeled by a continuous-time transfer function, which includes the models of the actuator and sensor.[2]

The progression of signals in Fig. 9-1 goes as follows. The analog output $y(t)$ is fed back and subtracted from the analog reference $r(t)$ to produce the analog error $e(t)$. The error $e(t)$ is converted into the digital error $e(kT)$, which is fed to the computer. The output of the computer is the digital control $u(kT)$, which is converted by the ADC into the analog control $u(t)$, which takes the form of a staircase continuous-time function.

In Fig. 9-2, the analog output $y(t)$ is converted into a digital signal $y(kT)$ and then subtracted from the digital reference $r(kT)$ to produce the digital error $e(kT)$. The summation

---

[2]The model could also be a state-space model but we will not consider state-space models in this chapter.

9-2 Discrete-Time Systems

node and the computer are boxed together in a dashed box to indicate that the calculation $e(kT) = r(kT) - y(kT)$ is usually performed in the same computer that implements the control algorithm.

In the analysis and design of digital control systems, the effect of quantization is ignored. Digital signals are treated as discrete-time signals whose values are defined continuously. The effect of quantization can be accounted for by treating it as a perturbation of the system's coefficients and by the noise that corrupts the signals. An example of such an examination is given in Section 9-9.

9-2 Discrete-Time Systems

A *discrete-time system* is a system whose inputs and outputs are discrete-time signals. A linear Single-Input–Single-Output (SISO) discrete-time system is modeled by the difference equation

$$y(k+n) = -a_{n-1} y(k+n-1) - a_{n-2} y(k+n-2) - \cdots - a_0 y(k) \\ + b_n u(k+n) + b_{n-1} u(k+n-1) + \cdots + b_0 u(k) , \quad (9.1)$$

where u is the input, y is the output, and k is the time variable ($k = 0, 1, 2, \ldots$). The present output $y(k+n)$ is a linear combination of the n past outputs, $y(k+n-1)$ to $y(k)$, the present input $u(k+n)$, and the n past inputs, $u(k+n-1)$ to $u(k)$. The equation is said to be of order n. The system can be represented by a transfer function using the z-transform, which plays a role for discrete-time systems analogous to the role played by the Laplace transform for continuous-time systems.

z-Transform

For the discrete-time signal $x(k)$, $k = 0, 1, 2, \ldots$, the z-transform is defined by

$$X(z) = \mathcal{Z}\{x(k)\} = \sum_{k=0}^{\infty} x(k) z^{-k} , \quad (9.2)$$

where z is a complex variable. The infinite series is convergent over a region of the form $|z| \geq r_0$ (region of convergence).

Properties of the z-transform

(a) Linearity:
$$\mathcal{Z}\{a_1 x_1(k) + a_2 x_2(k)\} = a_1 X_1(z) + a_2 X_2(z) . \quad (9.3)$$

(b) Time Shift by n:
$$\mathcal{Z}\{x(k+n)\} = z^n \left[X(z) - \sum_{i=0}^{n-1} x(i) z^{-i} \right]. \quad (9.4)$$

Also, if $x(i) = 0$ for $0 \leq i \leq n-1$, $\quad \rightarrow \quad \mathcal{Z}\{x(k+n)\} = z^n X(z)$,
and if $x(i) = 0$ for $i < 0$, $\quad \rightarrow \quad \mathcal{Z}\{x(k-n)\} = z^{-n} X(z)$.

(c) Scaling in the z-domain:
$$\mathcal{Z}\{a^{-k}x(k)\} = X(az) \, . \qquad (9.5)$$

(d) Convolution:
$$\mathcal{Z}\left\{\sum_{\ell=0}^{\infty} x_1(\ell)\, x_2(k-\ell)\right\} = X_1(z)\, X_2(z) \, . \qquad (9.6)$$

(e) Final Value Theorem: If all poles of $(z-1)\,X(z)$ are inside the unit circle, then
$$\lim_{k\to\infty} x(k) = \lim_{z\to 1}(z-1)\,X(z) \, . \qquad (9.7)$$

Example 9-1: z-Transform of an Exponential Function

Find the z-transform of $x(k) = a\lambda^k$, where λ is a complex number.

Solution:
$$X(z) = \sum_{k=0}^{\infty} a\lambda^k z^{-k} = a\sum_{k=0}^{\infty}\left(\frac{\lambda}{z}\right)^k = \frac{a}{1-(\lambda/z)} = \frac{az}{z-\lambda}, \quad \text{if } |z| > |\lambda| \, .$$

Transfer functions

The transfer function of a discrete-time system is the ratio of the z-transform of the output to the z-transform of the input, assuming zero initial conditions. Let $u(k)$ and $y(k)$ be the input and output, respectively. Then,
$$U(z) = \mathcal{Z}\{u(k)\}, \qquad Y(z) = \mathcal{Z}\{y(k)\} \, .$$

The transfer function $G(z)$ is defined by
$$Y(z) = G(z)\, U(z) \, .$$

Consider the system
$$y(k+n) = -a_{n-1}\, y(k+n-1) - a_{n-2}\, y(k+n-2) - \cdots - a_0 y(k)$$
$$+ b_n\, u(k+n) + b_{n-1}\, u(k+n-1) + \cdots + b_0 u(k) \, .$$

Take the z-transform of both sides assuming zero initial conditions:
$$z^n Y(z) = -a_{n-1}\, z^{n-1} Y(z) - a_{n-2}\, z^{n-2} Y(z) - \cdots - a_0 Y(z)$$
$$+ b_n z^n U(z) + b_{n-1} z^{n-1} U(z) + \cdots + b_0\, U(z) \, ,$$

9-2. DISCRETE-TIME SYSTEMS

which can be rewritten as

$$(z^n + a_{n-1}z^{n-1} + a_{n-2}\, z^{n-2} + \cdots + a_0)\, Y(z) = (b_n\, z^n + b_{n-1}\, z^{n-1} + \cdots + b_0)\, U(z)\;,$$

leading to the transfer function

$$G(z) = \frac{Y(z)}{U(z)} = \frac{b_n\, z^n + b_{n-1}\, z^{n-1} + \cdots + b_0}{z^n + a_{n-1}z^{n-1} + a_{n-2}\, z^{n-2} + \cdots + a_0}\;. \tag{9.8}$$

The transfer function can be written from the difference equation by inspection, and vice versa.

Example 9-2: Finding the Transfer Function from the Difference-Equation Model

Find the transfer function of a system represented by the difference equation

$$y(k+2) = \tfrac{1}{2}\, y(k+1) - \tfrac{1}{4}\, y(k) + u(k+1)\;.$$

Solution: Rewrite the equation as

$$y(k+2) - \tfrac{1}{2}\, y(k+1) + \tfrac{1}{4}\, y(k) = u(k+1)\;,$$

which gives

$$G(z) = \frac{z}{z^2 - \tfrac{1}{2}\, z + \tfrac{1}{4}}\;.$$

Example 9-3: Finding the Difference-Equation Model from the Transfer Function

Find the difference-equation model of a system represented by the transfer function

$$G(z) = \frac{0.96z + 0.92}{z^2 - 0.9z + 0.81}\;.$$

Solution:

$$\frac{Y(z)}{U(z)} = G(z) = \frac{0.96z + 0.92}{z^2 - 0.9z + 0.81}\;,$$

which can be rearranged as

$$(z^2 - 0.9z + 0.81)\, Y(z) = (0.96z + 0.92)\, U(z)\;,$$

and then inverse z-transformed into

$$y(k+2) - 0.9\, y(k+1) + 0.81\, y(k) = 0.96\, u(k+1) + 0.92\, u(k)\;.$$

Impulse response

Taking the inverse z-transform of $Y(z) = G(z)U(z)$ yields the convolution summation

$$y(k) = \sum_{\ell=0}^{\infty} g(\ell) u(k-\ell),$$

where

$$u(k) = \mathcal{Z}^{-1}\{U(z)\}, \qquad y(k) = \mathcal{Z}^{-1}\{Y(z)\}, \qquad g(k) = \mathcal{Z}^{-1}\{G(z)\}.$$

Consider the impulse input:

$$u(k) = \delta(k) = \begin{cases} 1, & k = 0 \\ 0, & k > 0 \end{cases} \quad \longrightarrow \quad y(k) = g(k).$$

The function $g(k)$ is called the *impulse response*.

Poles and zeros

For the rational transfer function $G(z) = \frac{N(z)}{D(z)}$, the roots of $N(z) = 0$ are the zeros of $G(z)$ and the roots of $D(z) = 0$ are the poles of $G(z)$.

Stability

The system is stable if for every bounded input, the output is bounded.

Theorem 9.1 *A discrete-time system represented by a rational proper transfer function $G(z)$ is stable if and only if all poles of $G(z)$ are inside the unit circle; that is, $|p_i| < 1$ for $i = 1, \ldots, n$, which is equivalent to $\sum_{k=0}^{\infty} |g(k)| < \infty$, where $g(k) = \mathcal{Z}^{-1}\{G(z)\}$.*

Proof of sufficiency for the special case when $G(z)$ has distinct poles:[3]

Write $G(z) = z \frac{G(z)}{z}$ and expand $\frac{G(z)}{z}$ using partial-fraction expansion:

$$G(z) = \frac{k_1 z}{z - p_1} + \cdots + \frac{k_n z}{z - p_n} = \sum_{i=1}^{n} \frac{k_i z}{z - p_i}. \tag{9.9}$$

Take the inverse z-transform using Example 9-1:

$$g(k) = \mathcal{Z}^{-1}\{G(z)\} = \mathcal{Z}^{-1}\left\{\sum_{i=1}^{n} \frac{k_i z}{z - p_i}\right\} = \sum_{i=1}^{n} k_i p_i^k,$$

$$\sum_{k=0}^{\infty} |g(k)| = \sum_{k=0}^{\infty} \left|\sum_{i=1}^{n} k_i p_i^k\right| \leq \sum_{i=1}^{n} |k_i| \sum_{k=0}^{\infty} |p_i|^k,$$

---

[3] A complete proof can be carried out by steps analogous to the proofs of Theorem 3.1 and 3.2.

9-2. DISCRETE-TIME SYSTEMS

$$|p_i| < 1 \quad \rightarrow \quad \sum_{k=0}^{\infty} |p_i|^k = \frac{1}{1-|p_i|} < \infty \quad \rightarrow \quad \sum_{k=0}^{\infty} |g(k)| < \infty.$$

The output $y(k)$ is given by $y(k) = \sum_{\ell=0}^{\infty} g(\ell) u(k-\ell)$. For a bounded input, $|u(k-\ell)| \leq K_u$. Therefore,

$$|y(k)| = \left| \sum_{\ell=0}^{\infty} g(\ell) u(k-\ell) \right| \leq \sum_{\ell=0}^{\infty} |g(\ell)| |u(k-\ell)| \leq K_u \sum_{\ell=0}^{\infty} |g(\ell)|,$$

which shows that the output is bounded.

Steady-state output to step inputs

Let $G(z)$ be a stable transfer function and

$$u(k) = r\, 1(k) = \begin{cases} r, & k \geq 0, \\ 0, & k < 0, \end{cases}$$

where $1(k)$ is the unit step function. In the z-domain,

$$U(z) = \frac{rz}{z-1} \quad \rightarrow \quad Y(z) = G(z) \frac{rz}{z-1}.$$

All poles of $(z-1)Y(z) = rz\, G(z)$ are inside the unit circle. By the final value theorem,

$$y_{ss} \stackrel{\text{def}}{=} \lim_{k \to \infty} y(k) = \lim_{z \to 1}(z-1)\, Y(z) = \lim_{z \to 1} rz\, G(z) = G(1)\, r.$$

The function $G(1)$ is called the *dc gain*.

Example 9-4: Steady-State Output to a Step Input

Show that

$$G(z) = \frac{0.96z + 0.92}{z^2 - 0.9z + 0.81}$$

is stable and find the steady-state output to a unit step input.

Solution:

$$z^2 - 0.9z + 0.81 = 0 \quad \rightarrow \quad p_{1,2} = 0.45 \pm j0.7794,$$

$$|p_{1,2}| = 0.9 \quad \rightarrow \quad G(z) \text{ is stable},$$

$$y_{ss} = G(1) = 2.0659.$$

Frequency response

Let $G(z)$ be a stable transfer function and, for convenience, assume that its poles $p_1,\ldots,p_n$ are distinct. Let $u(k) = A\,\cos(k\omega T)$, where $2\pi/(\omega T)$ is an integer; $u(k)$ is a sinusoidal function of k with period $2\pi/(\omega T)$:

$$u(k) = \frac{A}{2}\left[e^{jk\omega T} + e^{-jk\omega T}\right],$$

$$U(z) = \frac{A}{2}\left[\frac{z}{z - e^{j\omega T}} + \frac{z}{z - e^{-j\omega T}}\right],$$

and

$$Y(z) = G(z)\,U(z) = G(z)\,\frac{A}{2}\left[\frac{z}{z - e^{j\omega T}} + \frac{z}{z - e^{-j\omega T}}\right].$$

By partial-fraction expansion,

$$Y(z) = \frac{A}{2}\,G\!\left(e^{j\omega T}\right)\frac{z}{z - e^{j\omega T}} + \frac{A}{2}\,G\!\left(e^{-j\omega T}\right)\frac{z}{z - e^{-j\omega T}} + \sum_{i=1}^{n}\frac{k_i z}{z - p_i},$$

which translates into

$$y(k) = \frac{A}{2}\left[G\!\left(e^{j\omega T}\right)e^{jk\omega T} + G(e^{-j\omega T})e^{-jk\omega T}\right] + \sum_{i=1}^{n} k_i p_i^k.$$

Write $G\!\left(e^{j\omega T}\right) = M e^{j\psi}$, where $M = \left|G\!\left(e^{j\omega T}\right)\right|$ and $\psi = \angle G\!\left(e^{j\omega T}\right)$. It can be shown that $G\!\left(e^{-j\omega T}\right) = M e^{-j\psi}$, and

$$\frac{A}{2}\left[G\!\left(e^{j\omega T}\right)e^{jk\omega T} + G\!\left(e^{-j\omega T}\right)e^{-jk\omega T}\right] = \frac{A}{2}M\left[e^{j(k\omega T + \psi)} + e^{-(jk\omega T + \psi)}\right]$$

$$= AM\cos(k\omega T + \psi).$$

Hence,

$$y(k) = AM\cos(k\omega T + \psi) + \sum_{i=1}^{n} k_i p_i^k,$$

$$|p_i| < 1 \quad\longrightarrow\quad \lim_{k\to\infty}\sum_{i=1}^{n} k_i p_i^k = 0 \quad\longrightarrow\quad \lim_{k\to\infty}\left[y(k) - AM\cos(k\omega T + \psi)\right] = 0.$$

The steady-state output to a sinusoidal input is a sinusoidal signal of the same frequency with the magnitude and phase modified by the magnitude and phase of $G\!\left(e^{j\omega T}\right)$:

$$G\!\left(e^{j\omega T}\right) = G(z)\big|_{z = e^{j\omega T}} \text{ is the frequency response of } G(z).$$

Here, $G\!\left(e^{j\omega T}\right)$ is a periodic function of ω with period $2\pi/T$. Therefore, the frequency response of discrete-time systems is applicable only for

$$-\frac{\pi}{T} \leq \omega \leq \frac{\pi}{T}.$$

9-3 Discrete-Time Equivalent of Digital Control System

A digital control system consists of the interconnection of continuous-time components and discrete-time components. To analyze and design such a system, we derive a discrete-time equivalent of the system in which all signals are defined in discrete time. Thus, continuous-time signals are represented by their samples at the sampling points.

The starting point is to derive a discrete-time equivalent of a continuous-time system preceded by zero-order hold. Figure 9-4 shows a typical setup where a digital computer computes the discrete-time control signal $u(kT)$, which is defined at the sampling points kT, for $k = 0, 1, 2, \ldots$. The control signal $u(kT)$ is converted by the analog-to-digital converter into a continuous-time control signal $u(t)$, defined by

$$u(t) = u(kT), \quad kT \leq t < (k+1)T \; .$$

The signal $u(t)$ is a staircase continuous-time function where $u(t)$ is kept constant over the sampling period at its value at the beginning of the period; that is why the analog-to-digital converter is said to be zero-order hold. Given the transfer function $\tilde{G}(s)$ of the continuous-time system, the goal is to derive a discrete-time transfer function $G(z)$ that shows how the discrete-time samples of the output, $y(kT)$, are produced by the input $u(kT)$.

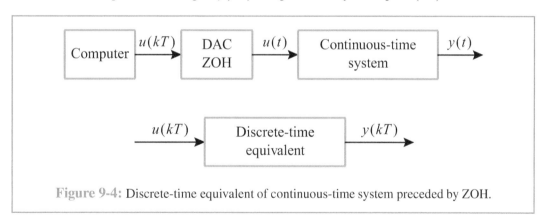

Figure 9-4: Discrete-time equivalent of continuous-time system preceded by ZOH.

The derivation starts by finding a mathematical model of how a staircase continuous-time function is related to the sampled continuous-time function. Let $f(t)$ be a continuous-time function and $\bar{f}(t)$ be the staircase continuous-time function defined by

$$\bar{f}(t) = f(kT), \quad kT \leq t < (k+1)T, \quad k = 0, 1, 2, \ldots$$

The function $\bar{f}(t)$ is generated from $f(t)$ by sampling and zero-order hold, as shown in Fig. 9-5. We plan to find a mathematical model of the block Sample/ZOH in the figure. We start by taking the Laplace transform of $\bar{f}(t)$:

$$\bar{F}(s) = \mathcal{L}\{\bar{f}(t)\} = \int_0^\infty \bar{f}(t) e^{-st} \, dt$$

$$= \sum_{k=0}^\infty \int_{kT}^{kT+T} f(kT) e^{-st} \, dt = \sum_{k=0}^\infty f(kT) \int_{kT}^{kT+T} e^{-st} \, dt \; . \quad (9.10)$$

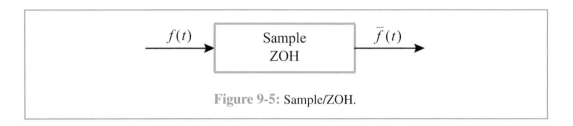

Figure 9-5: Sample/ZOH.

By computing the integral

$$\int_{kT}^{kT+T} e^{-st}\,dt = \left(-\frac{1}{s}e^{-st}\right)\Big|_{t=kT}^{t=kT+T} = \frac{e^{-kTs}(1-e^{-Ts})}{s}$$

and substituting the result in the expression for $\bar{F}(s)$, we obtain

$$\bar{F}(s) = \sum_{k=0}^{\infty} f(kT)\frac{e^{-kTs}(1-e^{-Ts})}{s} = \frac{(1-e^{-Ts})}{s}\sum_{k=0}^{\infty} f(kT)e^{-kTs}.$$

Next, we define

$$G_{\text{ZOH}}(s) = \frac{(1-e^{-Ts})}{s}, \quad F^{\star}(s) = \sum_{k=0}^{\infty} f(kT)e^{-kTs},$$

and rewrite the expression for $\bar{F}(s)$ as

$$\bar{F}(s) = G_{\text{ZOH}}(s)\,F^{\star}(s).$$

This equation represents a continuous-time system whose transfer function is $G_{\text{ZOH}}(s)$ with input $F^{\star}(s)$ and output $\bar{F}(s)$. Transfer function $G_{\text{ZOH}}(s)$ is called the *transfer function of the zero-order hold*. We already know that $\bar{F}(s)$ is the Laplace transform of $\bar{f}(t)$. What is $f^{\star}(t)$ whose Laplace transform is $F^{\star}(s)$?

The inverse Laplace transform of $F^{\star}$ is given by

$$f^{\star}(t) = \mathcal{L}^{-1}\{F^{\star}(s)\} = \mathcal{L}^{-1}\left\{\sum_{k=0}^{\infty} f(kT)e^{-kTs}\right\}$$

$$= \sum_{k=0}^{\infty} f(kT)\,\mathcal{L}^{-1}\{e^{-kTs}\} = \sum_{k=0}^{\infty} f(kT)\,\delta(t-kT). \quad (9.11)$$

The time-domain function $f^{\star}(t)$ consists of a train of impulses at the sampling points. The strength of the impulse at kT is $f(kT)$. Next, we define an ideal sampler that generates $f^{\star}(t)$ from $f(t)$.

9-3. DIGITAL CONTROL SYSTEMS

Figure 9-6: Ideal (Impulse) sampler.

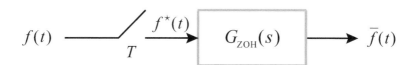

Figure 9-7: Representation of Sample/ZOH by ideal sampler followed by $G_{\text{ZOH}}(s)$.

Ideal (or impulse) sampler:

Figure 9-6 shows the block diagram symbol for an (artificial) process that generates $f^\star(t)$ from $f(t)$. With this definition of the ideal sampler, the (physical) Sample/Zero-Order Hold process can be represented (mathematically) by the block diagram of Fig. 9-7.

Now, the output $y(t)$ of a continuous-time transfer function $\tilde{G}(s)$, preceded by zero-order hold, is represented by the block diagram of Fig. 9-8.

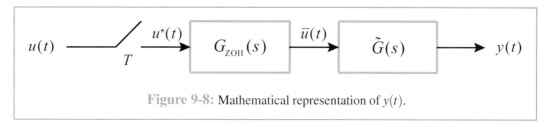

Figure 9-8: Mathematical representation of $y(t)$.

Let $P(s) = \tilde{G}(s) G_{\text{ZOH}}(s)$, which leads to

$$Y(s) = P(s)\, U^\star(s) = P(s) \sum_{\ell=0}^{\infty} u(\ell T)\, e^{-\ell T s} = \sum_{\ell=0}^{\infty} u(\ell T)\, P(s) e^{-\ell T s} \qquad (9.12\text{a})$$

and

$$y(t) = \mathcal{L}^{-1}\{Y(s)\} = \sum_{\ell=0}^{\infty} u(\ell T) \mathcal{L}^{-1}\left\{P(s) e^{-\ell T s}\right\}. \qquad (9.12\text{b})$$

Next, we let $p(t) = \mathcal{L}^{-1}\{P(s)\}$, which means that

$$p(t - \ell T) = \mathcal{L}^{-1}\left\{P(s)\, e^{-\ell T s}\right\}$$

and

$$y(t) = \sum_{\ell=0}^{\infty} u(\ell T)\, p(t-\ell T)\,.$$

The samples $y(kT)$ of $y(t)$ are given by

$$y(kT) = \sum_{\ell=0}^{\infty} u(\ell T)\, p(kT-\ell T)\,.$$

The right-hand side of this equation is a convolution summation of the discrete-time signals $u(kT)$ and $p(kT)$. Taking the z-transform and using the property that the z-transform of the convolution summation of two signals is the product of their z-transforms, we obtain

$$Y(z) = G(z)\,U(z), \quad \text{where } G(z) = \mathcal{Z}\{p(kT)\}\,.$$

$G(z)$ is the transfer function of the discrete-time equivalent of $\tilde{G}(s)$ preceded by ZOH.

The calculation of $G(z)$ can be simplified as follows:

$$P(s) = \tilde{G}(s)\,G_{\text{ZOH}}(s) = \tilde{G}(s)\left[\frac{1-e^{-Ts}}{s}\right] = \frac{\tilde{G}(s)}{s} - e^{-Ts}\frac{\tilde{G}(s)}{s}\,.$$

Next, set

$$q(t) = \mathcal{L}^{-1}\left\{\frac{\tilde{G}(s)}{s}\right\},$$

which leads to

$$p(t) = \mathcal{L}^{-1}\{P(s)\} = q(t) - q(t-T) \quad \longrightarrow \quad p(kT) = q(kT) - q(kT-T)\,.$$

$$\text{Let } Q(z) = \mathcal{Z}\{q(kT)\} \quad \longrightarrow \quad G(z) = Q(z) - z^{-1}Q(z) = (1-z^{-1})\,Q(z)\,.$$

The calculation of $G(z)$ is summarized by the process

$$\frac{\tilde{G}(s)}{s} \xrightarrow{\mathcal{L}^{-1}} q(t) \xrightarrow{T} q(kT) \xrightarrow{\mathcal{Z}} Q(z) \qquad (9.13\text{a})$$

and

$$G(z) = (1-z^{-1})\,Q(z)\,. \qquad (9.13\text{b})$$

Let us introduce the notation

$$H(z) = \mathcal{Z}\left\{\mathcal{L}^{-1}\{H(s)\}\right\},$$

which combines taking the inverse Laplace transform and the z transform. The notation $\mathcal{Z}\{\mathcal{L}^{-1}\{(H(s)\}\}$ says that we first take the inverse Laplace transform of $H(s)$ to obtain $h(t)$, then sample $h(t)$ at the sampling points to obtain $h(kT)$, and follow up by taking the z-transform of $h(kT)$. This calculation is facilitated by a combined Laplace/z-transform table whose first column is $H(s)$, its second column is $h(kT)$, and the third column is $H(z)$; see Appendix C. The second and third hcolumns provide a z-transform table.

9-3. DIGITAL CONTROL SYSTEMS

With this notation, the calculation of the discrete-time equivalent $G(z)$ of the continuous-time $\tilde{G}(s)$, preceded by zero-order hold, is summarized by

$$G(z) = \left(\frac{z-1}{z}\right) \mathcal{Z}\left\{\mathcal{L}^{-1}\left\{\frac{\tilde{G}(s)}{s}\right\}\right\}. \tag{9.14}$$

Example 9-5: Discrete-Time Equivalent of a Continuous-Time Transfer Function Preceded by Zero-Order Hold

For each of the following transfer functions $\tilde{G}(s)$, preceded by zero-order hold, find the equivalent transfer function $G(z)$.

(a) $\tilde{G}(s) = \dfrac{1}{s^2}$

(b) $\tilde{G}(s) = \dfrac{a}{s+a}$

Solution:

(a)

$$\tilde{G}(s) = \frac{1}{s^2}, \quad \frac{\tilde{G}(s)}{s} = \frac{1}{s^3},$$

$$\frac{1}{s^3} \rightarrow \frac{z(z+1)T^2}{2(z-1)^3},$$

$$G(z) = \left(\frac{z-1}{z}\right)\frac{z(z+1)T^2}{2(z-1)^3} = \frac{T^2(z+1)}{2(z-1)^2}.$$

$\tilde{G}(s)$ has two poles at $s = 0$; $G(z)$ has two poles at $z = 1$. The poles of $G(z)$ are mapped from the poles of $\tilde{G}(s)$ using $z = e^{sT}$.

(b)

$$\tilde{G}(s) = \frac{a}{s+a}, \quad \frac{\tilde{G}(s)}{s} = \frac{a}{s(s+a)},$$

$$\frac{a}{s(s+a)} \rightarrow \frac{(1-e^{-aT})z}{(z-1)(z-e^{-aT})},$$

$$G(z) = \left(\frac{z-1}{z}\right)\frac{(1-e^{-aT})z}{(z-1)(z-e^{-aT})} = \frac{(1-e^{-aT})}{z-e^{-aT}}.$$

The pole $s = -a$ of $\tilde{G}(s)$ maps into the pole $z = e^{-aT}$ of $G(z)$. The mapping function is $z = e^{sT}$.

The observations made in the example about the poles mapping are true in general.

▶ The transfer functions $\tilde{G}(s)$ and $G(z)$ have the same number of poles. The poles of $\tilde{G}(s)$ map into the poles of $G(z)$ using the mapping function $z = e^{sT}$. ◀

The calculation of $G(z)$ from $\tilde{G}(s)$ is done in MATLAB using the command "c2d". If G1 is the MATLAB name for $\tilde{G}$ and T is the sampling period, then

$$G = c2d(G1,T,'zoh') .$$

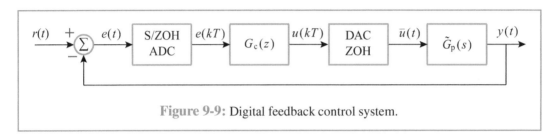

Figure 9-9: Digital feedback control system.

Consider the digital feedback control system of Fig. 9-9. The plant is represented by the continuous-time transfer function $\tilde{G}_p(s)$. The controller is represented by the discrete-time transfer function $G_c(z)$. The input-output relationship of the controller is

$$U(z) = G_c(z)\, E(z) .$$

The error signal $e(t)$ is given by

$$e(t) = r(t) - y(t) \quad \longrightarrow \quad e(kT) = r(kT) - y(kT)$$
$$\longrightarrow \quad E(z) = R(z) - Y(z)$$

and

$$G_p(z) = \left(\frac{z-1}{z}\right) \mathcal{Z}\left\{\mathcal{L}^{-1}\left\{\frac{\tilde{G}_p(s)}{s}\right\}\right\} .$$

The discrete-time equivalent is shown in Fig. 9-10. The closed-loop transfer function is

$$\frac{Y(z)}{R(z)} = \frac{G_p(z)\, G_c(z)}{1 + G_p(z)\, G_c(z)} . \tag{9.15}$$

Example 9-6: Closed-Loop Transfer Function of the Discrete-Time Equivalent System

A digital control system has the plant transfer function

$$\tilde{G}_p(s) = \frac{a}{s+a}$$

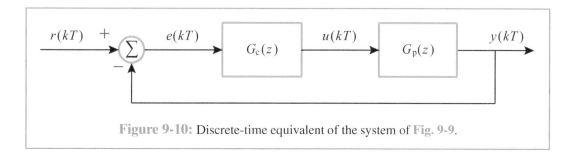

Figure 9-10: Discrete-time equivalent of the system of Fig. 9-9.

and the controller transfer function $G_c(z) = K$. Find the closed-loop transfer function $Y(z)/R(z)$.

Solution: From Example 9-5,

$$G_p(z) = \frac{(1-e^{-aT})}{z-e^{-aT}}$$

and

$$\frac{Y(z)}{R(z)} = \frac{G_p(z)\,G_c(z)}{1+G_p(z)\,G_c(z)}$$

$$= \frac{K\dfrac{(1-e^{-aT})}{z-e^{-aT}}}{1+K\dfrac{(1-e^{-aT})}{z-e^{-aT}}} = \frac{K(1-e^{-aT})}{z-e^{-aT}+K(1-e^{-aT})}.$$

9-4 Stability of Digital Control Systems

Consider the feedback control system of Fig. 9-10. The system is stable if all poles of the closed-loop transfer function

$$\frac{G_p(z)\,G_c(z)}{1+G_p(z)\,G_c(z)}$$

are inside the unit circle; that is, the absolute value of all poles is less than one. The poles of the closed-loop transfer function are the roots of the characteristic equation

$$1+G_p(z)\,G_c(z) = 0.$$

Methods for the determination of stability:

(a) Calculation of the closed-loop poles.

(b) Root locus method.

(c) Routh-Hurwitz criterion (using the bilinear transformation).

(d) Nyquist criterion.

(e) Jury test (not covered).[4]

---
[4]See Section 5.4 of [9].

9-4.1 Calculation of the Closed-Loop Poles

The closed-loop poles can be determined by manually calculating the roots of the characteristic equation or by using the MATLAB commands "*pole*" or "*roots*".

Example 9-7: Determination of Stability by Calculation of the Closed-Loop Poles

---

Determine the stability of a digital control system where the continuous-time plant $\tilde{G}_p(s)$ and the discrete-time PI controller $G_c(z)$ are given by

$$\tilde{G}_p(s) = \frac{1}{s+1}, \qquad G_c(z) = 1 + \frac{0.1}{z-1}.$$

The sampling period is $T = 0.01$ s.

Solution:

$$G_p(z) = \left(\frac{z-1}{z}\right) \mathcal{Z}\left\{\mathcal{L}^{-1}\left\{\frac{\tilde{G}(s)}{s}\right\}\right\} = \frac{1-e^{-T}}{z-e^{-T}} = \frac{1-e^{-0.01}}{z-e^{-0.01}}$$

and

$$G_c(z) = 1 + \frac{0.1}{z-1} = \frac{z-0.9}{z-1}.$$

Find the roots of

$$1 + \frac{(1-e^{-0.01})(z-0.9)}{(z-e^{-0.01})(z-1)} = 0,$$

$$(z-e^{-0.01})(z-1) + (1-e^{-0.01})(z-0.9) = 0,$$

$$z^2 - 2e^{-0.01}z + 1.9e^{-0.01} - 0.9 = 0 \quad \longrightarrow \quad z = 0.99 \pm j0.0299,$$

$$|z| = 0.9905 < 1 \quad \longrightarrow \quad \text{the system is stable.}$$

9-4.2 Root Locus Method

Let $G_p(z)G_c(z) = KL(z)$. Construct the locus of the roots of $1 + KL(z) = 0$ in the z-plane, as K changes from zero to infinity, and determine the range of K for which the locus is inside the unit circle. The rules for constructing the root locus are the same as those used in Section 4-2.

Example 9-8: Determination of Stability from the Root Locus: First-Order Case

---

Determine the stability of a digital control system where the discrete-time equivalent of the plant $G_p(z)$ and the discrete-time controller $G_c(z)$ are given by

$$G_p(z) = \frac{1-a}{z-a}, \qquad 0 < a < 1, \qquad G_c(z) = K \geq 0.$$

9-4. STABILITY OF DIGITAL CONTROL SYSTEMS

Solution:
$$G_p(z)\, G_c(z) = \frac{K(1-a)}{z-a} = KL(z), \quad \text{where } L(z) = \frac{(1-a)}{z-a}.$$

There is one branch of the root locus that starts at $z = a < 1$ and moves along the real axis towards $-\infty$. The root locus crosses the unit circle at $z = -1$. Hence,

$$K_{max}|L(z)|_{z=-1} = 1,$$

$$\frac{K_{max}(1-a)}{|-1-a|} = 1 \quad \rightarrow \quad K_{max} = \frac{1+a}{1-a}.$$

The system is stable for $K < K_{max}$.

Example 9-9: Determination of Stability from the Root Locus: Second-Order Case

---

Determine the stability of a digital control system where the discrete-time equivalent of the plant $G_p(z)$ and the discrete-time controller $G_c(z)$ are given by

$$G_p(z) = \frac{1-e^{-1}}{z-e^{-1}}, \quad G_c(z) = \frac{K(z-0.9)}{z-1}, \quad K \geq 0.$$

Solution:
$$G_p(z)\, G_c(z) = \frac{K(1-e^{-1})(z-0.9)}{(z-e^{-1})(z-1)} = KL(z), \quad \text{where } L(z) = \frac{(1-e^{-1})(z-0.9)}{(z-e^{-1})(z-1)}.$$

$L(z)$ has two poles at $z = e^{-1}$ and $z = 1$ and one zero at $z = 0.9$. The zero is located between the two poles because $e^{-1} = 0.368 < 0.9$. There are two branches of the root locus, the first branch starts at $z = 1$ and terminates at $z = 0.9$ and the second branch starts at $z = e^{-1}$ and moves along the real axis towards $-\infty$. Because at $K = 0$ there is a root at $z = 1$, K must be positive for stability. The second branch crosses the unit circle at $z = -1$. Hence,

$$K_{max}|L(z)|_{z=-1} = 1.$$

$$\left|\frac{K_{max}(1-e^{-1})(-1-0.9)}{(-1-e^{-1})(-1-1)}\right| = 1 \quad \rightarrow \quad K_{max} = \frac{2(1+e^{-1})}{1.9(1-e^{-1})} = 2.2778.$$

The system is stable for $0 < K < 2.2778$.

Example 9-10: Determination of Stability from the Root Locus: Second-Order Case

---

Determine the stability of a digital control system where the discrete-time equivalent of the plant $G_p(z)$ and the discrete-time controller $G_c(z)$ are given by

$$G_p(z) = \frac{1-e^{-1}}{z-e^{-1}}, \quad G_c(z) = \frac{K(z+1)}{z-1}, \quad K \geq 0.$$

Solution:

$$G_p(z)\, G_c(z) = \frac{K(1-e^{-1})(z+1)}{(z-e^{-1})(z-1)}\;.$$

The MATLAB-constructed root locus is shown in Fig. 9-11. The root locus crosses the unit circle at $K = 1.01$. The system is stable for $0 < K < 1.01$.

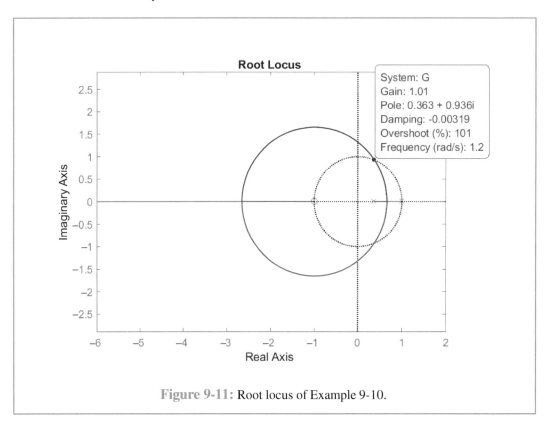

Figure 9-11: Root locus of Example 9-10.

Bilinear transformation:

The bilinear transformation allows us to map stability determination and frequency-response analysis from the discrete-time domain (z-plane) to a continuous-time domain (w-plane), where

$$w = c\left(\frac{z-1}{z+1}\right), \quad c > 0 \quad \Longleftrightarrow \quad z = \frac{1+w/c}{1-w/c},$$

$$w = \sigma + jv \quad \rightarrow \quad z = \frac{\left(1+\frac{\sigma}{c}\right) + j\frac{v}{c}}{\left(1-\frac{\sigma}{c}\right) - j\frac{v}{c}} \quad \rightarrow \quad |z|^2 = \frac{\left(1+\frac{\sigma}{c}\right)^2 + \left(\frac{v}{c}\right)^2}{\left(1-\frac{\sigma}{c}\right)^2 + \left(\frac{v}{c}\right)^2},$$

$$\sigma = 0 \quad \Longleftrightarrow \quad |z| = 1,$$

$$\sigma < 0 \quad \Longleftrightarrow \quad |z| < 1.$$

9-4. STABILITY OF DIGITAL CONTROL SYSTEMS

The bilinear transformation maps the unit circle in the *z*-plane into the imaginary axis in the *w*-plane and maps the interior of the unit circle in the *z*-plane into the left half of the *w*-plane.

Frequency response in the *z*-plane:

$$z = e^{j\omega T}, \quad -\frac{\pi}{T} \leq \omega \leq \frac{\pi}{T},$$

Frequency response in the *w*-plane:

$$w = jv, \quad -\infty \leq v \leq \infty,$$

$$w = jv \rightarrow z = \frac{1+j\frac{v}{c}}{1-j\frac{v}{c}} \rightarrow \angle z = 2\tan^{-1}\left(\frac{v}{c}\right),$$

$$z = e^{j\omega T} \rightarrow \angle z = \omega T \rightarrow \omega T = 2\tan^{-1}\left(\frac{v}{c}\right),$$

$$v = c \tan\left(\frac{\omega T}{2}\right),$$

$$c = \frac{2}{T} \rightarrow \boxed{v = \frac{2}{T} \tan\left(\frac{\omega T}{2}\right)}, \quad \frac{\omega T}{2} \ll 1 \rightarrow v \approx \omega.$$

- For stability analysis, set $c = 1$.
- For mapping the frequency response, set $c = 2/T$.

9-4.3 Routh-Hurwitz Criterion

To determine whether or not the roots of a polynomial equation $F(z) = 0$ are inside the unit circle, apply the Routh-Hurwitz criterion to determine if the roots of

$$F(z)|_{z=(1+w)/(1-w)} = 0$$

are in the left-half plane.

Example 9-11: Determination of Stability by the Routh-Hurwitz Criterion

---

Determine the stability of a digital control system where the discrete-time equivalent of the plant $G_p(z)$ and the discrete-time controller $G_c(z)$ are given by

$$G_p(z) = \frac{1-e^{-0.01}}{z-e^{-0.01}}, \quad G_c(z) = \frac{K(z-0.9)}{z-1}.$$

Solution:

$$1 + \frac{K(1-e^{-0.01})(z-0.9)}{(z-e^{-0.01})(z-1)} = 0,$$

$$(z - e^{-0.01})(z - 1) + K(1 - e^{-0.01})(z - 0.9) = 0,$$

$$z \rightarrow \frac{1+w}{1-w},$$

$$\left(\frac{1+w}{1-w} - e^{-0.01}\right)\left(\frac{1+w}{1-w} - 1\right) + K\left(1 - e^{-0.01}\right)\left(\frac{1+w}{1-w} - 0.9\right) = 0,$$

$$2w\left[(1 + e^{-0.01})w + 1 - e^{-0.01}\right] + K\left(1 - e^{-0.01}\right)(1-w)(1.9w + 0.1) = 0,$$

$$(3.9801 - 0.0189K)w^2 + (0.0199 + 0.0179K)w + 0.001K = 0.$$

By the Routh-Hurwitz criterion, all coefficients must have the same sign. Since $K \geq 0$, the system is stable for

$$0 < K < \frac{3.9801}{0.0189} = 210.6.$$

9-4.4 Nyquist Criterion

The Nyquist criterion is derived using Cauchy's principle of the argument. Let

$$G_p(z)\, G_c(z) = KG(z), \quad K > 0.$$

Draw a clockwise contour in the z-plane that encircles the outside of the unit circle. **Figure 9-12** shows how this contour is constructed. Starting at $z = 1$ where $\omega = 0$, move counter clockwise along the upper half of the unit circle until you reach the point $z = -1$ where $\omega = \pi/T$. Move along a horizontal line infinitesimally close to the real axis until you reach a circle of radius R; this will be at the point $z = -R$. Complete one clockwise revolution around this circle until you come back to the point $z = -R$. Move along a horizontal line infinitesimally close to the real axis until you reach the point $z = -1$. Finally move clockwise along the lower half of the unit circle until you reach the stating point $z = 1$. Now let $R \to \infty$ so that the contour encircles the outside of the unit circle clockwise. This is the *Nyquist contour*.

If $G(z)$ has a pole at $z = 1$, make an infinitesimally small indentation (outside the unit circle) around $z = 1$[5]:

$$z = 1 + \varepsilon e^{j\phi}, \quad \varepsilon \to 0, \quad \phi : -\frac{\pi}{2} \to 0 \to \frac{\pi}{2}.$$

Map the Nyquist contour into the G-plane. This is the *Nyquist plot*. The important part of the Nyquist plot is mapping the motion on the unit circle, which is the frequency response of $G(e^{j\omega T})$ as ω changes from 0 to π/T. The frequency response as ω changes form $-\pi/T$ to 0 is the mirror image of the frequency response from 0 to π/T. The frequency response can be calculated and plotted by the MATLAB command "*nyquist*". The infinite-radius circle maps into a single point, which is the origin when $G(z)$ is strictly proper. The motions on the horizontal lines from $z = -1$ to $-\infty$ and back map into segments of the Nyquist plot on the real axis that reach the origin. These segments do not interfere with counting the number of

---

[5]Small indentations should also be made around any pole on the unit circle, but we shall not encounter such cases.

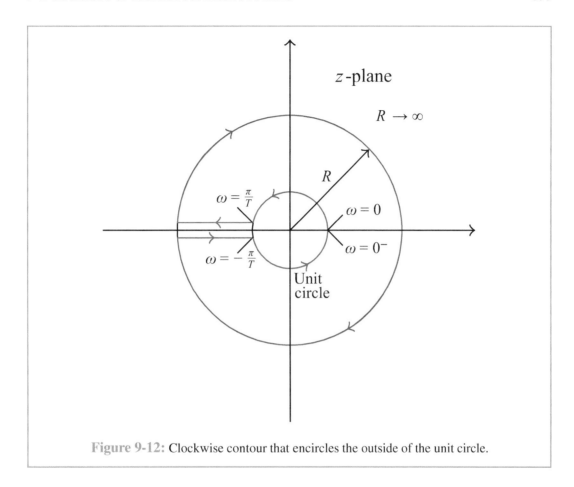

Figure 9-12: Clockwise contour that encircles the outside of the unit circle.

encirclements of the critical point. Therefore, we do not pay attention to them. If there is a small infinitesimal indentation around $z = 1$, it maps into an infinite-radius arc that closes the Nyquist plot.

Let:

$P =$ number of poles of $G(z)$ outside the unit circle,

$Z =$ number of zeros of $1 + KG(z)$ outside the unit circle, and

$N =$ number of counterclockwise encirclements of the point $-\frac{1}{K} + 0j$ by the Nyquist plot.

The system is stable when $Z = 0$.

Nyquist criterion: $Z = P - N$

Example 9-12: Determination of Stability by the Nyquist Criterion

Determine the stability of a digital control system where the continuous-time plant $\tilde{G}_p(s)$ and the discrete-time controller $G_c(z)$ are given by

$$\tilde{G}_p(s) = \frac{1}{(s+1)(s+2)(s+3)}, \qquad G_c(z) = K \geq 0.$$

The sampling period is $T = 0.01$ s.

Solution: Using MATLAB, the discrete-time equivalent of the plant is

$$G_p(z) = \frac{1.6419 \times 10^{-7}(z+0.264)(z+3.677)}{(z-0.99)(z-0.9802)(z-0.9704)},$$

and the Nyquist plot of $G_p(z)$ is shown in Fig. 9-13. Zooming on the intersection with the real axis, it is determined that the Nyquist plot intersects the real axis at -0.0169. Because $G_p(z)$ has no poles outside the unit circle; that is, $P = 0$, the system is stable when $N = 0$. This is the case if

$$-\frac{1}{K} < -0.0169 \quad \longleftrightarrow \quad \frac{1}{K} > 0.0169 \quad \longleftrightarrow \quad K < \frac{1}{0.0169} = 59.17.$$

The system is stable for $K < 59.17$.

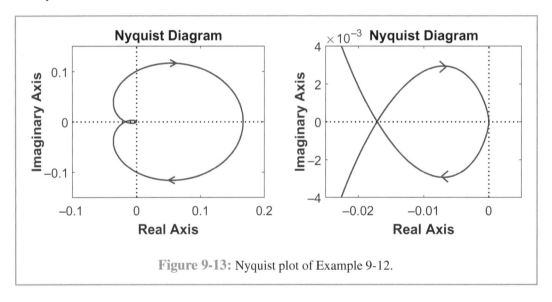

Figure 9-13: Nyquist plot of Example 9-12.

Example 9-13: Determination of Stability by the Nyquist Criterion: A Case with a Pole at $z = 1$

Determine the stability of a digital control system where the continuous-time plant $\tilde{G}_p(s)$ and the discrete-time controller $G_c(z)$ are given by

$$\tilde{G}_p(s) = \frac{1}{s(s+1)}, \qquad G_c(z) = K \geq 0.$$

9-4. STABILITY OF DIGITAL CONTROL SYSTEMS

The sampling period is $T = 0.1$ s.

Solution: Using MATLAB, the discrete-time equivalent of the plant is

$$G_p(z) = \frac{0.0048374(z+0.9672)}{(z-1)(z-0.9048)},$$

and the frequency response part of the Nyquist plot of $G_p(z)$ is shown in Fig. 9-14.

Indentation around $z = 1$:

$$z = 1 + \varepsilon e^{j\phi}, \qquad \varepsilon \to 0, \qquad \phi: -\frac{\pi}{2} \to 0 \to \frac{\pi}{2},$$

$$G_p(z) = \frac{0.0048374(1 + \varepsilon e^{j\phi} + 0.9672)}{(1 + \varepsilon e^{j\phi} - 1)(1 + \varepsilon e^{j\phi} - 0.9048)} \approx \frac{0.0048374 \times 1.9672 e^{-j\phi}}{0.0952\varepsilon},$$

$$|G_p| \to \infty, \qquad -\phi: \frac{\pi}{2} \to 0 \to -\frac{\pi}{2}.$$

The mapping of the indentation is a clockwise infinite-radius arc that closes the Nyquist plot on the right side. Zooming on the intersection with the real axis, it is determined that the Nyquist plot intersects the real axis at -0.0493. Because $G_p(z)$ has no poles outside the unit circle, that is, $P = 0$, the system is stable when $N = 0$. This is the case if

$$-\frac{1}{K} < -0.00493 \quad \longleftrightarrow \quad \frac{1}{K} > 0.0493 \quad \longleftrightarrow \quad K < \frac{1}{0.0493} = 20.28.$$

The system is stable for $0 < K < 20.28$.

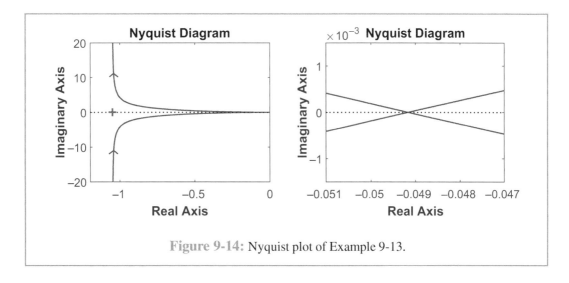

Figure 9-14: Nyquist plot of Example 9-13.

Stability margins

The gain and phase margins are defined as in the continuous-time case.

- The Gain Margin is the gain increase that makes the system marginally stable.
- The Phase Margin is the phase decrease that makes the system marginally stable.

They can be calculated using the MATLAB command "*margin*".

Example 9-14: Stability Margins for a Digital Control System

Find the gain and phase margins of the system of Example 9-13 when $K = 5$.

Solution: The MATLAB statement

$$\text{margin}(5*Gp)$$

produces Fig. 9-15.

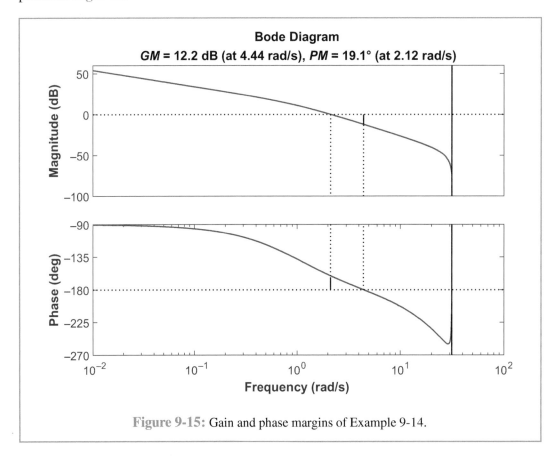

Figure 9-15: Gain and phase margins of Example 9-14.

9-5 Steady-State Error

Consider the feedback control system of **Fig. 9-10**. Suppose the closed-loop system is stable. Set $G(z) = G_p(z)\, G_c(z)$. The z-transform of the tracking error is given by:

$$E(z) = \frac{1}{1+G(z)} R(z) \,. \tag{9.16}$$

Type of feedback system:

The system is type $N \geq 0$ if $G(z)$ has N poles at $z = 1$.

Step input

$$r(kT) = A1(k) = A \text{ for } k \geq 0, \quad \longrightarrow \quad R(z) = \frac{Az}{z-1},$$

$$e(\infty) = \lim_{k \to \infty} e(k) = \lim_{z \to 1}[(z-1)\,E(z)] = \lim_{z \to 1}\left[(z-1)\,\frac{Az}{[1+G(z)](z-1)}\right],$$

or

$$e(\infty) = \begin{cases} \frac{A}{1+G(1)} & \text{if } N = 0, \\ 0 & \text{if } N \geq 1, \end{cases}$$

For $N = 0$,

$$K_p \stackrel{\text{def}}{=} G(1), \qquad \textit{(position error constant)}$$

$$e(\infty) = \frac{A}{1+K_p}\,.$$

Ramp input

$$r(kT) = A\,kT \text{ for } k \geq 0, \quad \longrightarrow \quad R(z) = \frac{ATz}{(z-1)^2},$$

$$e(\infty) = \lim_{k \to \infty} e(k) = \lim_{z \to 1}[(z-1)\,E(z)] = \lim_{z \to 1}\left[(z-1)\,\frac{ATz}{[1+G(z)](z-1)^2}\right],$$

or

$$e(\infty) = \begin{cases} \infty & \text{if } N = 0 \\ \dfrac{AT}{(z-1)G(z)|_{z=1}} & \text{if } N = 1 \\ 0 & \text{if } N \geq 2. \end{cases}$$

For $N = 1$,

$$K_v \stackrel{\text{def}}{=} \frac{1}{T}(z-1)\,G(z)|_{z=1}, \qquad \textit{(velocity error constant)}$$

$$e(\infty) = \frac{A}{K_v}\,.$$

> **Example 9-15: Steady-State Error**

Consider a digital control system where the discrete-time equivalent of the plant $G_p(z)$ and the discrete-time controller $G_c(z)$ are given by

$$G_p(z) = \frac{1-a}{z-a}, \quad 0 < a < 1, \quad G_c(z) = K \geq 0.$$

Find the steady-state error to a unit step input. Can we make the error arbitrarily small by choosing K large enough? If the answer is NO, how would you make the error arbitrarily small?

Solution: Determine the range of K for stability. The characteristic equation is

$$1 + \frac{K(1-a)}{(z-a)} = 0 \quad \longleftrightarrow \quad z - a + K(1-a) = 0.$$

There is one closed-loop pole at $z = a - K(1-a)$. For stability

$$-1 < z < 1 \quad \longleftrightarrow \quad -1 < a - K(1-a) < 1 \quad \longleftrightarrow \quad -1 < K < \frac{1+a}{1-a}.$$

The lower bound on K is satisfied because $K \geq 0$. Hence, the system is stable for $K < \frac{1+a}{1-a}$.

Find the steady-state error to a unit step input:

$$G(z) = G_p(z)\, G_c(z) = \frac{K(1-a)}{z-a}, \quad \text{(Type Zero)}$$

$$K_p = G(1) = K \quad \longrightarrow \quad e(\infty) = \frac{1}{1+K}.$$

The error cannot be made arbitrarily small by choosing a large value for K because K is limited by $K < (1+a)/(1-a)$ for stability. The error can be reduced by using a PI controller. The transfer function of the PI controller is:

$$G_c(z) = K\left[1 + \frac{b}{z-1}\right], \quad K > 0, \quad b > 0.$$

Hence,

$$G(z) = G_p(z)\, G_c(z) = \frac{K(z-1+b)(1-a)}{(z-1)(z-a)}. \quad \text{(Type One)}$$

Design K and b such that the system is stable. The steady-state error to a unit step input is zero. We can calculate the steady-state error to a ramp input:

$$K_v = \frac{1}{T}(z-1)\, G(z)\big|_{z=1} = \frac{1}{T}\frac{K(z-1)(z-1+b)(1-a)}{(z-1)(z-a)}\bigg|_{z=1},$$

$$K_v = \frac{Kb}{T}.$$

The steady-state error to a unit ramp input is given by

$$e(\infty) = \frac{1}{K_v} = \frac{T}{Kb}.$$

9-6 Sample-Frequency Selection

An important step in the design of digital control systems is the choice of the sampling period or the sampling frequency, which are defined as follows:

$$T = \text{sampling period (s)},$$

$$f_s = \frac{1}{T} = \text{sampling frequency in Hz},$$

$$\omega_s = \frac{2\pi}{T} = \text{sampling frequency in rad/s}.$$

The choice of the sampling frequency is a compromise between performance and cost. To achieve the performance of an analog controller, T should be as small as possible, but reducing T increases the cost because we need a faster computer with a larger wordlength (more bits).

To see the need for more bits when T is small, consider the analog PI controller $G_c(s) = K_p + (K_I/s)$, which can be implemented digitally as $G_c(z) = K_p + K_I(Tz/z-1)$. As T decreases, more bits are needed to represent the coefficient $K_I T$; otherwise it would be rounded off to zero.

9-6.1 Sampling Theory

A fundamental result in digital signal processing is the sampling theory, which we summarize here.

(a) Spectrum of a sampled signal:

Let $f(t)$ be a continuous-time signal and $F(\omega)$ be its Fourier Transform:

$$F(\omega) = \mathcal{F}[f(t)] = \int_{-\infty}^{\infty} f(t)\, e^{-j\omega t}\, dt.$$

Pass $f(t)$ through an ideal sampler to obtain

$$f^*(t) = \sum_{k=-\infty}^{\infty} f(kT)\, \delta(t-kT).$$

Compute the Fourier transform of $f^\star(t)$:

$$F^*(\omega) = \mathcal{F}[f^*(t)] = \int_{-\infty}^{\infty} \sum_{k=-\infty}^{\infty} f(kT)\,\delta(t-kT)\,e^{-j\omega t}\,dt$$

$$= \int_{-\infty}^{\infty} \sum_{k=-\infty}^{\infty} f(t)\,\delta(t-kT)\,e^{-j\omega t}\,dt$$

$$= \int_{-\infty}^{\infty} f(t) \sum_{k=-\infty}^{\infty} \delta(t-kT)\,e^{-j\omega t}\,dt\,.$$

Here, $\sum_{k=-\infty}^{\infty} \delta(t-kT)$ is a periodic signal of period T. It can be represented by its Fourier series:

$$\sum_{k=-\infty}^{\infty} \delta(t-kT) = \sum_{\ell=-\infty}^{\infty} c_\ell e^{j\omega_s \ell t}, \qquad \omega_s = \frac{2\pi}{T}\,,$$

where

$$c_\ell = \frac{1}{T} \int_{-T/2}^{T/2} \sum_{k=-\infty}^{\infty} \delta(t-kT)\,e^{-j\omega_s \ell t}\,dt = \frac{1}{T}\,.$$

Hence,

$$\sum_{k=-\infty}^{\infty} \delta(t-kT) = \frac{1}{T} \sum_{\ell=-\infty}^{\infty} e^{j\omega_s \ell t}\,.$$

Substitution of this expression in $F^\star(\omega)$ yields

$$F^*(\omega) = \int_{-\infty}^{\infty} f(t) \frac{1}{T} \sum_{\ell=-\infty}^{\infty} e^{j\omega_s \ell t} e^{-j\omega t}\,dt\,,$$

$$= \frac{1}{T} \sum_{\ell=-\infty}^{\infty} \underbrace{\int_{-\infty}^{\infty} f(t)\,e^{-j(\omega-\omega_s \ell)t}\,dt}_{F(\omega-\omega_s \ell)}\,,$$

$$= \frac{1}{T} \sum_{\ell=-\infty}^{\infty} F(\omega - \omega_s \ell)\,. \tag{9.17}$$

When a continuous-time signal $f(t)$ is ideally sampled, the frequency spectrum of the sampled signal $f^\star(t)$ is a periodic repetition of the spectrum of $f(t)$, multiplied by the factor $1/T$. The period of the periodic repetition in the frequency domain is the sampling frequency ω_s. This is illustrated in Fig. 9-16. The figure shows the phenomenon of *aliasing*, where the repeated components of the spectrum overlap. Aliasing distorts information because the spectrum of the original signal cannot be recovered from the spectrum of the sampled signal.

If $f(t)$ is a band-limited signal, that is,

$$|F(\omega)| = 0, \quad \text{for } |\omega| > \omega_{\max}\,,$$

9-6. SAMPLE-FREQUENCY SELECTION

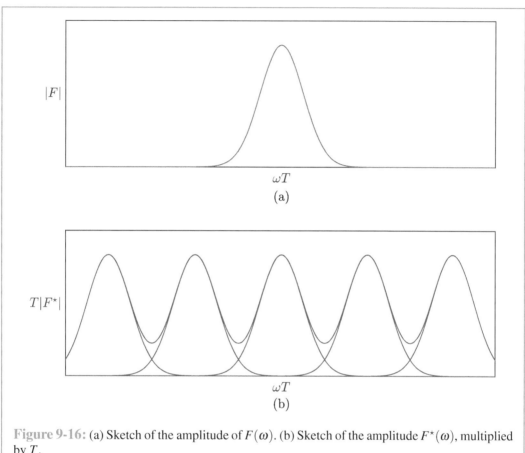

Figure 9-16: (a) Sketch of the amplitude of $F(\omega)$. (b) Sketch of the amplitude $F^\star(\omega)$, multiplied by T.

choosing $\omega_s > 2\omega_{max}$ ensures that there is no aliasing in the spectrum of the sampled signal. Then, the original signal can be recovered by passing the sampled signal through an ideal low-pass filter with a cutoff frequency of $\omega_s/2$.

(b) Shannon's sampling theorem:

A band-limited signal with maximum frequency ω_{max} can be uniquely determined from its samples if $\omega_s > 2\,\omega_{max}$.

9-6.2 Antialiasing Filter (Prefilter)

Physical signals are not band-limited. However, in many cases the energy (information) content of a signal is concentrated within a limited frequency range. In such cases, we may treat the signal as band-limited. To reduce the error due to aliasing, the signal is pre-filtered (before sampling) by a low-pass filter having a cutoff frequency $\omega_s/2$. Figure 9-17 shows the effect of the antialiasing filter in eliminating aliasing.

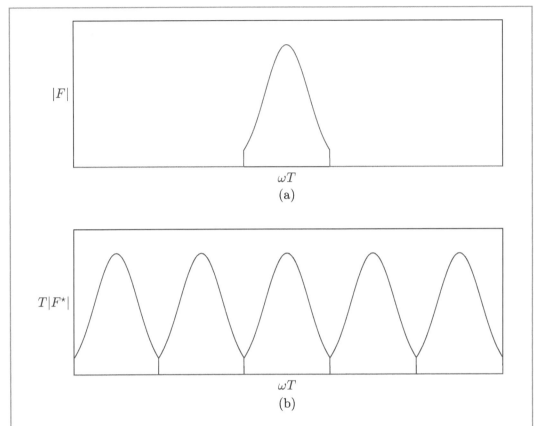

Figure 9-17: (a) Sketch of the amplitude of $F(\omega)$ after passing through an antialiasing filter. (b) Sketch of the amplitude of $F^*(\omega)$, multiplied by T.

The sampling theorem requires that $\omega_s > 2\,\omega_{max}$, but what is ω_{max} for a control system? It is usually set to be the bandwidth of the closed-loop transfer function, ω_{BW}, because for a well-designed control system, ω_{BW} will cover the bandwidth of the reference signal and will be adequate to ensure good dynamic response. Hence, the sampling theorem requirement becomes $\omega_s > 2\omega_{BW}$. This choice, however, is usually not adequate. We need to choose ω_s much higher than $2\omega_{BW}$ to achieve acceptable performance.

9-6.3 Methods for Choosing ω_s

(a) Effect of sampling on smoothness of the transient response.

(b) Effect of ZOH delay on stability.

(c) Effect of ZOH delay on tracking and disturbance rejection.

9-6. SAMPLE-FREQUENCY SELECTION

Smoothness of transient response

Fit 5 to 10 samples within the rise time of the step response:

$$5 \leq \frac{T_r}{T} \leq 10.$$

What is the relationship to ω_{BW}? From the relations of Fig. 6-39, we have

$$T_r \approx \frac{2.16\zeta + 0.6}{\omega_n}, \quad \omega_{BW} \approx (1.83 - 1.16\zeta)\omega_n,$$

$$\longrightarrow \quad T_r \omega_{BW} \approx (2.16\zeta + 0.6)(1.83 - 1.16\zeta),$$

$$\frac{T_r}{T} \approx \frac{\omega_s}{2\pi} \times \frac{(2.16\zeta + 0.6)(1.83 - 1.16\zeta)}{\omega_{BW}}.$$

For $0.3 \leq \zeta \leq 0.75$,

$$\frac{(2.16\zeta + 0.6)(1.83 - 1.16\zeta)}{2\pi} \approx \frac{1}{3}.$$

Hence,

$$\frac{T_r}{T} \approx \frac{\omega_s}{3\,\omega_{BW}},$$

and

$$5 \leq \frac{T_r}{T} \leq 10 \quad \longrightarrow \quad 15 \leq \frac{\omega_s}{\omega_{BW}} \leq 30.$$

Delay Due to ZOH

The effect of sampling and zero order hold on the analog control system can be approximated by the transfer function $(1 - e^{-sT})/(sT)$, as shown in Fig. 9-18. The transfer function of ZOH is $(1 - e^{-sT})/s$ and the factor $1/T$ is due to sampling.

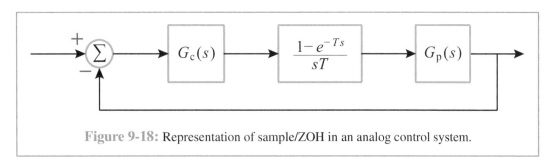

Figure 9-18: Representation of sample/ZOH in an analog control system.

For a sufficiently small sampling period T, the transfer function $(1 - e^{-sT})/(sT)$ can be approximated by a time delay equal to $T/2$ because

$$\frac{1 - e^{-sT}}{sT} = \frac{1 - (1 - sT + s^2T^2/2 - \cdots)}{sT} \approx 1 - \frac{sT}{2}$$

and

$$e^{-sT/2} \approx 1 - \frac{sT}{2} \quad \longrightarrow \quad \frac{1-e^{-sT}}{sT} \approx e^{-sT/2}.$$

Time delay undermines stability because it reduces the phase margin:

$$\text{Reduction in PM} = \frac{T\omega_c}{2} \times \frac{180}{\pi} \quad (\text{deg}),$$

where ω_c is the crossover frequency. For 5° to 10° reduction in *PM*, *T* should satisfy

$$0.175 \leq T\omega_c \leq 0.35.$$

What is the relationship to ω_{BW}?

$$\omega_{BW} \approx 1.6\,\omega_c \quad \longrightarrow \quad T\omega_c \approx \frac{2\pi}{\omega_s} \times \frac{\omega_{BW}}{1.6}.$$

Therefore,

$$11.25 \leq \frac{\omega_s}{\omega_{BW}} \leq 22.5.$$

Effect of ZOH delay on tracking and disturbance rejection

Holding the control signal constant over the sampling period limits the system's ability to react to changes in the reference or disturbance signals. Numerical investigation show that the sampling frequency should satisfy[6]

$$\omega_s \geq 20\omega_{BW}$$

in order to preserve the tracking and disturbance rejection properties of the analog control system.

Example 9-16: Sample Period Selection

A feedback control system is expected to have a settling time of no more than 100 ms and an overshoot of no more than 20%.

(a) What is the sampling period *T* if the reduction in the phase margin is to be no more than 10°?

(b) What is the sampling period *T* if there are to be 10 samples per rise time?

---

[6]See [12].

Solution: For 20% overshoot, set $\zeta = 0.5$. Hence,

$$T_s = \frac{4}{\zeta \omega_n} = 0.1 \quad \rightarrow \quad \omega_n = 80 \text{ rad/s},$$

$$T_r \approx \frac{2.16\zeta + 0.6}{\omega_n} = 0.021 \text{ s},$$

and

$$\omega_c \approx \frac{(3.35 - 2\zeta)\omega_n}{3} = 62.67 \text{ rad/s}.$$

(a)

$$\frac{T}{2} \times \omega_c \times \frac{180}{\pi} = 10,$$

$$T = \frac{10 \times 2 \times \pi}{180 \times 62.67} = 0.0056 \text{ s} = 5.6 \text{ ms}.$$

(b)

$$T = \frac{T_r}{10} = \frac{0.021}{10} = 2.1 \text{ ms}.$$

9-7 Design by Emulation

The controller is designed in continuous time and converted into a discrete-time controller by approximation.

Design steps:

(1) Design a continuous-time controller $\tilde{G}_c(s)$ to meet the design specifications.

(2) Choose the sampling period.

(3) Approximate the continuous-time controller $\tilde{G}_c(s)$ by a discrete-time controller $G_c(z)$.

(4) Evaluate the performance of the digital controller by analysis and/or simulation of the discrete-time equivalent of the closed-loop system.

Common approximations of continuous-time transfer functions

- Forward Difference Method (FDM)

$$G_c(z) = \tilde{G}_c(s)\big|_{s=(z-1)/T}.$$

- Backward Difference Method (BDF)

$$G_c(z) = \tilde{G}_c(s)\big|_{s=(z-1)/(Tz)}.$$

- Tustin (Trapezoidal or Bilinear Transformation (BT)) Method

$$G_c(z) = \tilde{G}_c(s)\big|_{s=(2/T)(z-1)/(z+1)} .$$

- ZOH Method

$$G_c(z) = \left(\frac{z-1}{z}\right) \mathcal{Z}\left\{\mathcal{L}^{-1}\left\{\frac{\tilde{G}_c(s)}{s}\right\}\right\} .$$

The ZOH method is based on approximating the input by a staircase continuous-time function that is constant in-between the sampling points. The other three methods are derived from numerical integration methods. Consider the integral

$$y(t) = \int x(\tau)\, d\tau ,$$

whose Laplace transform is

$$Y(s) = \frac{1}{s} X(s) .$$

Integration over one sampling period is given by

$$y[(k+1)T] = y(kT) + \int_{kT}^{(k+1)T} x(\tau)\, d\tau .$$

FDM:
$$x(\tau) \approx x(kT) .$$
$$y[(k+1)T] = y(kT) + T\, x(kT) .$$

Taking the z-transform yields

$$(z-1)\, Y(z) = T\, X(z) \quad \longleftrightarrow \quad Y(z) = \frac{T}{z-1} X(z) .$$

Comparison of the s- and z-transfer functions shows that

$$\frac{1}{s} \mapsto \frac{T}{z-1} \quad \longleftrightarrow \quad s \mapsto \frac{z-1}{T} .$$

BDM:
$$x(\tau) \approx x[(k+1)T] ,$$
$$y[(k+1)T] = y(kT) + Tx[(k+1)T] .$$

Take the z-transform:

$$(z-1)\, Y(z) = Tz\, X(z) \quad \longleftrightarrow \quad Y(z) = \frac{Tz}{z-1} X(z) ,$$

$$\frac{1}{s} \mapsto \frac{Tz}{z-1} \quad \longleftrightarrow \quad s \mapsto \frac{z-1}{Tz} .$$

9-7. DESIGN BY EMULATION

Tustin:

$$x(\tau) \approx \frac{1}{2}\{x(kT) + x[(k+1)T]\},$$

$$y[(k+1)T] = y(kT) + \frac{T}{2}\{x(kT) + x[(k+1)T]\}.$$

Take the z-transform.

$$(z-1)Y(z) = \frac{T}{2}(z+1)X(z) \quad \longleftrightarrow \quad Y(z) = \frac{T}{2}\left(\frac{z+1}{z-1}\right)X(z),$$

$$\frac{1}{s} \mapsto \frac{T}{2}\left(\frac{z+1}{z-1}\right) \quad \longleftrightarrow \quad s \mapsto \frac{2}{T}\left(\frac{z-1}{z+1}\right).$$

The Tustin and ZOH conversions can be calculated using the MATLAB command "*c2d*". If G1 is the MATLAB name for $\tilde{G}_c(s)$, G2 is the MATLAB name for $G_c(z)$, and T is the sampling period, then

- G2 = c2d(G1,T,'tustin') provides the Tustin approximation, and
- G2 = c2d(G1,T,'zoh') provides the ZOH approximation.

Note that the use of ZOH conversion here is different from our earlier use. Earlier we used the ZOH conversion to find the discrete-time equivalent of the continuous-time plant driven by the staircase continuous-time control signal coming from DAC/ZOH. That calculation was exact. In the current case, the continuous-time controller is driven by the error signal, which is not a staircase continuous-time function. Therefore, the use of the ZOH conversion is based on approximating the error signal by a staircase continuous-time function.

Example 9-17: Design by Emulation

Given the plant transfer function

$$\tilde{G}_p(s) = \frac{0.2}{(s+1)(s+0.2)},$$

design a discrete-time controller to meet the specifications:

- Steady-state error to step inputs $\leq 1\%$,
- Overshoot $\leq 20\%$, and
- Settling time < 10 s.

Solution: We start with the design of a continuous-time controller $\tilde{G}_c(s)$, which we take to be a PI controller:
$$\tilde{G}_c(s) = K_p + \frac{K_I}{s} = K_p \frac{(s+b)}{s}, \qquad b = \frac{K_I}{K_p}.$$

The PI controller increases the type of the system to Type 1; hence, the steady-state error to step inputs is zero. Take $b = 0.2$ to cancel the pole at -0.2. The compensated system has poles at 0 and -1. Hence, the complex part of the root locus is a vertical line at $s = -0.5$. For a pair of complex poles,
$$\zeta \omega_n = 0.5 \quad \longrightarrow \quad T_s = \frac{4}{\zeta \omega_n} = \frac{4}{0.5} = 8\text{ s} < 10\text{ s}.$$

For 20% overshoot, set $\zeta = 0.5$. The desired poles are $-\frac{1}{2} \pm j \frac{1}{2}\sqrt{3}$. Apply the magnitude criterion to find K_p.
$$1 = \left.\frac{0.2 K_p}{s(s+1)}\right|_{s=-(1/2)+j(1/2)\sqrt{3}} \quad \longrightarrow \quad K_p = 5,$$
$$\tilde{G}_c(s) = \frac{5(s+0.2)}{s} = 5 + \frac{1}{s}.$$

The closed-loop system has the following properties:
- Step response: $T_r = 1.639$ s, $T_s = 8.0759$ s, $PO = 16.3\%$.
- Stability margins: $GM = \infty$, $PM = 51.8°$ at $\omega_c = 0.786$ rad/s.

Choice of T:
$$T_r = 1.639\text{ s}, \qquad 5 \leq \frac{T_r}{T} \leq 10 \quad \longrightarrow \quad 0.1639 \leq T \leq 0.328\text{ s}.$$

Try $T = 0.2$ s and $T = 0.3$ s. We use the Tustin method to convert the continuous-time controller into a discrete-time controller:
$$G_c(z) = \left[5 + \frac{1}{s}\right]_{s=(2/T)(z-1)/(z+1)} = 5 + \frac{T}{2}\left(\frac{z+1}{z-1}\right).$$

The discrete-time equivalent of the plant is
$$G_p(z) = \left(\frac{z-1}{z}\right) \mathcal{Z}\left\{\mathcal{L}^{-1}\left\{\frac{\tilde{G}_p(s)}{s}\right\}\right\}$$
$$= \left(\frac{z-1}{z}\right) \mathcal{Z}\left\{\mathcal{L}^{-1}\left\{\frac{0.2}{s(s+1)(s+0.2)}\right\}\right\}.$$

For $T = 0.2$ s,
$$G_p(z) = \frac{0.0036959(z+0.9231)}{(z-0.9608)(z-0.8187)} \quad \text{and} \quad G_c(z) = \frac{5.1(z-0.9608)}{(z-1)}.$$

9-7. DESIGN BY EMULATION

For $T = 0.3$ s

$$G_p(z) = \frac{0.0079989(z+0.887)}{(z-0.9418)(z-0.7408)} \quad \text{and} \quad G_c(z) = \frac{5.15(z-0.9417)}{(z-1)}.$$

Figure 9-19 compares the step response of the digital control system with the step response of the analog control systems, and Table 9-1 compares properties of the two systems. As expected, the performance of the digital controller is closer to the performance of the analog controller for smaller T.

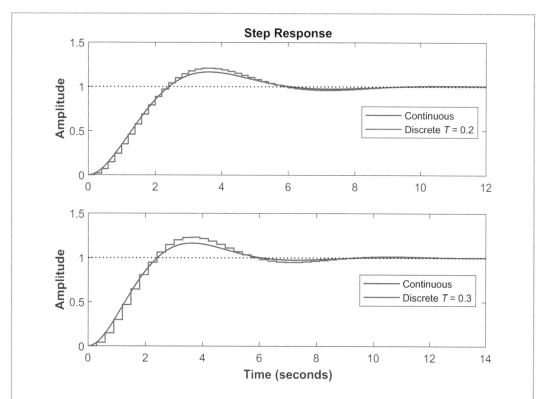

Figure 9-19: Comparison of the step responses of the analog and digital control systems of Example 9-17.

Table 9-1: Comparison of analog control and digital control properties of Example 9-17.

Property	Analog	Digital ($T = 0.2$ s)	Digital ($T = 0.3$ s)
T_r (s)	1.639	1.6	1.5
T_s (s)	8.0759	8.4	8.7
PO	16.3%	20.58%	22.94%
GM (dB)	∞	20.4	17
PM (deg)	51.8	47.4	45.2

9-8 Discrete-Time Design

The controller is designed in the discrete-time domain.

Design steps:

(1) Choose T based on the design specifications (larger T relative to the emulation method).

(2) Find the discrete-time equivalent of the plant

$$G_p(z) = \left(\frac{z-1}{z}\right) \mathcal{Z}\left\{\mathcal{L}^{-1}\left\{\frac{\tilde{G}_p(s)}{s}\right\}\right\}.$$

(3) Design the controller $G_c(z)$ to meet the design specifications.

(4) Evaluate the performance of the digital controller by analysis and/or simulation.

We shall consider a time-domain design using the root locus method and a frequency-domain design using Bode plots.

9-8.1 Root Locus Design

Design procedure:

Step 1: Construct the root locus of $G_p(z)\,G_c(z) = KL(z)$ as K varies from zero to infinity.

Step 2: Choose a point on the root locus that meets the design specifications.

Step 3: Find K at the chosen point by using the MATLAB plot of the root locus or by applying the magnitude criterion

$$|KL(z)|_{\text{at desired root}} = 1.$$

Design Specifications:

- Steady-state error: Requirement on the steady-state error or the error constant (K_p, K_v, etc.).

- Transient response: Choose a pair of dominant poles in the s-plane that meet the requirements of the transient response. Map the poles into the z-plane using $z = e^{sT}$.

$$s = -\zeta\omega_n \pm j\omega_d, \quad \zeta < 1, \quad \omega_d = \omega_n\sqrt{1-\zeta^2},$$
$$z = e^{(-\zeta\omega_n \pm j\omega_d)T} = e^{-\zeta\omega_n T} e^{\pm j\omega_d T},$$
$$z = re^{\pm j\theta} \quad \rightarrow \quad r = e^{-\zeta\omega_n T}, \quad \theta = \omega_d T.$$

9-8. DISCRETE-TIME DESIGN

Compensation:

Design $G_c(z)$ to reshape the root locus.

- Use a lead compensator to stabilize the system or improve the transient response by shifting the root locus to the left.
- Use a lag or PI compensator to increase the error constant $(K_p, K_v, \ldots)$ or increase the type of the feedback system; hence reducing the steady-state error.

$$G_c(z) = \frac{K(z-a)}{(z-b)}, \quad -1 \leq a,b \leq 1,$$

$a > b \rightarrow$ lead, $\quad a < b \rightarrow$ lag, $\quad b = 1 \rightarrow$ PI.

Example 9-18: Root Locus Design

Given the plant transfer function

$$\tilde{G}_p(s) = \frac{0.2}{(s+1)(s+0.2)},$$

design a discrete-time controller to meet the specifications:

- Steady-state error to step inputs $\leq 1\%$,
- Overshoot $\leq 20\%$, and
- Settling time < 10 s.

Solution: For 20% overshoot, set $\zeta = 0.5$.

$$T_s = \frac{4}{\zeta \omega_n} = \frac{4}{0.5 \omega_n} < 10 \quad \rightarrow \quad \omega_n > \frac{4}{0.5 \times 10} = 0.8 \text{ rad/s}.$$

Design for $\omega_n = 1$ rad/s.

$$T_r = \frac{2.16\zeta + 0.6}{\omega_n} = 1.68 \text{ s},$$

$$5 \leq \frac{T_r}{T} \leq 10 \quad \rightarrow \quad 0.187 \leq T \leq 0.336.$$

Set $T = 0.3$ s.

$$G_p(z) = \frac{0.0079989(z+0.887)}{(z-0.9418)(z-0.7408)}.$$

Use a PI controller (zero steady-state error):

$$G_c(z) = K \frac{(z-a)}{(z-1)}.$$

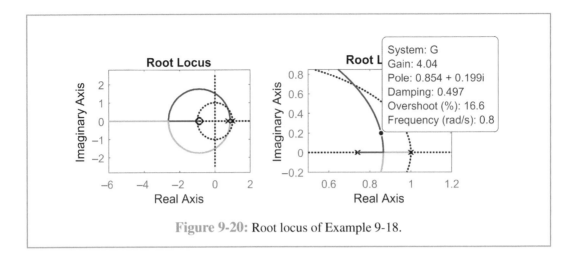

Figure 9-20: Root locus of Example 9-18.

Take $a = 0.9418$ to cancel the pole at 0.9418. Construct the root locus of

$$K\left(\frac{z-0.9418}{z-1}\right)G_p(z) = K\frac{0.0079989(z+0.887)}{(z-1)(z-0.7408)}.$$

The root locus is shown in Fig. 9-20. The point on the root locus that corresponds to $\zeta = 0.497$ has gain $4.04 \approx 4$. Thus, the controller is given by

$$G_c(z) = \frac{4(z-0.9418)}{(z-1)}.$$

The properties of the discrete-time design are compared with the properties of the emulation designs of Example 9-17 in Table 9-2. For the same sampling period $T = 0.3$ s, the discrete-time design has smaller overshoot and larger stability margins compared with the emulation design, while satisfying the settling time requirement.

Table 9-2: Comparison of the discrete-time design of Example 9-18 with the emulation designs of Example 9-17.

Property	Analog	Emulation $T = 0.2$ s	Emulation $T = 0.3$ s	Discrete design $T = 0.3$ s
T_r (s)	1.639	1.6	1.5	2.1
T_s (s)	8.0759	8.4	8.7	9.3
PO	16.3%	20.58%	22.94%	16.3%
GM (dB)	∞	20.4	17	19.2
PM (deg)	51.8	47.4	45.2	51.5

9-8.2 Frequency-Domain Design

The frequency response of $G(e^{j\omega T})$ is not a rational function of ω. Design using Bode plots will be more complicated than the continuous-time case. Therefore, we use the bilinear transformation to convert the discrete-time design into a continuous-time design.

Bilinear Transformation:

$$w = \frac{2}{T}\left(\frac{z-1}{z+1}\right) \quad \longleftrightarrow \quad z = \frac{1+Tw/2}{1-Tw/2},$$

Frequency υ in the w-domain: $w = j\upsilon$,

Frequency ω in the z-domain: $z = e^{j\omega T}$,

$$\upsilon = \frac{2}{T}\tan\left(\frac{\omega T}{2}\right).$$

Bode plot design procedure:

(1) Choose T based on the design specifications.

(2) Find the discrete-time equivalent of the plant.

$$G_p(z) = \left(\frac{z-1}{z}\right)\mathcal{Z}\left\{\mathcal{L}^{-1}\left\{\frac{\tilde{G}_p(s)}{s}\right\}\right\}.$$

(3) Transform the plant to the w-domain.

$$G'_p(w) = G_p(z)\big|_{z=(1+Tw/2)/(1-Tw/2)}.$$

(4) Design a controller $G'_c(w)$ in the (continuous-time) w-domain to meet the design specifications. Make sure to map frequencies (e.g., crossover frequency) using the mapping $\upsilon = (2/T)\tan(\omega T/2)$.

(5) Transform the controller to the (discrete-time) z-domain.

$$G_c(z) = G'_c(w)\big|_{w=(2/T)(z-1)/(z+1)}$$

(6) Evaluate the performance of the digital controller by analysis and/or simulation.

The transfer function transformations are done in MATLAB as follows:

$$\tilde{G}_p(s) \to G_p(z): \quad \text{Gp = c2d(tildeGp,T,'zoh')},$$
$$G_p(z) \to G'_p(w): \quad \text{primeGp = d2c(Gp,'tustin')}.$$
$$G'_c(w) \to G_c(z): \quad \text{Gc = c2d(primeGc,T,'tustin')}.$$

Example 9-19: Bode Plot Design for Desired PM and ω_c

Given the plant transfer function

$$\tilde{G}_p(s) = \frac{0.2}{s(0.4s+1)},$$

design a discrete-time controller to have $PM = 50°$ at $\omega_c \approx 6$ rad/s.

Solution: Choose T to limit the loss in the phase margin to $5°$.

$$\frac{T\omega_c}{2} \times \frac{180}{\pi} = 5 \quad \rightarrow \quad T = \frac{2\pi \times 5}{6 \times 180} = 0.029 \approx 0.03,$$

$$G_p(z) = \frac{0.00021948(z+0.9753)}{(z-1)(z-0.9277)},$$

$$G'_p(w) = \frac{0.2(1-0.015w)(1+1.8748 \times 10^{-4}w)}{w(1+0.4w)}.$$

Map crossover frequency:

$$v_c = \frac{2}{0.03} \tan\left(\frac{6 \times 0.03}{2}\right) = 6.016 \approx 6.$$

Design for PM = 50° at 6 rad/s.

$$\angle G'_p(j6) = -\tan^{-1}(0.015 \times 6) + \tan^{-1}(1.8748 \times 10^{-4} \times 6) - 90° - \tan^{-1}(0.4 \times 6)$$
$$= -162.46°.$$

Uncompensated phase margin at 6 rad/s = $180 - 162.46 = 17.54°$.

Use a lead compensator:

$$G'_c(w) = K \frac{1+\tau w}{1+\beta \tau w}, \quad \beta < 1,$$

$$\beta = \frac{1-\sin\phi_{max}}{1+\sin\phi_{max}}, \quad \omega_{max} = \frac{1}{\tau\sqrt{\beta}},$$

$$\phi_{max} = 50 - 17.54 = 32.46° \quad \rightarrow \quad \beta = \frac{1-\sin 32.46}{1+\sin 32.46} \approx 0.3,$$

$$\frac{1}{\tau\sqrt{\beta}} = 6 \quad \rightarrow \quad \tau = \frac{1}{6\sqrt{0.3}} \approx 0.3.$$

Choose K such that $|G'_p(w) G'_c(w)|_{w=j6} = 1$:

$$\left| K \frac{(1+j0.3 \times 6)0.2(1-j0.015 \times 6)(1+j1.8748 \times 10^{-4} \times 6)}{(1+j0.09 \times 6)j6(1+j0.4 \times 6)} \right| = 1,$$

9-8. DISCRETE-TIME DESIGN

$$K \approx 42.88 \quad \rightarrow \quad G'_c(w) = \frac{42.88(1+0.3w)}{1+0.09w},$$

$$G_c(z) = \frac{128.64(z-0.9048)}{(z-0.7143)}.$$

MATLAB check: $PM = 50.1°$ at $\omega_c = 5.98$ rad/s.

Example 9-20: Bode Plot Design for Desired PM, ω_c, and K_v

Given the plant transfer function

$$\tilde{G}_p(s) = \frac{0.2}{s(0.4s+1)},$$

design a discrete-time controller to meet the design specifications:

- $PM = 45°$ at $\omega_c \approx 6$ rad/s and
- $K_v \geq 20$.

Solution: Based on the previous example calculations, set $T = 0.03$ s.

Anticipating the use of a lag compensator to increase K_v, design for $PM = 50°$ at $v_c \approx 6$ rad/s.

$$G_p(z) = \frac{0.00021948(z+0.9753)}{(z-1)(z-0.9277)},$$

$$G'_p(w) = \frac{0.2(1-0.015w)(1+1.8748 \times 10^{-4}w)}{w(1+0.4w)}.$$

Use the same lead compensator of the previous example to increase the phase margin to 50°.

$$G'_c(w) = \frac{42.88(1+0.3w)}{1+0.09w},$$

$$K_v = w\, G'_c(w)\, G'_p(w)\big|_{w=0} = 42.88 \times 0.2 \approx 8.576 < 20.$$

Use a lag compensator to increase K_v by the factor $20/8.576 = 2.33$.

$$G'_{c2}(w) = \frac{K_1 \alpha(\tau_1 w + 1)}{\alpha \tau_1 w + 1}, \quad \alpha > 1.$$

Take

$$\frac{1}{\tau_1} = 0.1 v_c \quad \rightarrow \quad \tau_1 = \frac{1}{0.1 \times 6} = 1.6667,$$

$$\alpha \geq 2.33; \quad \text{take } \alpha = 2.5.$$

Take $K_1 = 1$. Lead-Lag compensator:

$$G'_c(w) = \frac{42.88(1+0.3w)}{1+0.09w} \times \frac{2.5(1+1.6667w)}{1+2.5 \times 1.6667w},$$

$$G_c(z) = \frac{129.33(z-0.9048)(z-0.9822)}{(z-0.7143)(z-0.9928)}.$$

MATLAB Check: $PM = 46.6°$ at 6 rad/s.

$$K_v = \frac{1}{T}(z-1)\,G_p(z)\,G_c(z)\Big|_{z=1} = 21.2957 > 20.$$

9-9 Quantization

Numbers in digital computers are represented by a finite number of bits; so, there is only a finite number of values that a number can take. Other values are rounded off to the nearest available values. This process is known as quantization. Quantization affects digital control systems in two ways:

- *coefficient quantization*, and
- *signal quantization*.

The effect of coefficient quantization can be studied by analysis and/or simulation of the closed-loop control system after all the coefficients of the controller are quantized.

Signal quantization is studied by treating quantization as noise and studying the effect of the noise on the performance of the control system. This is the subject of this section. Let x be a real number and x_q be its quantized value. x_q can be represented as $x_q = x + e$, as shown below, where e is quantization noise.

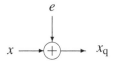

We study signal quantization under the following assumptions:

- $|x| \leq 1$.

- Fixed-point, twos-complement number system.

- One sign bit plus ℓ magnitude bits. Wordlength $= \ell + 1$. The distance between quantization levels $q = 2^{-\ell}$.

- Round-off: x is approximated by the nearest quantization level:

$$|x - x_q| \leq \frac{q}{2} = 2^{-(\ell+1)}$$

- No overflow.

9-9. PROBLEMS

Signal Quantization Analysis

- Identify points of signal quantization:

  (1) ADC: Quantization takes place when an analog signal is converted into a digital signal.

  (2) Coefficient multiplication nodes: When two numbers, each represented by ℓ magnitude bits plus a sign bit, are multiplied, the product will have 2ℓ magnitude bits, which is rounded off to a number with ℓ magnitude bits.

- At each point, introduce quantization noise e_i.

- Find the transfer function $H_i(z)$ from e_i to y.

Example 9-21: Analysis of Quantization Noise

Consider the digital control system of Fig. 9-9, where the controller transfer function is

$$G_c(z) = \frac{K(z-a)}{(z-b)} = K\left[1 + \frac{\alpha}{(z-b)}\right], \quad \text{where } \alpha = b - a.$$

Suppose the controller is implemented by the difference equation

$$\eta(k+1) = b\eta(k) + e(k), \quad u(k) = K[\alpha\,\eta(k) + e(k)],$$

$$\eta(k+1) = b\eta(k) + e(k) \;\rightarrow\; z\eta = b\eta + E \;\rightarrow\; \eta = z^{-1}(b\eta + E),$$

$$u(k) = K[\alpha\,\eta(k) + e(k)] \;\rightarrow\; U = K(\alpha\eta + E).$$

The foregoing equations define the arithmetic operations that take place in the computer: addition, multiplication, and delay by one cycle. A schematic diagram of the controller is shown in Fig. 9-21, together with four sources of quantization noise. The first noise is added after the ADC converter, which is at the input of the controller. The other three noise signals are introduced at the outputs of the gains b, α, and K, since the output of each gain is the multiplication of the input signal by the gain.

Compute the transfer function from each noise input to the controller output u.

$$\eta = z^{-1}(E + E_1 + E_2 + b\eta),$$

$$(1 - bz^{-1})\eta = z^{-1}(E + E_1 + E_2),$$

$$\eta = \frac{1}{(z-b)}(E + E_1 + E_2),$$

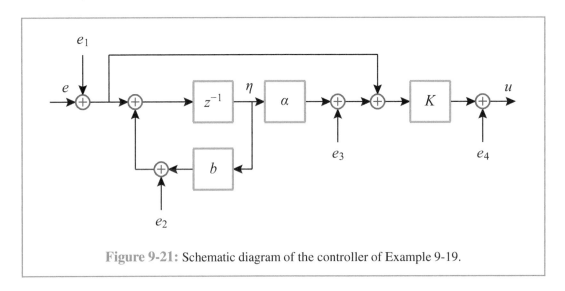

Figure 9-21: Schematic diagram of the controller of Example 9-19.

$$U = E_4 + K(E + E_1 + E_3 + \alpha\eta)$$
$$= E_4 + K(E + E_1 + E_3) + \frac{K\alpha}{z-b}(E + E_1 + E_2)$$
$$= \underbrace{K\left(1 + \frac{\alpha}{z-b}\right)}_{G_c(z)}(E + E_1) + \frac{K\alpha}{z-b}E_2 + KE_3 + E_4$$
$$= G_c(z)(E + E_1) + \frac{K\alpha}{z-b}E_2 + KE_3 + E_4.$$

Now put the controller inside the feedback control system, as shown in Fig. 9-22, and compute the transfer function from each noise input to the output y. Notice the scaling factor $V_{\max}$ in the block diagram. We assumed that numbers inside the computer are limited to ± 1. If the voltage signals at the input of the ADC and the output of the DAC are limited to $\pm V_{\max}$, we multiply the signal by $1/V_{\max}$ as it enters the computer and multiply it by $V_{\max}$ as it leaves the computer.

The transfer function from each input e_i to the output y is given by the transfer function of the forward path from e_i to y, divided by $[1 + G_p(z) G_c(z)]$.

$$H_1 = \frac{Y}{E_1} = \frac{V_{\max}G_pG_c}{1 + G_pG_c}, \qquad H_2 = \frac{Y}{E_2} = \frac{V_{\max}G_p}{1 + G_pG_c} \cdot \frac{K\alpha}{(z-b)},$$
$$H_3 = \frac{Y}{E_3} = \frac{KV_{\max}G_p}{1 + G_pG_c}, \qquad H_4 = \frac{Y}{E_4} = \frac{V_{\max}G_p}{1 + G_pG_c}.$$

Output error due to round-off:

Assume the closed-loop system is stable. There are two approaches to calculate the output error:

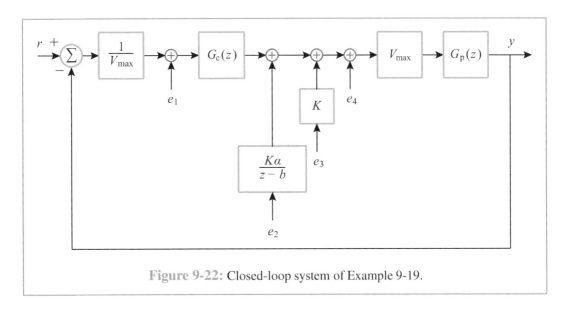

Figure 9-22: Closed-loop system of Example 9-19.

- *worst-case analysis*, and
- *stochastic analysis*.

Worst-case analysis (maximum error):

The maximum round-off error is $q/2 = 2^{-(\ell+1)}$.

$$Y_i(z) = H_i(z)\, E_i(z)\,,$$

$$y_i(k) = \sum_{\ell=0}^{\infty} h_i(\ell)\, e_i(k-\ell), \quad \text{where } h_i(k) = \mathcal{Z}^{-1}[H_i(z)]\,,$$

$$|y_i(k)| \leq \sum_{\ell=0}^{\infty} |h_i(\ell)|\, |e_i(k-\ell)| \leq \frac{q}{2} \sum_{\ell=0}^{\infty} |h_i(\ell)|\,,$$

$$|y(k)| \leq |y_1| + |y_2| + \cdots \leq \frac{q}{2} \left[\sum_{\ell=0}^{\infty} |h_1(\ell)| + \sum_{\ell=0}^{\infty} |h_2(\ell)| + \cdots \right].$$

Stochastic analysis (root-mean-square error):

Model the round-off error e as a random variable, uniformly distributed over $[-q/2,\, q/2]$:

$$\text{Mean value of } e = E\{e\} = 0, \quad \text{variance of } e = \frac{q^2}{12}\,.$$

Model $e(k)$, for $k = 0, 1, 2, \ldots$, as a sequence of independent identically distributed random variables:

$$y_i(k) = \sum_{\ell=0}^{\infty} h_i(\ell) \, e_i(k-\ell) \,,$$

$$E\{y_i(k)\} = \sum_{\ell=0}^{\infty} h_i(\ell) \, \underbrace{E\{e_i(k-\ell)\}}_{0} = 0 \,,$$

$$\sigma_i^2 = \text{variance of } y_i(k) = E\{y_i^2(k)\}$$

$$= E\left\{ \sum_{\ell=0}^{\infty} h_i(\ell) \, e_i(k-\ell) \sum_{m=0}^{\infty} h_i(m) \, e_i(k-m) \right\}$$

$$= \sum_{\ell=0}^{\infty} \sum_{m=0}^{\infty} h_i(\ell) \, h_i(m) E\{e_i(k-\ell) \, e_i(k-m)\} \,,$$

$$E\{e_i(k-\ell) \, e_i(k-m)\} \begin{cases} q^2/12 & \text{for } \ell = m \,, \\ 0 & \text{otherwise} \,, \end{cases}$$

$$\sigma_i^2 = \frac{q^2}{12} \sum_{\ell=0}^{\infty} h_i^2(\ell) \,.$$

Assuming $e_1, e_2, \ldots$ are independent,

$$\sigma^2 = E\{y^2(k)\} = \frac{q^2}{12} \left[\sum_{\ell=0}^{\infty} h_1^2(\ell) + \sum_{\ell=0}^{\infty} h_2^2(\ell) + \cdots \right] \,.$$

σ is the root mean square (rms) of y.

Calculation of $\sum_{i=0}^{\infty} |h(i)|$ and $\sum_{i=0}^{\infty} h^2(i)$ using MATLAB:

- Define the discrete-time transfer function $H(z)$.

- Calculate the impulse response using

$$h = \text{impulse}(H) \,.$$

- Calculate the L_1 and L_2 norms of h using

$$L1 = \text{norm}(h,1), \quad L2 = \text{norm}(h,2).$$

-

$$\sum_{i=0}^{\infty} |h(i)| = L_1 \,, \qquad \sum_{i=0}^{\infty} h^2(i) = L_2^2 \,.$$

9-9. PROBLEMS

> Example 9-22: Output Error Due to Quantization

Consider a digital control system where the plant transfer function is

$$\tilde{G}_p(s) = \frac{20}{s(s+2)}, \quad \text{the digital controller is } G_c(z) = \frac{2(z-0.9)}{(z-0.6)},$$

and the sampling period $T = 0.05$ s. Assume the controller is implemented as

$$\eta(k+1) = 0.6[\eta(k) - e(k)], \quad u(k) = \eta(k) + 2e(k).$$

(a) Verify the stability of the system.

(b) Draw a block diagram of the discrete-time closed-loop system and show all sources of quantization noise.

(c) Find the transfer function from each quantization-noise input to the output.

(d) Find the maximum and rms output error due to round-off. Assume 16-bit wordlength, $V_{max} = 10$ V, and no overflow.

Solution:

(a)

$$G_p(z) = \frac{z-1}{z} \mathcal{Z}\left\{\mathcal{L}^{-1}\left\{\frac{20}{s^2(s+2)}\right\}\right\} = \frac{0.024187(z+0.9672)}{(z-1)(z-0.9048)}.$$

Characteristic equation:

$$1 + G_p(z) G_c(z) = 0,$$

$$z^3 - 2.4564z^2 + 2.0509z - 0.5850 = 0.$$

The roots are 0.8975 and $0.7794 \pm 0.2103j$ (absolute value 0.8073). All roots are inside the unit circle. The system is stable.

(b) The realization of the controller and the noise inputs are shown in Fig. 9-23. No noise input is added after the gain 2 (why?).

$$\eta = z^{-1}[E_2 + 0.6(\eta - E - E_1)],$$

$$(1 - 0.6z^{-1})\eta = z^{-1}[E_2 - 0.6E - 0.6E_1],$$

$$\eta = \frac{1}{z - 0.6}[-0.6E - 0.6E_1 + E_2],$$

$$U = \eta + 2(E + E_1) = \frac{1}{z - 0.6}[-0.6E - 0.6E_1 + E_2] + 2(E + E_1)$$

$$= \left(2 - \frac{0.6}{z - 0.6}\right)(E + E_1) + \frac{1}{z - 0.6} E_2$$

$$= G_c(z)(E + E_1) + \frac{1}{z - 0.6} E_2.$$

The closed-loop block diagram is shown in Fig. 9-24.

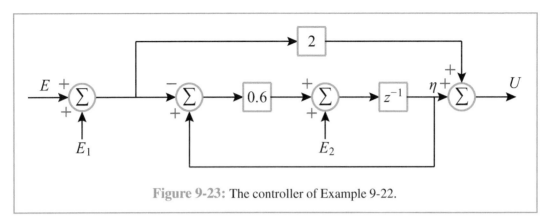

Figure 9-23: The controller of Example 9-22.

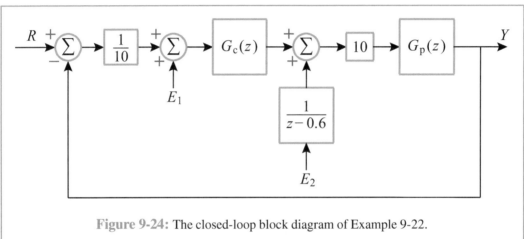

Figure 9-24: The closed-loop block diagram of Example 9-22.

(c)

$$H_1 = \frac{Y}{E_1} = \frac{10G_pG_c}{1+G_pG_c} = \frac{0.48374(z-0.9)(z+0.9672)}{(z-0.8975)(z^2-1.559z+0.6518)},$$

$$H_2 = \frac{Y}{E_2} = \left(\frac{1}{z-0.6}\right)\frac{10G_p}{1+G_pG_c} = \frac{0.24187(z+0.9672)}{(z-0.8975)(z^2-1.559z+0.6518)}.$$

(d) $q = 2^{-15}$.

$$\text{Maximum error} = \frac{q}{2}\left[\text{L1 norm of }(h_1) + \text{L1 norm of }(h_2)\right] = 0.0189,$$

$$\text{RMS error} = \sqrt{\frac{q^2}{12}\left\{[\text{L2 norm of }(h_1)]^2 + [\text{L2 norm of }(h_2)]^2\right\}} = 0.002.$$

Summary

Concepts

- Digital control allows for flexibility in implementation, higher accuracy, cost-effectiveness, improved user interface, and reduction of signal distortion.
- The basic components of a digital control system are: sample, ZOH (zero-order hold), ADC (analog to digital converter), digital computer, DAC (digital to analog converter), and plant.
- The z-transform has properties analogous to the Laplace transform.
- A discrete-time transfer function is stable if and only if all its poles are strictly inside the unit circle of the complex plane.
- When a continuous-time transfer function is preceded by zero-order hold, the samples of the output and input are related by a discrete-time transfer function, called the discrete-time equivalent.
- A digital control system is stable when all closed-loop poles are strictly inside the unit circle of the complex plane.
- Stability of the digital control system can be determined by calculating the roots of the characteristic equation, application of the Routh-Hurwitz criterion after transforming the characteristic equation into a continuous-time domain, application of the Nyquist criterion in the discrete-time domain, or drawing the root locus in the z-plane.
- Gain and phase margins are defined analogous to their definitions for continuous-time control systems.
- Steady-state errors to step and ramp references are expressed in terms of the error constants K_p and K_v, respectively.
- Sampling frequency needs to be much higher than twice the maximum frequency, which is suggested by the sampling theorem.
- Sampling frequency is chosen to smooth the transient response, reduce the phase margin loss due to zero-order-hold delay, or reduce the effect of zero-order-hold delay on tracking and disturbance rejection.
- The frequency spectrum of a sampled signal exhibits aliasing due to overlap of the repeated components of the spectrum of the continuous-time signal.
- An anti-aliasing filter is used to reduce the error due to aliasing.
- A controller can be designed in the continuous-time domain and converted into a discrete-time controller by one of several approximation methods; the process is called emulation. The approximation requires the sampling period to be sufficiently small.
- A digital controller can be designed in the discrete-time domain either by the root locus method in the z-plane or Bode plots in the w-domain.
- Effect of quantization can be studied by inserting quantization noise at points of signal quantization.

Mathematical Models

Frequency response of $G(z)$:	$G(e^{j\omega T})$
dc-gain of $G(z)$:	$G(1)$
Discrete-time equivalent of $\tilde{G}(s)$:	$G(z) = \left(\dfrac{z-1}{z}\right) \mathcal{Z}\left\{\mathcal{L}^{-1}\left\{\dfrac{\tilde{G}(s)}{s}\right\}\right\}$
Bilinear transformation:	$w = c(z-1)/(z+1), \quad c > 0$
	$z = (1 + w/c)/(1 - w/c)$
Frequency mapping for $c = 2/T$:	$v = (2/T)\tan(\omega T/2)$
Nyquist criterion:	$Z = P - N$
Position error constant:	$K_p = G(1)$
Velocity error constant:	$K_v = (1/T)(z-1)G(z)\big|_{z=1}$
Sampling period for smooth response:	$5 \leq T_r/T \leq 10$
Phase margin loss due to ZOH delay:	$(T\omega_c/2) \times (180/\pi)$ deg
Compensator:	$G_c(z) = K(z-a)/(z-b),$
	$\quad -1 \leq a,b \leq 1$
Lead:	$a > b$
Lag:	$a < b$
PI:	$b = 1$
Distance between quantization level:	$q = 2^{-\ell}, \ \ell =$ number of magnitude bits
Wordlength:	$\ell + 1$
Impulse response of $H_i(z)$:	$h_i(k)$
Maximum error due to quantization:	$\dfrac{q}{2}\left[\displaystyle\sum_{\ell=0}^{\infty}|h_1(\ell)| + \sum_{\ell=0}^{\infty}|h_2(\ell)| + \cdots\right]$
Mean square error due to quantization:	$\dfrac{q^2}{12}\left[\displaystyle\sum_{\ell=0}^{\infty}h_1^2(\ell) + \sum_{\ell=0}^{\infty}h_2^2(\ell) + \cdots\right]$

Important Terms Provide definitions or explain the meaning of the following terms:

ADC	ideal sampler	sampling frequency
aliasing	impulse response	sampling period
antialiasing filter	impulse sampler	sampling theorem
BDM	lag compensator	signal quantization
bilinear transformation	lead compensator	spectrum of sampled signal
Bode plots	maximum error	stability
characteristic equation	Nyquist contour	staircase function
coefficient quantization	Nyquist criterion	stochastic analysis
DAC	Nyquist plot	transfer function
dc gain	phase margin	Tustin method
difference equation	PI controller	type of feedback system
discrete-time equivalent	poles	unit step function
emulation	quantization	worst-case analysis
error constant	root locus method	zeros
FDM	root-mean-square error	ZOH
frequency response	roundoff	z-transform
gain margin	sampling	

PROBLEMS

9.1 For each of the following discrete-time systems with input u and output y:

  (i) Find the transfer function.

  (ii) Find the impulse response.

  (iii) Determine if the system is stable.

  (iv) Find the steady-state output to a unit step input.

  (v) Find the steady-state output when $u(k) = \cos(k\pi/4)$ for all $k \geq 0$.

  (vi) By solving the difference equation recursively, find $y(k)$ for $k \geq 0$ when the input is a unit step and $y(k) = 0$ for $k < 0$.

  (a) $y(k+1) = 0.5y(k) + u(k)$

  (b) $y(k+2) = 1.5y(k+1) - 0.54y(k) + 1.4u(k+1) + u(k)$

  (c) $y(k+3) = \frac{1}{4}y(k+2) + \frac{1}{4}y(k+1) - \frac{1}{16}y(k) + u(k+3) + \frac{1}{2}u(k+2) - \frac{1}{9}u(k+1)$

  (d) $y(k+3) = \frac{1}{2}y(k+2) + \frac{1}{16}y(k+1) - \frac{1}{32}y(k) + u(k+2) + \frac{1}{2}u(k+1) - \frac{1}{4}u(k)$

9.2 For each of the following cases:

  (i) Compute by hand and table look-up the discrete transfer function $G(z)$ when the given continuous transfer function $\tilde{G}(s)$ is preceded by ZOH. Set $T = 0.1$ s.

  (ii) Verify your answer using MATLAB. (Use the command "c2d" with "zoh").

(iii) Find the poles of $G(z)$ and show how they are related to the poles of $\tilde{G}(s)$.

(a) $\tilde{G}(s) = \dfrac{10}{(s+2)}$, (b) $\tilde{G}(s) = \dfrac{5(s+1)}{s+2}$, (c) $\tilde{G}(s) = \dfrac{s+1}{s(s+2)}$,

(d) $\tilde{G}(s) = \dfrac{5}{(s+4)}$, (e) $\tilde{G}(s) = \dfrac{6(s+1)}{s+3}$, (f) $\tilde{G}(s) = \dfrac{2(s+1)}{s(s+3)}$.

9.3 Consider a digital control system with continuous plant $\tilde{G}_p(s)$, discrete controller $G_c(z)$, and sampling period $T = 0.01$ s, where

$$\tilde{G}_p(s) = \dfrac{20}{s+10} \quad \text{and} \quad G_c(z) = K \geq 0.$$

(a) Find the discrete transfer function $G_p(z)$.

(b) Sketch the root locus in the z-plane as K varies from zero to infinity.

(c) Find the range of K for stability.

(d) Calculate the steady-state error to a unit step input, as function K.

9.4 Consider a digital control system where the discrete-time equivalent of the plant is $G_p(z)$ and the controller is $G_c(z) = K \geq 0$. For each of the following cases of $G_p(z)$:

(i) Sketch the root locus as K varies from zero to infinity.

(ii) From the root locus, determine the range of K for stability.

(iii) Using the Routh-Hurwitz criterion, determine the range of K for stability.

(iv) Determine the type of the feedback system and the corresponding error constant.

(a) $G_p(z) = \dfrac{1}{z-1}$

(b) $G_p(z) = \dfrac{a}{z-1}$, $a > 0$

(c) $G_p(z) = \dfrac{a(z-c)}{z(z-b)}$, $a > 0$, $0 < b < 1$, and $0 < c < b$

(d) $G_p(z) = \dfrac{z-0.7}{z(z-0.9)}$

9.5 For each of the following digital control systems with continuous plant $\tilde{G}_p(s)$, discrete controller $G_c(z) = K \geq 0$, and sampling period $T = 0.1$ s:

(i) Determine the range of K for stability using the Nyquist criterion.

(ii) Find the gain and phase margins and the corresponding crossover frequencies when $K = 1$.

(a) $\tilde{G}_p(s) = \dfrac{10}{s(s+1)(s+10)}$

(b) $\tilde{G}_p(s) = \dfrac{10}{s(s+2)(s+10)}$

(c) $\tilde{G}_p(s) = \dfrac{10}{(s+1)(s+2)(s+4)}$

9.6 Consider a digital control system where

$$G_p(z) = \dfrac{z-0.8}{(z-1)(z-1.2)} \quad \text{and} \quad G_c(z) = K > 0.$$

Use the Nyquist criterion to find the range of K for stability.

9.7 Consider the digital control system of a dc motor where the motor's transfer function from the input voltage to the speed is given by

$$\dfrac{20}{(0.1s+1)}$$

and a PI controller is implemented digitally by the transfer function

$$0.1\left(\dfrac{2Tz}{z-1}\right).$$

The sampling period is $T = 0.01$ s.
  (a) Find the discrete transfer function of the closed-loop system.
  (b) Is the system stable?
  (c) Find the type of the feedback system and the corresponding error constant.
  (d) Find the steady-state error to a unit step input.

9.8 A block diagram of a typical motion control system [33] is shown in Fig. P9.8.

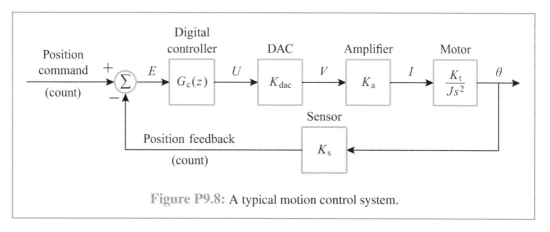

Figure P9.8: A typical motion control system.

The motor position θ is measured by optical encoder whose output is binary. The binary output of the sensor is subtracted from the position command, which is also binary. The position error is fed into a digital computer that implements the control algorithm. The binary output of the computer is fed into a digital-to-analog converter (DAC) whose output is a voltage signal. The voltage signal is fed into a current source amplifier that generates the electric current that runs the motor.

The amplifier's gain $K_a = 2$ A/V. The motor's torque constant $K_t = 0.1$ Nm/A and its total moment of inertial $J = 2 \times 10^{-4}$ kg·m². The position sensor is an absolute encoder with 12 bits of binary output. The sensor output changes between 0 and $2^{12} - 1 = 4095$. Thus, the sensor gain is $K_s = 4096/(2\pi) = 652$ counts/rad. The sampling period is 1 ms.

The digital controller, with input e and output u, implements the difference equations

$$f(k) = 0.15e(k) + f(k-1),$$
$$u(k) = 20e(k) + 200[e(k) - e(k-1)] + f(k),$$

which realize a PID controller.

The DAC has 14-bit resolution and its output signal ranges between -10 V and 10 V. Noting that the DAC input varies over the range ± 8192 counts and its output varies over the range ± 10 V, the DAC gain is $K_{dac} = 10/8192 = 1.22 \times 10^{-3}$ V/count.

(a) Find the discrete-time equivalent of the system.

(b) Assuming that the controller is multiplied by a gain K, draw the root locus in the z-plane as K varies from 0 to ∞. Show that the system is stable for $K_1 < K < K_2$ and find K_1 and K_2.

(c) Find the phase margin and the corresponding crossover frequency.

9.9 Consider a digital control system with continuous plant $\tilde{G}_p(s)$, discrete controller $G_c(z)$, and sampling period $T = 0.1$ s, where

$$\tilde{G}_p(s) = \frac{1}{s} \quad \text{and} \quad G_c(z) = K \geq 0.$$

(a) Compute by hand and lookup table the discrete transfer function $G_p(z)$.

(b) Sketch the root locus in the z-plane as K varies from zero to infinity.

(c) Find the range of K for stability.

(d) Calculate the steady-state error to a unit step input, as function K.

9.10 Consider a digital control system where

$$G_p(z) = \frac{z+1}{(z-1)(z-0.6)}, \qquad G_c(z) = K \geq 0,$$

and $T = 0.1$ s. Use the Nyquist criterion to find the range of K for stability.

9.11 What is the antialiasing filter and why is it used?

9.12 A feedback control system is expected to have a settling time of no more than 100 ms and an overshoot of no more than 5%.

(a) What is the sampling period T if the reduction in the phase margin is to be no more than $5°$?

(b) What is the sampling period T if there are to be 10 samples per rise time?

PROBLEMS

9.13 A dc motor is modeled by

$$\tilde{G}_p(s) = \frac{200}{s(s+10)}.$$

(a) Using the Bode plot method in the s-plane, design an analog compensator such that the phase margin is 60° at 10 rad/s.

(b) Choose the sampling period such that the phase-margin drop is limited to 10°.

(c) Convert the analog compensator into a digital one using the Tustin method. Check the design specifications in the z-domain.

9.14 A dc motor is modeled by

$$\tilde{G}_p(s) = \frac{500}{s(s+12.5)}.$$

(a) Using the root locus method in the s-plane, design an analog compensator such that the step response has 5% overshoot and 0.4 s settling time.

(b) Choose the sampling period to fit 10 samples within the rise time of the step response.

(c) Convert the analog compensator into a digital one using the Tustin method. Check the design specifications in the z-domain.

9.15 An analog control system with the controller

$$\tilde{G}_c(s) = \frac{8(s+1)}{s+4}$$

has the step-response characteristics shown in the table below:

Rise time	Settling time	Overshoot	Undershoot	Peak	Peak time
0.7597	2.1082	4.3210	0	1.0432	1.5658

(a) Choose the sampling period.

(b) Convert the analog controller into a digital controller using the Tustin Method.

9.16 A continuous-time plant is modeled by

$$\tilde{G}_p(s) = \frac{1}{s(s+1)}.$$

(a) Using the root locus method in the s-plane, design an analog controller such that the step response has 5% overshoot and 1 s settling time.

(b) Choose the sampling period.

(c) Convert the analog controller into a digital controller using the Tustin Method.

(d) Find the overshoot and settling time of the step response of the digital control system.

9.17 In a digital control system, the discrete-time equivalent of the plant is

$$G_p(z) = \frac{0.1(z+1)}{z(z-1)}$$

and the controller is $G_c(z) = K$.
  (a) Draw the root locus of the system as K varies from 0 to ∞.
  (b) Choose K such that the closed-loop poles have damping ratio $\zeta \approx 0.7$.
  (c) Design a compensator such that the closed-loop poles have $\zeta \approx 0.7$ and natural frequency $\omega_n \approx 0.4\pi/T$.

9.18 Consider the digital control of a dc motor, modeled by

$$\tilde{G}_p(s) = \frac{10}{s(s+10)}.$$

The sampling period is $T = 0.01$ s.
  (a) Using the root locus method in the z-plane, design a digital compensator such that the step response has 20% overshoot and 0.4 s settling time.
  (b) Using the w-plane method, design a digital compensator to achieve 50-degree phase margin at 16 rad/s.

9.19 In a digital control system, the discrete transfer function of the plant is given by

$$G_p(z) = \frac{10^{-3}(z+0.966)}{(z-1)(z-0.9)}.$$

Using the root locus method, design a compensator $G_c(z)$ such that the closed-loop system has a pair of dominant complex poles at $z = 0.9 \pm j0.1$.

9.20
  (a) What is the difference between a discrete-time signal and a digital signal?
  (b) In fixed-point quantization analysis, why do we not add quantization noise after addition of two signals?
  (c) What is the maximum quantization round-off error for a system where the wordlength is: (i) 16 bits, (ii) 32 bits, (iii) 64 bits?
  (d) Why do we not need to add quantization noise after multiplication by a coefficient that is a power of 2?
  (e) In the quantization analysis of a feedback control system, why do we pre-multiply the controller by $1/V_{max}$ and post-multiply by V_{max}? What is V_{max}?

9.21 In a digital control system, the discrete-time equivalent of the plant is

$$G_p(z) = \frac{1}{z - 0.4}$$

and the controller is

$$G_c(z) = \frac{0.2}{z - 1}.$$

(a) Verify the stability of the closed-loop system.

(b) Draw a block diagram of the closed-loop system and show all sources of quantization noise.

(c) Find the transfer function from each quantization-noise input to the output.

(d) Find the maximum and rms output errors due to round-off. Assume 32-bit wordlength, $V_{max} = 15$, and no overflow.

9.22 Consider a digital control system where the plant transfer function is

$$\tilde{G}_p(s) = \frac{10}{s+1},$$

the sampling period T is chosen such that $e^{-T} = 0.9$, and the digital controller is implemented by

$$\eta(k+1) = \eta(k) + \tfrac{1}{2} e(k), \qquad u(k) = \eta(k) + e(k).$$

(a) Verify the stability of the system.

(b) Draw a block diagram of the discrete-time closed-loop system and show all sources of quantization noise.

(c) Find the transfer function from each quantization noise input to the output.

(d) Find the maximum and rms output errors due to round-off assuming 8-bit wordlength, $V_{max} = 15$ V, and no overflow.

9.23 Repeat Problem 9.22 if the plant is

$$\tilde{G}_p(s) = \frac{10}{s(s+2)}$$

and the controller is

$$\eta(k+1) = 0.7\eta(k) - 0.6e(k), \qquad u(k) = \eta(k) + 3e(k).$$

In part (d), assume 16-bit wordlength and $V_{max} = 10$ V.

9.24 Repeat Problem 9.22 if the plant is

$$\tilde{G}_p(s) = \frac{1}{s(s+1)},$$

the controller is

$$\eta(k+1) = 0.6\eta(k) - 0.4e(k), \qquad u(k) = \eta(k) + 2e(k),$$

and $T = 0.1$ s. In part (d), assume 16-bit wordlength and $V_{max} = 10$ V.

Appendix A
Review of Laplace Transform

The (one-sided) Laplace transform of $f(t)$ is defined by

$$F(s) = \mathcal{L}\{f(t)\} = \int_{0^-}^{\infty} f(t)\, e^{-st}\, dt\ ,$$

where s is a complex variable. The integral is defined from 0^-; that is, just before $t = 0$, to include the impulse function $\delta(t)$. The Laplace transform exists if the integral exists. If $f(t)$ does not grow faster that an exponential function, that is, $|f(t)| \leq Ke^{\alpha t}$, for some positive constants K and α, then, for $s = \sigma + j\omega$,

$$\left| \int_{0^-}^{\infty} f(t)\, e^{-st}\, dt \right| \leq \int_{0}^{\infty} Ke^{-(\sigma-\alpha)t}\, dt < \infty, \qquad \text{for } \sigma > \alpha\ .$$

In general, the Laplace transform exists over a region of convergence of the form $\text{Re}[s] \geq \sigma_0$.

Properties of the Laplace transform

- Linearity:
$$\mathcal{L}\{\alpha_1\, f_1(t) + \alpha_2\, f_2(t)\} = \alpha_1\, F_1(s) + \alpha_2\, F_2(s)$$

- Time delay:
$$\mathcal{L}\{f(t-\tau)\} = e^{-s\tau}\, F(s)$$

- Time scaling:
$$\mathcal{L}\{f(at)\} = \frac{1}{a}\, F\left(\frac{s}{a}\right)$$

- Frequency shift:
$$\mathcal{L}\{e^{-at} f(t)\} = F(s+a)$$

- Differentiation:
$$\mathcal{L}\left\{\frac{df}{dt}\right\} = sF(s) - f(0^-)\ ,$$

$$\mathcal{L}\left\{\frac{d^n f}{dt^n}\right\} = s^n F(s) - \sum_{k=1}^{n} s^{n-k} f^{(k-1)}(0^-)$$

- Integration:
$$\mathcal{L}\left\{\int_{0^-}^{t} f(\tau)\, d\tau\right\} = \frac{1}{s}\, F(s)$$

- Convolution:

$$\mathcal{L}\left\{\int_0^t f_1(\tau)\, f_2(t-\tau)\, d\tau\right\} = F_1(s)\, F_2(s)$$

- Final value theorem: If all the poles of $sF(s)$ have negative real parts,

$$\lim_{t\to\infty} f(t) = \lim_{s\to 0} s\, F(s)\,.$$

- Initial value theorem: If $f(t)$ is continuous or has step discontinuity at $t = 0$,

$$f(0^+) = \lim_{s\to\infty} s\, F(s).$$

Inverse Laplace transform using partial fraction expansion

Pairs of functions $f(t)$ and their Laplace transforms $F(s)$ are listed in Laplace transform tables. The table can be used to find the inverse Laplace transform of a function $F(s)$. However, the table has a limited number of pairs. To find the inverse Laplace transform of a function $F(s)$, which does not exist in the table, we use partial fraction expansion to represent $F(s)$ as the sum of terms that are available in the table. The process is illustrated by the following examples.

Example A-1:

Find the inverse Laplace transform of

$$F(s) = \frac{1}{s^3 + 12s^2 + 44s + 48}\,.$$

Solution:

$$F(s) = \frac{1}{(s+2)(s+4)(s+6)} = \frac{K_1}{s+2} + \frac{K_2}{s+4} + \frac{K_3}{s+6}\,.$$

$$K_1 = (s+2)\,F(s)|_{s=-2} = \frac{1}{8}\,,$$

$$K_2 = (s+4)\,F(s)|_{s=-4} = \frac{-1}{4}\,,$$

$$K_3 = (s+6)\,F(s)|_{s=-6} = \frac{1}{8}\,,$$

$$F(s) = \frac{1}{8}\left(\frac{1}{s+2} - \frac{2}{s+4} + \frac{1}{s+6}\right),$$

$$f(t) = \mathcal{L}^{-1}\{F(s)\} = \frac{1}{8}\left(e^{-2t} - 2e^{-4t} + e^{-6t}\right),\quad t \geq 0.$$

Example A-2:

Find the inverse Laplace transform of

$$F(s) = \frac{1}{s^3 + 2s^2 + 2s + 1}.$$

Solution:

$$F(s) = \frac{1}{(s+1)(s^2+s+1)} = \frac{K_1}{s+1} + \frac{K_2 s + K_3}{s^2+s+1},$$

$$K_1 = (s+1)F(s)|_{s=-1} = 1,$$

$$K_1(s^2+s+1) + (K_2 s + K_3)(s+1) = 1,$$

Coefficient of $s^2 = K_1 + K_2 = 0 \;\; \longrightarrow \;\; K_2 = -1,$

Coefficient of $s^0 = K_1 + K_3 = 1 \;\; \longrightarrow \;\; K_3 = 0,$

$$F(s) = \frac{1}{s+1} - \frac{s}{s^2+s+1} = \frac{1}{s+1} - \frac{s+\frac{1}{2}}{(s+\frac{1}{2})^2 + \frac{3}{4}} + \frac{\frac{1}{\sqrt{3}}\left(\frac{\sqrt{3}}{2}\right)}{(s+\frac{1}{2})^2 + \frac{3}{4}},$$

$$f(t) = \mathcal{L}^{-1}\{F(s)\} = e^{-t} - e^{-t/2}\cos\left(\frac{\sqrt{3}}{2}t\right) + \frac{1}{\sqrt{3}}e^{-t/2}\sin\left(\frac{\sqrt{3}}{2}t\right), \;\; t \geq 0.$$

The partial fraction expansion can be computed using the "residue" command of MATLAB.

Solution of differential equations

Steps:

- Compute the Laplace transform of both sides of the equation.
- Compute the Laplace transform of the solution from the algebraic equation that results from the previous step.
- Compute the inverse Laplace transform of the solution.

Example A-3:

Solve the differential equation

$$\ddot{y} + 3\dot{y} + 2y = 1(t), \quad y(0) = -1, \quad \dot{y}(0) = 1.$$

Solution:

$$s^2 Y(s) - sy(0) - \dot{y}(0) + 3[s Y(s) - y(0)] + 2Y(s) = \frac{1}{s},$$

$$(s^2 + 3s + 2) Y(s) + s - 1 + 3 = \frac{1}{s},$$

$$Y(s) = \frac{-s^2 - 2s + 1}{s(s+1)(s+2)} = \frac{K_1}{s} + \frac{K_2}{s+1} + \frac{K_3}{s+2},$$

$$K_1 = s Y(s)|_{s=0} = \frac{1}{2},$$
$$K_2 = (s+1) Y(s)|_{s=-1} = -2,$$
$$K_3 = (s+2) Y(s)|_{s=-2} = \frac{1}{2},$$

$$Y(s) = \frac{\frac{1}{2}}{s} - \frac{2}{s+1} + \frac{\frac{1}{2}}{s+2},$$

$$y(t) = \mathcal{L}^{-1}\{Y(s)\} = \frac{1}{2} - 2e^{-t} + \frac{1}{2} e^{-2t}, \quad t \geq 0.$$

Appendix B
Elements of Matrix Analysis

It is assumed that the reader is familiar with the definition of matrices, their addition, multiplication and transposition, diagonal, triangular, square and identity matrices, and the determinant of a matrix.

An n-dimensional column vector is defined by

$$v = \begin{bmatrix} x_1 \\ x_2 \\ \vdots \\ x_n \end{bmatrix},$$

where x_1 to x_n are real variables, and its transpose v^T is a row vector.

Linear independence and rank:

A group of vector $v_1, \ldots, v_k$ are linearly independent if no vector in the group can be represented as a linear combination of the other vectors. This is equivalent to saying that $\sum_{i=1}^{k} a_i v_i = 0$ if and only if $a_i = 0$ for $i = 1, \ldots k$. For an $n \times m$ matrix M,

$$\text{rank}(M) = r \iff M \text{ has } r \text{ linearly indepedent columns,}$$
$$\iff M \text{ has } r \text{ linearly indepedent rows.}$$

The rank of a matrix is invariant to elementary row or elementary column operations. Elementary row operations are:

- Exchange any two rows.
- Multiply any row by a non-zero constant.
- Add a multiple of one row to another row.

Elementary column operations are defined similarly. The rank of a matrix can be computed in MATLAB using the command "*rank*".

Nonsingular matrix:

An $n \times n$ matrix M is nonsingular if there is an $n \times n$ matrix M^{-1} such that $MM^{-1} = M^{-1}M = I$.

$$\text{An } n \times n \text{ matrix } M \text{ is nonsingular} \iff \text{rank}(M) = n \iff \det(M) \neq 0.$$

The inverse matrix is given by

$$M^{-1} = \frac{1}{\det(M)} \text{ adjoint}(M).$$

The adjoint of a square matrix A is the transpose of the cofactor matrix, which is a square matrix whose (i,j)th entry is defined by

$$c_{ij} = (-1)^{i+j} \det(A_{ij}),$$

where A_{ij} is the submatrix of A obtained from A by removing the ith row and jth column. For a 2×2 matrix,

$$\begin{bmatrix} a & b \\ c & d \end{bmatrix}^{-1} = \frac{1}{(ad-bc)} \begin{bmatrix} d & -b \\ -c & a \end{bmatrix}.$$

In MATLAB, the inverse of a matrix is computed using the command "*inv*".

If A, B, and C are square matrices of the same dimension, $A = BC$, and A is nonsingular, then both B and C are nonsingular because $\det(A) = \det(B)\det(C)$.

Eigenvalues and eigenvectors:

For an $n \times n$ square matrix A, the n-dimensional vector v and the scalar λ are the eigenvector and eigenvalue of A, respectively, if

$$Av = \lambda v.$$

This equation can be rewritten as

$$(\lambda I - A)v = 0,$$

which shows that the columns of $(\lambda I - A)$ are linearly dependent because a linear combination of them equals zero. Therefore,

$$\det(\lambda I - A) = 0.$$

This is a polynomial equation of degree n, called the *characteristic equation* of A. It has n roots, which are the eigenvalues of A. The eigenvalues of a real matrix could be complex, but complex eigenvalues will be in conjugate pairs because the polynomial $\det(\lambda I - A)$ has real coefficients.

In MATLAB, the eigenvalues and eigenvector of a matrix are computed using the command "*eig*".

Appendix C
Laplace and *z*-Transform Tables

Table C.1: Laplace transform table.

$f(t),\ t \geq 0$	$F(s)$
$\delta(t)$	1
$1(t)$	$\dfrac{1}{s}$
t	$\dfrac{1}{s^2}$
t^m	$\dfrac{m!}{s^{m+1}}$
e^{-at}	$\dfrac{1}{(s+a)}$
$\sin \omega t$	$\dfrac{\omega}{s^2+\omega^2}$
$\cos \omega t$	$\dfrac{s}{s^2+\omega^2}$

The entries of the table can be extended using the properties of the Laplace transform. For example, using the frequency-shift property,

$$\mathcal{L}\{te^{-at}\} = \frac{1}{(s+a)^2},$$

$$\mathcal{L}\{e^{-at}\sin(\omega t)\} = \frac{\omega}{(s+a)^2+\omega^2}.$$

Table C.2: Combined Laplace transform–z-transform table.

Laplace transform	Discrete time	z-transform
$\dfrac{1}{s}$	$1(kT)$	$\dfrac{z}{z-1}$
$\dfrac{1}{s^2}$	kT	$\dfrac{zT}{(z-1)^2}$
$\dfrac{1}{s^3}$	$\dfrac{1}{2}(kT)^2$	$\dfrac{z(z+1)T^2}{2(z-1)^3}$
$\dfrac{1}{(s+a)}$	e^{-akT}	$\dfrac{z}{(z-e^{-aT})}$
$\dfrac{1}{(s+a)^2}$	kTe^{-akT}	$\dfrac{e^{-aT}zT}{(z-e^{-aT})^2}$
$\dfrac{a}{s(s+a)}$	$1-e^{-akT}$	$\dfrac{(1-e^{-aT})z}{(z-1)(z-e^{-aT})}$
$\dfrac{a}{s^2(s+a)}$	$kT - \dfrac{1}{a}(1-e^{-akT})$	$\dfrac{zT}{(z-1)^2} - \dfrac{(1-e^{-aT})z}{a(z-1)(z-e^{-at})}$
$\dfrac{\omega}{s^2+\omega^2}$	$\sin(\omega kT)$	$\dfrac{\sin(\omega T)z}{z^2-2\cos(\omega T)z+1}$
$\dfrac{s}{s^2+\omega^2}$	$\cos(\omega kT)$	$\dfrac{z[z-\cos(\omega T)]}{z^2-2\cos(\omega T)z+1}$

Appendix D
MATLAB Tutorial

This appendix gives a brief MATLAB tutorial on how MATLAB can be used to do the computations that we encounter throughout the book. We assume that the reader is familiar with the basics of MATLAB and the tutorial targets commands that are available in the *control system toolbox*. It is highly recommended that the readers consult one of the available online tutorials for control system analysis using MATLAB.[1]

D-1 Transfer Function Definition and Manipulation

```
% Define transfer functions.
% There are two ways to define a transfer function.
% Let G(s) = 4(s+2)/[(s+1)(s+3)(s+5)] = (4s+8)/(s^3+9s^2+23s+15).
      G = tf([4 8],[1 9 23 15]);
      H = zpk(-2,[-1 -3 -5],4);
% Find poles, zeros, and DC gain of a transfer function.
      pole(G)
      zero(G)
      dcgain(G)
% Find the roots of a polynomial; example: s^3+9s^2+23s+15.
      roots([1 9 23 15])
% Calculate the series, parallel, or feedback connection of two transfer functions.
      H1 = tf(1,[1 1]), H2 = tf(4,[1 2]);
% Series Connection
      Hs = H1*H2;
% Parallel Connection
      Hp = H1+H2;
% Feedback Connection
      Hf = feedback(H1,H2);
% Calculate and plot the step response.
      wn=5; zeta=0.5; G = tf(wn^2 ,[1 2*zeta*wn wn^2]);
      step(G)
% The response appears in the graphics window.
      stepinfo(G)
% The characteristics of the response are displayed.
% Calculate and plot the response to other inputs.
% The input is a ramp signal in this example.
      t = 0:0.01:1; u = 2*t;
      y = lsim(G,u,t);
      plot(t,u,t,y,'r'),xlabel('Time'),ylabel('Output'),legend('Input','Output')
```

---

[1] See, for example, https://ctms.engin.umich.edu/CTMS/index.php?aux=Home

% The time delay transfer function e^{-Ts} is defined as follows.
   G = tf(1,1,'InputDelay',T);

D-2 Root Locus Analysis

Plotting the root locus requires the commands "*rlocus*" and "*rlcofind*" from the *control system toolbox*. The use of "*rlocus*" of a given transfer function opens the graphics window with the root locus plot. Clicking on a point on the root locus opens a box that lists the value of *K* and the poles at the chosen point. For a pair of complex poles, it gives the damping ratio ζ and the natural frequency ω_n, for which the pair of complex poles are represented as $-\zeta\omega_n \pm j\omega_n\sqrt{1-\zeta^2}$. It also gives the percentage overshoot corresponding to ζ, which is calculated as $100 \times exp(-\zeta\pi/\sqrt{1-\zeta^2})$. This formula gives the exact overshoot for a second-order transfer function with no zero, but it is not the overshoot for other transfer functions.

Another method to find *K* and the poles for a point on the root locus is to use the command "*rlocfind*", which allows you to select a point on the root locus using a pointer that appears in the window. When you select a point on the locus, MATLAB displays the value of *K* and all the corresponding closed-loop poles. The selected point appears as a red plus sign. Even if the selected point is not exactly on the locus, MATLAB selects the closest point on the locus.

Another command that is useful in design is the "*sgrid*" command. In design calculations it is sometimes required to choose a point on the complex part of the root locus that has a desired value of ζ or ω_n. The "*sgrid*" command generates a grid of curves of points that have the same ζ or ω_n. Points in the complex plan that have the same ζ form a line originating at the origin that makes an angle ϕ with the negative real axis where $\cos\phi = \zeta$. For the angle that appears in the grid to be equal to the physical angle, the two axes of the root locus must have the same scale. This can be done by using the command "*axis equal*".
   G = zpk([ ],[-1 -2 -3],1);
   rlocus(G)
% The root locus appears in the graphics window.
   [K,poles] = rlocfind(G)
   Select a point in the graphics window
% K and the poles are displayed.

D-3 Error Constants

% Calculation of K_p.
   G0 = tf(12,[1 6 11 6]);
   Kp = dcgain(G0);
% Calculation of K_v.
   G1 =zpk([ ],[0 -1 -2],10);
% to calculate Kv, multiply G1 by s then apply minreal to the
% product to perform pole-zero cancelation of s. Finally calculate
% the dc gain of the product.
   H = zpk(0,[ ],1);
   G1m = minreal(H*G1);

```
        Kv = dcgain(G1m);
% Calculation of $K_a$.
        G2 = zpk([ ],[0 0 -2],6);
% Calculation of Ka is similar to calculation of Kv except that we multiply by $s^2$.
        H2 = zpk([0 0],[ ],1);
        G2m = minreal(H2*G2);
        Ka = dcgain(G2m);
```

D-4 Frequency Response

The polar plot of a transfer function can be computed using the command *"nyquist"*. The plot appears in the graphics window. The bode plots are generated by the command *"bode"* and a grid is added with the command *"grid"*. You can use *"bode"* to compute the magnitude and phase of a transfer function and plot them yourself. This is useful when you want to do more with the plots, like overlaying the asymptotic plots over the exact plots. Finally, the command *"margin"* is used compute the gain and phase margins and their corresponding frequencies. The result appears in the graphics window.

```
        G = zpk([ ],[-1 -10 -100],10000);
% Polar plot
        nyquist(G)
% Bode plots
        bode(G)
        grid
% In the following lines it is shown how to compute the
% magnitude and phase curves and plot them yourself.
% The calculation is done from 0.01 rad/s to 10000 rad/s.
        [M,P,w] = bode(G,0.01,10000);
        Magnitude = 20*log10(squeeze(M(1,:,:)));
        Phase = squeeze(P(1,:,:));
        subplot(2,1,1)
        semilogx(w,Magnitude)
        subplot(2,1,2)
        semilogx(w,Phase)
% Stability margins
        margin(G)
```

D-5 State-Space Calculations

```
% Define the state-space model.
        A = [0 1 0;1 0 0;0 0 -2]; B = [0;1;1];
        C = [1 0 1]; D = 0;
% Check stability.
        eig(A)
```

% Test controllability and observability.
        CM = ctrb(A,B); rank(CM)
        OM = obsv(A,C); rank(OM)
% Find the transfer function.
        [num,den] = ss2tf(A,B,C,D);
        G = tf(num,den);
% Find a state-space model of G.
        [AA,BB,CC,DD] = tf2ss(num,den);
% The model {AA,BB,CC,DD} could be different from {A,B,C,D}
% Recall that the state-space model is not unique.
% Compute K to assign the eigenvalues of (A-BK) at -3, -2 + j, -2 -j.
        p = [-3;-2+i;-2-i];
% You can use one of the following two commands.
        K = acker(A,B,p);
        K = place(A,B,p);
% Design observer gain with observer eigenvalues at -10, -10, -10.
        p = [-10;-10;-10];
        L = (acker(A',C',p))';
% Feedforward gain.
        N = - 1/(C*inv(A-B*K)*B);
% Construct the closed-loop system.
        Acl = [A-B*K, B*K; zeros(3,3), A-L*C];
        Bcl = [B*N;zeros(3,1)]; Ccl = [C,zeros(1,3)]; Dcl = D;
% Closed-loop transfer function.
        [Num,Den] = ss2tf(Acl,Bcl,Ccl,Dcl);
        Gcl = tf(Num,Den);

D-6 Digital Control

% Define a discrete-time transfer function.
        T = 1;
        G=tf([0 1 0],[1 -1/2 1/4],T);
% Find poles, zero, and dcgain.
        pole(G)
        zero(G)
        dcgain(G)
% Check stability.
        abs(pole(G))
% Discrete equivalent of continuous plant preceded by ZOH.
        T = 0.1;
        TildeGp = tf(1,[1 1]);
        Gp = c2d(TildeGp,T,'zoh');
% Define a discrete controller.
        Gc = tf([1 -0.9],[1 -1],T);

```matlab
% Closed-loop transfer function
    Gcl = feedback(Gp*Gc,1);
% Root locus.
    rlocus(Gp*Gc)
% Frequency response.
    nyquist(Gp*Gc)
    bode(Gp*Gc)
    margin(Gp*Gc)
% Design by emulation.
% Define a continuous controller.
    TildeGc = zpk(-0.2,0,5);
    T = 0.2;
% Convert the continuous controller into a discrete controller
% using one of two methods.
    Gc = c2d(TildeGc,T,'tustin');
    Gc = c2d(TildeGc,T,'zoh');
% Design in the w-domain requires three transformations.
    Gp = c2d(TildeGp,T,'zoh');
    primeGp = d2c(Gp,'tustin');
    Gc = c2d(primeGc,T,'tustin');
% Quantization noise; see Example 9.22.
    T = 0.05;
    q = 2^(-15);
    H1 = zpk([0.9 -0.9672],0.8975,0.48374,T)*tf(1,[1 -1.559 0.6518],T);
    H2 = zpk(-0.9672,0.8975,0.24187,T)*tf(1,[1 -1.559 0.6518],T);
    h1 = impulse(H1); h2 = impulse(H2);
    L11 = norm(h1,1);L12 = norm(h2,1);
    L21 = norm(h1,2);L22 = norm(h2,2);
    MaximumError = (q/2)*(L11+L12);
    rmsError = sqrt((q^2/12)*(L21^2 + L22^2));
```

Appendix E
Symbols

$\equiv$	identically equal
$\approx$	approximately equal
$\stackrel{\text{def}}{=}$	defined as
$<\;(>)$	less (greater) than
$\leq\;(\geq)$	less (greater) than or equal to
$\ll\;(\gg)$	much less (greater) than
$\forall$	for all
$\rightarrow$	tends to
$\Longrightarrow$	implies
$\longleftrightarrow$	equivalent to, if and only if
$\sum$	summation
$\prod$	product
$\lvert a \rvert$	the absolute value of a scalar a
$\lVert x \rVert$	the norm of a vector x
$\lVert A \rVert$	the induced norm of a matrix A
$\det(A)$ or $\lvert A \rvert$	determinant of a matrix A
$A^{\mathrm{T}}\;(x^{\mathrm{T}})$	the transpose of a matrix A (a vector x)
$\angle \theta$	angle of θ
max	maximum
min	minimum
j	$\sqrt{-1}$
$\mathrm{Re}[z]$	the real part of a complex variable z
$\mathrm{Im}[z]$	the imaginary part of a complex variable z
$\frac{\partial f}{\partial x}$	the Jacobian matrix
$\dot{y}$	the first derivative of y with respect to time
$\ddot{y}$	the second derivative of y with respect to time
$\dddot{y}$	the third derivative of y with respect to time
$y^{(i)}$	the ith derivative of y with respect to time
[xx]	see reference number xx in the bibliography

$1(t)$	unit step function
$1(k)$	discrete-time unit step function
ζ	damping ratio
ω_n	undamped natural frequency
ω_d	damped natural frequency
T_r	rise time of the step response
T_p	peak time of the step response
T_s	settling time of the step response
PO	percent overshoot of the step response
ω_p	peak frequency of the frequency response
$M_{p\omega}$	ratio of the peak of the frequency response to the dc gain
ω_{BW}	bandwidth of the frequency response
GM	gain margin
PM	phase margin
ω_c	crossover frequency
ω_{pc}	phase crossover frequency
ϕ_{max}	maximum phase of the lead compensator
ω_{max}	frequency at which the phase of the lead compensator is maximum
K_p	position error constant
K_v	velocity error constant
K_a	acceleration error constant

Bibliography

[1] T. Andersson, N. Persson, A. Fattouh, and M. C. Ekstrom. A loop shaping method for stabilising a riderless bicycle. In *European Conference on Mobile Robots*, pages 1–6, 2019.

[2] P. J. Anstaklis and A. N. Michel. *Linear Systems Primer*. Birkhäuser, Boston, 2007.

[3] K. J. Astrom and T. Hagglund. PID control. In W. S. Levine, editor, *The Control Handbook*, pages 198–209. CRC Press, Boca Raton, FL, 1996.

[4] K. J. Astrom and B. Wittenmark. *Computer-Controlled Systems*. Prentice-Hall, Upper Saddle River, NJ, third edition, 1997.

[5] W. L. Bialkowski. Control of pulp and paper making procss. In W. S. Levine, editor, *The Control Handbook*, pages 1219–1242. CRC Press, Boca Raton, FL, 1996.

[6] A. Cinar. Artificial pancreas systems: An introduction to the special issue. *Control Systems Magazine*, 38:26–29, 2018.

[7] J. A. Cook, J. W. Grizzle, and J. Sun. Engine control. In W. S. Levine, editor, *The Control Handbook*, pages 1261–1274. CRC Press, Boca Raton, FL, 1996.

[8] R. C. Dorf and R. H. Bishop. *Modern Control Systems*. Pearson, Boboken, NJ, thirteenth edition, 2017.

[9] M. S. Fadali and A. Visioli. *Digital Control Engineering: Analysis and Design*. Elsevier, Boston, third edition, 2019.

[10] A. Ferrante. A simple proof of the Routh test. *IEEE Trans. Automat. Contr.*, 44:1306–1309, 1999.

[11] G. F. Franklin, J. D. Powell, and A. Emami-Naeini. *Feedback Control of Dynamic Systems*. Pearson, Upper Saddle River, NJ, sixth edition, 2010.

[12] G. F. Franklin, J. D. Powell, and M. L. Workman. *Digital Control of Dynamic Systems*. Addison Wesley, Reading, MA, third edition, 1990.

[13] H. Gao, W. Sun, and P. Shi. Robust sampled-data h_∞ control for vehicle active suspension systems. *IEEE Trans. Contr. Syst. Tech.*, 18:238–245, 2010.

[14] E. Garone, S. Di Cairano, and I. V. Kolmanovsky. Reference and command governors for systems with constraints: A survey of their theory and applications. *Automatica*, 75:306–328, 2017.

[15] M. Green and D. J. N. Limebeer. *Linear Robust Control*. Prentice Hall, Englewood Cliffs, NJ, 1995.

[16] D. W. Gu, P. H. Petkov, and M. M. Konstantinov. *Robust Control Design with MATLAB*. Springer, London, 2013.

[17] T. Hagglund and K. J. Astrom. Automatic tuning of PID controllers. In W. S. Levine, editor, *The Control Handbook*, pages 817–826. CRC Press, Boca Raton, FL, 1996.

[18] T. Hu and Z. Lin. *Control Systems with Actuator Saturation: Analysis and Design*. Springer, New York, 2001.

[19] L. M. Huyett, E. Gassau, H. C. Zisser, and F. J. Doyle III. Design and evaluation of a robust PID controller for a fully implantable artificial pancreas. *Ind. Eng. Chem. Res.*, 54:10311–10321, 2015.

[20] K. Keesman. *System Identification: An Introduction*. Springer, London, second edition, 2011.

[21] H. K. Khalil. *Nonlinear Systems*. Prentice Hall, Upper Saddle River, NJ, third edition, 2002.

[22] H. K. Khalil. *Nonlinear Control*. Pearson, Boston, 2015.

[23] H. Kwakernaak and R. Sivan. *Linear Optimal Control Systems*. Wiley-Interscience, New York, 1972.

[24] F. L. Lewis, D. L. Vrabie, and V. L. Syrmos. *Optimal Control*. Wiley, Hoboken, NJ, 2012.

[25] J. S. Lim, J. R. Ryoo, Y. I. Lee, and S. Y. Son. Design of a fixed-order controller for the track following control of optical disc drives. *IEEE Trans. Contr. Syst. Tech.*, 20:205–213, 2012.

[26] L. Ljung. *System Identification: Theory for the User*. Printice-Hall, Upper Saddle River, NJ, second edition, 1999.

[27] G. Meinsma. Elementary proof of the Routh-Hurwitz test. *Syst. Contr. Lett.*, 25:237–242, 1995.

[28] N. S. Nise. *Control Systems Engineering*. Wiley, Hoboken, NJ, eighth edition, 2019.

[29] K. Ogata. *Modern Control Engineering*. Pearson, Boboken, NJ, fifth edition, 2010.

[30] M. A. Pai. *Power System Stability Analysis by the Direct Method of Lyapunov*. North-Holland, Amsterdam, 1981.

[31] C. L. Philips, H. T. Nagle, and A. Chakrabortty. *Digital Control Systems: Analysis and Design*. Pearson, Upper Saddle River, NJ, fourth edition, 2014.

[32] W. J. Rugh. *Linear System Theory*. Prentice-Hall, Upper Saddle River, NJ, second edition, 1996.

[33] J. Tal. Motion control systems. In W. S. Levine, editor, *The Control Handbook*, pages 1382–1386. CRC Press, Boca Raton, FL, 1996.

[34] K. van Heusden, E. Gassau, H. C. Zisser, D. E. Seborg, and F. J. Doyle III. Control relevant models for glucose control using a priori patient characteristics. *IEEE Trans. Biomed. Eng.*, 59:1839–1849, 2012.

[35] K. Zhou and J. C. Doyle. *Essential of Robust Control*. Prentice Hall, Englewood Cliffs, NJ, 1998.

[36] K. Zhou, J. C. Doyle, and K. Glover. *Robust and Optimal Control*. Prentice Hall, Upper Saddle River, NJ, 1996.

Index

Note: Page numbers in *italics* refer to occurrences in the end-of-chapter problems.

acceleration error constant, 128
ADC, 379
admittance, 18
aliasing, 406
analog-to-digital converter, 379
angle criterion, 110
anti-windup scheme, 194, 199
antialiasing filter, 407
approximation of continuous-time transfer functions, 411
asymptotic Bode plots, 220
automobile suspension, 29, 322, 342, *373*
    active, *373*

backward difference method (BDF), 411
band-limited signal, 406
bandwidth, 256
bilinear transformation, 396, 419
bilinear transformation (BT) method, 412
Bode plot design, discrete-time systems, 419
Bode plots, 217
    error constants, 231
    second-order transfer function, 228
bounded signal, 65
bounded-input–bounded-output stability, 66
break frequency, 220
break-in point, 116
breakaway point, 116

capacitor, 14
Cauchy's principle of the argument, 235
characteristic equation of a matrix, 443

characteristic equation of closed-loop system, 109
circuit realization of compensators, 163
closed-loop eigenvalues, 353
combined Laplace transform–z-transform table, 445
component with memory, 15
contour mapping, 234
control constraints, 140
control system
    closed-loop, 2
    features of a well-designed, 6
    open-loop, 2
controllability, 333
controllability Gramian, 336
controllability matrix, 336
controllable canonical form, 344
corner frequency, 220
critically damped, 83, 84
crossover frequency, 251
cruise control, 3
cruise control system, 149
current source, 15

DAC, 379
damped natural frequency, 85
damping ratio, 85
dc gain of the transfer function, 79
dc gain, discrete-time, 385
dc motor, 38, 109, 124, 138, 323, 343, *371*, *433*, *435*
dc-to-dc power converter, 327, *372*
design by emulation, 411
design by gain adjustment, 156, 280
design specifications
    frequency domain, 278

INDEX

time-domain, 155
difference equation, 381
digital control, 377
digital control system, 378
    basic components, 379
    stability of, 393
digital-to-analog converter, 379
discrete-time design, 416
discrete-time equivalent, 387
discrete-time system, 381
disturbance effect, 133
duality, 340
dynamic component, 15
dynamic response, 78
    periodic input, 81
    sinusoidal input, 80
    step input, 79

eigenvalues, 331, 443
    selection of, 355
eigenvectors, 443
elementary column operations, 442
elementary row operations, 442
estimator, 358

feedforward control, 361
forced response, 78
forward difference method (FDM), 411
free-body diagram, 25
frequency response, 80
    discrete-time systems, 386
frequency-domain approach, advantages of, 306
frequency-domain design, 277
    discrete-time systems, 419

gain margin, 250
    discrete-time systems, 402
gear, 35
gear ratio, 35

Hook's law, 24

ideal sampler, 389
idle speed control, *150*, *314*
impedance, 18

impulse response, 65, 384
impulse sampler, 389
inductor, 14
insulin pump, *152*, *316*
integral control, 363
integrator windup, 198
Internal Model Principle, *151*, *315*, *316*
inverted pendulum on a cart, *374*

Jacobian matrix, 324
jury test, 393

Kirchhoff's current law, 15
Kirchhoff's voltage law, 15

lag compensator, 179, 295
Laplace transform, 327, 438
Laplace transform table, 444
lead compensator, 165, 287
    phase relations, 289
lead-lag compensator, 187, 302
linear independence, 442
linearization, 50, 324

magnetic levitation system, *372*
magnitude criterion, 110
marginally stable transfer function, 68
mass, 25
MATLAB tutorial, 446
measurement noise, 139
memoryless component, 15
minimum-phase system, 91
modes, 332
monic polynomial, 110
motion control system, *433*

natural response, 78
Newton's second law, 25, 31
non-minimum-phase system, 92
non-unity feedback system, 131
nonsingular matrix, 442
Nyquist contour, 238
Nyquist criterion, 234
    discrete-time systems, 398
    gain adjustment, 240
    poles on the imaginary axis, 243

Nyquist plot, 239
Nyquist stability criterion, 240

observability, 338
observability Gramian, 339
observability matrix, 340
Observable Canonical Form, 347
observer, 358
Ohm's law, 14
op-amp, 21
open-loop eigenvalues, 353
open-loop transfer function, 110
operational amplifier, 21
optical disc drive, 314
output equation, 319
output error due to quantization
    stochastic analysis, 425
    worst-case analysis, 425
output feedback, 352
output feedback control, 359
overdamped, 83
overshoot, 87

Padé approximation, 13
parallel realization, 349, 350
parameter uncertainty, 136
peak frequency, 256
peak of the frequency response, 256
peak time, 87
pendulum, 51
percent overshoot, 87
phase criterion, 110
phase crossover frequency, 251
phase margin, 251
    discrete-time systems, 402
PI (Proportional-Integral), 109
PI controller, 299
PID controller, tuning, 194
polar plot, 211
poles and zeros, discrete-time, 384
poles, closed-loop, 109
position error constant, 127
    discrete-time systems, 403
proper transfer function, 12
proportional controller, 156

proportional derivative (PD) controller, 173
proportional-integral (PI) controller, 181
proportional-integral-derivative (PID) controller, 190
pulp and paper industry, *151, 315, 316*

quantization, 422
Quarter Amplitude Delay Method, 195

rank of a matrix, 442
relative stability, 249
resistor, 14
riderless bicycle, *150, 208, 275*
rise time, 82, 87
rocket, 333, 338
root locus design, discrete-time systems, 416
root locus method, 110
    discrete-time systems, 394
rotating mass, 31
rotational mechanical system, 31
Routh-Hurwitz criterion, 70
    discrete-time systems, 397

sample, 379
sampling frequency, 405
sampling frequency, methods for choosing, 408
sampling period, 379
sampling theorem, 407
satellite, *375*
sensitivity transfer function, 137
set point, 126
set-point regulation, 361
set-point weighting, 177, 185, 194
settling time, 82, 87
spectrum of a sampled signal, 405
spring, 24
stability margins, 250, 251
    discrete-time systems, 402
stability margins from Bode plots, 251
stability of digital control systems, 393
stability of transfer functions, 66
stability, discrete-time systems, 384
stability, state-space models, 330

staircase continuous-time function, 379
state equation, 319
    solution of, 327
state feedback, 352
state feedback control, 353
state-space design, 352
state-space model of a transfer function, 344
state-space model, linear, 320
state-space models, 318
static component, 15
steady-state error, 126
steady-state output to step inputs, discrete-time systems, 385
steady-state response, 79
step response, 81
    effect of zeros, 90
    first-order transfer function, 81
    second-order transfer function, 83
strictly proper transfer function, 12
superposition principle, 81

three-term controller, 194
time delay, 13, 65
time-delay systems, 233
time-domain–frequency-domain relations, 256, 260
torsional spring, 31
transfer function, 11
    discrete-time, 382
transfer function of a state-space model, 340
transfer functions, approximation of higher-order, 93
transient response, 79
transition matrix, 328
    properties of, 335
translational mechanical system, 24
trapezoidal method, 412
Tustin method, 412
type of feedback control system, 128
type of feedback system, discrete-time, 403

unbounded signal, 65
uncompensated system, 165
undamped natural frequency, 85
underdamped, 83, 85

unit step function, 65
    discrete-time, 385
unity feedback system, 127

velocity error constant, 128
    discrete-time systems, 404
viscous damper, 24, 31
voltage source, 15

z-transform, 381
z-transform table, 445
zero-order hold, 379
    transfer function of, 388
ZOH, 379
ZOH method, 412

CONTROL SYSTEMS: AN INTRODUCTION

Professor Hassan K. Khalil is Distinguished University Professor Emeritus of Electrical and Computer Engineering at Michigan State University. He joined MSU in 1978, shortly after completing a Ph.D. degree in electrical engineering at the University of Illinois, which was preceded by B.S. and M.S. degrees, also in electrical engineering, from Cairo University in Egypt. Professor Khalil has consulted for General Motors and Delco Products and has published over 120 journal papers on singular perturbation methods and nonlinear control. He is also the author of several highly cited books, including *High-Gain Observers in Nonlinear Feedback Control* (SIAM 2017), *Nonlinear Control* (Pearson 2015), and *Nonlinear Systems* (Macmillan 1992; Prentice Hall 1996 & 2002), and a coauthor of *Singular Perturbation Methods in Control: Analysis and Design* (Academic Press 1986; SIAM 1999).

Professor Khalil was named IEEE Fellow in 1989 and IFAC Fellow in 2007, and has received numerous international honors and awards, including the 1989 IEEE-CSS George S. Axelby Outstanding Paper Award, the 2000 AACC Ragazzini Education Award, the 2002 IFAC Control Engineering Textbook Prize, the 2004 AACC O. Hugo Schuck Best Paper Award, and the 2015 IEEE-CSS Bode Lecture Prize. At MSU he received the 1995 Distinguished Faculty Award and the 2020 Withrow Teaching Excellence Award. He has served as Associate Editor of the *IEEE Transactions on Automatic Control*, *Automatica*, and *Neural Networks*, and as Editor of *Automatica* for nonlinear systems and control. He was Program Chair of the 1988 ACC, and General Chair of the 1994 ACC.

ISBN 978-1-60785-826-3